Structure Analysis by Small-Angle X-Ray and Neutron Scattering

Structure Analysis by Small-Angle X-Ray and Neutron Scattering

L. A. Feigin and
D. I. Svergun
Institute of Crystallography
Academy of Sciences of the USSR
Moscow, USSR

Edited by
George W. Taylor
Princeton Resources
Princeton, New Jersey

Plenum Press • New York and London

Library of Congress Cataloging in Publication Data

Svergun, D. I.
Structure analysis by small-angle X-ray and neutron scattering.

Bibliography: p.
Includes index.
1. X-rays—Scattering. 2. Neutrons—Scattering. 3. Matter—Structure. 4. Small-angle scattering. I. Feigin, L. A. II. Taylor, George W. III. Title.
QC482.S3S8513 1987 539.7′222 87-25489
ISBN 0-306-42629-3

Printed in the United States of America

Preface

Small-angle scattering of X rays and neutrons is a widely used diffraction method for studying the structure of matter. This method of elastic scattering is used in various branches of science and technology, including condensed matter physics, molecular biology and biophysics, polymer science, and metallurgy. Many small-angle scattering studies are of value for pure science and practical applications.

It is well known that the most general and informative method for investigating the spatial structure of matter is based on wave-diffraction phenomena. In diffraction experiments a primary beam of radiation influences a studied object, and the scattering pattern is analyzed. In principle, this analysis allows one to obtain information on the structure of a substance with a spatial resolution determined by the wavelength of the radiation.

Diffraction methods are used for studying matter on all scales, from elementary particles to macro-objects. The use of X rays, neutrons, and electron beams, with wavelengths of about 1 Å, permits the study of the condensed state of matter, solids and liquids, down to atomic resolution. Determination of the atomic structure of crystals, i.e., the arrangement of atoms in a unit cell, is an important example of this line of investigation. Another line deals with the study of the structure of matter at the superatomic level, i.e., with a spatial resolution from 10 Å up to thousands and even several tens of thousands of angstroms. For this line of investigation the basic tool is small-angle scattering, mainly of X rays and neutrons.

In small-angle scattering research the wavelength used is, as a rule, on the order of several angstroms. As interatomic distances are of the same order of magnitude, the diffraction pattern, corresponding to the superatomic structure, lies in the small-angle region, and this is how the method received its name. The most important feature of the small-angle scattering method is its potential for analyzing the inner structure of disordered systems, and frequently the application of this method is a unique way to obtain direct structural information on systems with random arrangement of density inhomogeneities of colloid size, namely, 10^1–10^4 Å.

Small-angle scattering studies date from the classical works of A. Guinier, published in 1938. Later, P. Debay, G. Porod, O. Kratky, V. Luzzati, W. Beeman, P. Schmidt, and others developed the theoretical and experimental fundamentals of the method. The principles of designing small-angle scattering facilities were developed, and the potential for applying the method for determining general structural characteristics of various types of highly dispersed systems was shown.

New progress in refining the method of small-angle scattering began in the 1970s and is continuing today. This stage of work is characterized by new opportunities both for experimentation (powerful neutron beams, synchrotron X-ray sources, position-sensitive detectors) and for structural interpretation (contrast variation, isomorphous replacement, direct methods, analysis of characteristic functions). Certainly, the use of various computers is of great significance in advancing these two aspects of small-angle scattering applications. The result has been a gradual expansion of the range of studied objects and an increase in the spatial resolution in both directions. Thus, for example, the mechanisms of phase separation and the sizes and degrees of dispersion of dispersed structures of alloys, powders, and glasses; the specific features of the configuration of polymer chains in different aggregate states; and the geometrical and weight parameters (and, sometimes, three-dimensional structure) of biological macromolecules and their complexes in solution can all be subjected to analysis.

Currently, the small-angle scattering technique, with its well-developed experimental and theoretical procedures and wide range of studied objects, is a self-contained branch of the structural analysis of matter.

Despite the wide application of the small-angle scattering technique in solving various fundamental and applied problems, there are very few monographs on this subject. Only two books are known: the classical monograph *Small-Angle Scattering of X Rays* by A. Guinier and A. Fournet (1955) and *Small-Angle X-Ray Scattering*, edited by O. Glatter and O. Kratky (1982). The latter, in fact, is a collection of review articles. Both books deal with X-ray scattering and do not discuss small-angle diffraction of neutrons. As a result, general information on the small-angle scattering method is dispersed in individual original and review articles, and in the proceedings of several international small-angle scattering conferences (1965, 1973, and 1977). As a consequence, scientists and engineers applying the small-angle scattering technique to solve their particular problems frequently are not well informed on the possibilities of the method and may use it inefficiently.

The aim of this book is to present the current state of small-angle scattering knowledge and to show how this technique makes it possible to establish several physicochemical parameters (geometry and weight) of investigated objects, and to analyze their spatial structure. The main purpose of the authors is to demonstrate methods for obtaining structural information from experimental small-angle scattering data. The X-ray and neutron scattering methods are considered in parallel, and their advantages and potentials for combined application are discussed.

Part I deals with the basic theory of diffraction and scattering intensity calculations for systems of various degrees of ordering. The theory, which describes the relation of the elastic scattering pattern to the spatial distribution of matter within an object, is the same for all types of radiation. However, the concrete form and specific features of the diffraction phenomena for X rays and neutrons are determined by different characteristics of matter. Thus, Chapter 1 discusses general relationships for the amplitude and intensity of elastic scattering; methods of Fourier transforms are considered briefly as well as the basic features of the interaction of X rays and neutrons with matter. The main equations relating the scattering intensity to the structural parameters of various objects are derived in Chapter 2. Particle solutions, isolated particles, nonparticulate systems, and some other model systems are treated.

Part II discusses methods for obtaining structural information from small-angle scattering data for monodisperse systems. These methods are divided into three groups. The first one includes the calculation of general geometric and weight parameters of particles (invariants of the scattering curves) and modeling of their structure by the trial-and-error technique. These approaches (Chapter 3) are used mainly in the case of particles that can be treated as homogeneous. Chapter 4 deals with new methods for calculating the inner structure of a particle on the basis of novel experimental procedures, including contrast variation, isomorphous and isotopic replacement, and change of radiation type. Chapter 5 treats the so-called direct methods of interpretation, which permit one, if some additional *a priori* information on an object is available, to recover directly its structure, i.e., the scattering density distribution.

Part III deals with the application of the general theory of small-angle diffraction to polymers and inorganic substances. This branch of small-angle scattering research is gradually expanding, mainly because of applied and technological requirements. In the case of polymer systems (Chapter 6), some modifications of general methods of structural interpretation are necessary because of the great variety of structures of this kind. Here a model of the polymer chain forms the basis for the analysis. For polymers in solution or in an amorphous state, the characteristics of the coils are mainly determined; for crystal polymers the ways of packing the crystal and amorphous regions are investigated. Some applications of the small-angle scattering method to oriented objects are considered as well.

Inorganic materials require an even wider variety of models (Chapter 7) for data analysis. It should be noted that the most important procedures, such as determination of the phase composition of alloys or estimation of the pore sizes in ceramics, have been developed reliably, mainly for technological applications, and may be considered as routine procedures.

Part IV discusses the problems of small-angle scattering experimentation. The basic requirements for measuring devices and principles of design and construction are treated, including instrumentation for neutron beams and synchrotron sources (Chapter 8). Fundamental sources of instrumental distortions in small-angle scattering measurements are considered in Chapter 9, and a series of procedures for experimental data treatment is analyzed comprehensively. Both complicated methods of data analysis and relatively simple algorithms for minicomputers are presented.

ACKNOWLEDGMENTS. The authors wish to thank Professor B. Vainshtein for his continuous support in developing small-angle structural research at the Institute of Crystallography of the USSR Academy of Sciences. They are indebted to Professors P. B. Moore and G. W. Taylor for reading some chapters and offering valuable comments. The authors thank Dr. A. Ryskin for help with word processor use and also A. Semenyuk and T. Simagina for their assistance in preparing the manuscript for publication. They are also grateful to the staff of Plenum Press for their continued help and cooperation.

L. A. Feigin and D. I. Svergun

Moscow

Contents

PART I. SMALL-ANGLE SCATTERING AND THE STRUCTURE OF MATTER

1. Principles of the Theory of X-Ray and Neutron Scattering . . 3

1.1. Scattering of a Plane Wave by Matter 3
1.2. Fourier Transforms. Convolutions 6
1.3. Scattering by Simple Objects 10
1.3.1. Rectangular Parallelepiped 10
1.3.2. Homogeneous Thin Plate 12
1.3.3. Homogeneous Thin Rod 12
1.3.4. Inhomogeneous Parallelepiped 13
1.3.5. Periodic Set of Centers 13
1.3.6. Spherically Symmetric Body 14
1.4. Scattering of X Rays by Atoms 14
1.5. Scattering of Thermal Neutrons by Nuclei 17
1.6. Absorption of X Rays and Neutrons 22
1.7. Conclusion 24

2. General Principles of Small-Angle Diffraction 25

2.1. Scattering by Objects with Different Ordering 25
2.1.1. Single Crystals 26
2.1.2. One-Dimensional Periodic Structures 27
2.1.3. Cylindrically Symmetric Objects 28
2.1.4. Isotropic Systems 29
2.2. Small-Angle Scattering by Disperse Systems 30
2.2.1. Scattering Intensity by a Disordered Object . . . 30
2.2.2. Scattering at Low Angles 33
2.3. Particle Solutions 34

2.3.1. Monodisperse and Polydisperse Systems 34
2.3.2. Conception of Contrast 35
2.3.3. Concentrated and Dilute Systems 36
2.4. Isolated Particle 38
2.4.1. Debye Equation 38
2.4.2. Correlation Function 39
2.4.3. Uniform Particles 40
2.4.4. Asymptotic Behavior of Intensity. The Porod Invariant 45
2.4.5. Special Types of Particle 46
2.5. Nonparticulate Systems 50
2.5.1. Scattering Due to Statistical Fluctuations 50
2.5.2. Two-Phase and Multiphase Systems 52
2.6. Conclusion 55

PART II. MONODISPERSE SYSTEMS

3. Determination of the Integral Parameters of Particles . . . 59

3.1. Geometrical and Weight Invariants 59
3.1.1. Total Scattering Length and Radius of Gyration . . 60
3.1.2. Volume and Surface 61
3.1.3. Largest Dimension and Correlation Length . . . 62
3.1.4. Anisometric Particles 63
3.2. Information Content in Small-Angle Scattering Data . . . 63
3.2.1. General Approach 64
3.2.2. Number of Independent Parameters 65
3.3. Evaluation of the Invariants 68
3.3.1. Accuracy of Calculation of the Radius of Gyration . 68
3.3.2. Absolute Measurements. Molecular-Mass Determination 73
3.3.3. Possibilities of Homogeneous Approximation . . . 76
3.3.4. Estimate of the Largest Dimension 82
3.3.5. Evaluation of the Invariants of Anisometric Particles 83
3.3.6. List of the Basic Equations 87
3.4. Scattering by Particles of Simple Shape 90
3.5. Modeling Method 94
3.5.1. Demands on the Technique of Calculation 94
3.5.2. Subparticle Models 95
3.5.3. Spheres Method 95
3.5.4. Cube Method 97
3.5.5. Modeling in Real Space 98
3.6. Applications of Modeling 98
3.6.1. *Helix pomatia* Hemocyanin 99

3.6.2. Bacteriophage S_D 100
3.6.3. 30 S Ribosomal Subparticle 102
3.7. Conclusion 104

4. Interpretation of Scattering by Inhomogeneous Particles . . 107

4.1. Scattering by Inhomogenous Particles 107
4.1.1. Solvent Influence 108
4.1.2. General Equations for Intensity and Invariants . . 110
4.1.3. Spherically Symmetric Particles 112
4.1.4. Large-Angle Scattering 112
4.2. Contrast Variation 115
4.2.1. Basic Functions 115
4.2.2. Dependence of Invariants on Contrast 117
4.2.3. Contrasting Techniques 118
4.2.4. Applications of the Contrast-Variation Technique 122
4.3. Isomorphous-Replacement Methods 131
4.3.1. Heavy-Atom Labels 131
4.3.2. The Triangulation Method 132
4.4. Variation in the Applied Radiation 138
4.4.1. Combined Use of Various Types of Radiation . . 139
4.4.2. Anomalous (Resonant) Scattering 140
4.5. Conclusion 144

5. Direct Methods 147

5.1. One-Dimensional Density Distributions 148
5.2. Solving the One-Dimensional Sign Problem 150
5.2.1. Use of Correlation Functions 152
5.2.2. Box-Function Refinement 153
5.3. Multipole Theory of Small-Angle Diffraction 156
5.4. Determination of Multipole Components 164
5.4.1. Isometric Particles 164
5.4.2. Shape of Uniform Particles 165
5.4.3. Separation of Bessel Functions 167
5.4.4. Examples of Direct Structure Determination . . . 176
5.5. Conclusion 182

PART III. POLYMERS AND INORGANIC MATERIALS

6. Investigations of Polymer Substances 187

6.1. Models of Polymer Chains 188
6.1.1. Gaussian Chains 188

6.1.2. Persistent Chain 191
6.1.3. Perturbed Chains 194
6.1.4. Molecular Mass Distribution 195
6.2. Polymers in Solution and in the Amorphous State . . . 197
6.2.1. Polymers in Solution 197
6.2.2. Amorphous Polymers 201
6.3. Crystalline Polymers 203
6.3.1. Lamellar Model 203
6.3.2. Scattering by the Lamellar Stacks 205
6.3.3. Correlation Functions 207
6.3.4. Determination of Chain Folding 209
6.4. Anisotropic Systems 210
6.4.1. Oriented Amorphous Polymers 210
6.4.2. Fibrillar Systems 211
6.4.3. Lamellar Systems 214
6.5. Conclusion 217

7. Structural Studies of Inorganic Materials 219

7.1. Crystalline Materials 220
7.1.1. Defects in Single Crystals 220
7.1.2. Phase Separation in Alloys 225
7.2. Polydisperse Objects. Calculation of Size Distribution . . 230
7.2.1. Analytical Methods 231
7.2.2. Numerical Methods 236
7.3. Amorphous Solids and Liquids 240
7.3.1. Study of the Structure of Glasses 240
7.3.2. Concentration Fluctuations and Clusters 243
7.4. Conclusion 245

PART IV. INSTRUMENTATION AND DATA ANALYSIS

8. X-Ray and Neutron Instrumentation 249

8.1. Basic Designs of Instrumentation 249
8.1.1. Angular Resolution 250
8.1.2. Main Characteristics of Instruments 252
8.2. Laboratory X-Ray Instruments 255
8.2.1. X-Ray Tubes 255
8.2.2. X-Ray Detectors 256
8.2.3. Point Collimation System 257
8.2.4. Slit Collimation System 258
8.3. Synchrotron Radiation Instruments 262

8.3.1. Main Characteristics of Synchrotron Radiation . . 262
8.3.2. Monochromatization and Focusing of X Rays . . 263
8.3.3. Small-Angle Synchrotron Instruments 265
8.4. Small-Angle Neutron Scattering Apparatus 268
8.4.1. Neutron Sources, Monochromatization, Detectors . 268
8.4.2. Collimation Systems and Instruments 269
8.5. Conclusion 273

9. Data Treatment 275

9.1. General Scheme of Small-Angle Data Processing 276
9.1.1. Instability of Experimental Conditions 276
9.1.2. Additive Scattering Components 276
9.1.3. Influence of Beam and Detector Dimensions . . . 277
9.1.4. Beam Polychromaticity 279
9.1.5. Statistical Errors 280
9.1.6. General Expression for Experimental Intensity . . 280
9.2. Preliminary Data Processing 281
9.3. Experimental-Data Smoothing 283
9.3.1. Algebraic Polynomials 284
9.3.2. Spline Functions 288
9.3.3. Frequency Filtering Method 291
9.3.4. Problem of Optimum Smoothing 292
9.4. Collimation Corrections 295
9.4.1. Weighting Functions 295
9.4.2. Slit-Width Correction 296
9.4.3. Slit-Height Correction 298
9.5. Corrections for Polychromaticity 303
9.6. Termination Effects 305
9.7. Simultaneous Elimination of Various Distortions . . . 309
9.7.1. Iteration Methods 309
9.7.2. Orthogonal Expansions 310
9.7.3. Use of the Sampling Theorem 314
9.7.4. General Regularization Procedure 317
9.8. Conclusion 320

REFERENCES 321

INDEX 333

I

Small-Angle Scattering and the Structure of Matter

1

Principles of the Theory of X-Ray and Neutron Scattering

X rays (wavelengths from 0.5 to 2 Å), thermal neutrons (1–10 Å), and fast electrons (0.05–1 Å) are all used in diffraction studies of the three-dimensional structure of matter. Electron diffraction is not widespread in small-angle scattering investigations, which require a larger wavelength. Thus X rays and neutrons are used in practice only as a tool for structure determinations by the small-angle scattering technique.

In Section 1.1 we treat the scattering principles of pure elastic scattering by matter without discussing the nature of the incident radiation. This is followed by a minimum amount of mathematics (Section 1.2) and some applications involving calculation of the scattering amplitude and intensity for simple bodies and systems (Section 1.3). However, real scattering phenomena depend much on the nature of the primary radiation and Sections 1.4 and 1.5 deal respectively with the basis of X-ray and neutron scattering by matter. The absorption of these radiations is discussed in Section 1.6.

1.1. Scattering of a Plane Wave by Matter

We assume a plane monochromatic wave $A_0 \exp(i\mathbf{k}_0\mathbf{r})$ incident at the point scattering center O, which generates a secondary spherical wave (see Figure 1.1). Then, at some observation point, the resulting wave (Cowley, 1975) is given by

$$A_0 \exp(i\mathbf{k}_0\mathbf{r}) + (A_0 b/r)\exp(i\mathbf{kr}) \tag{1.1}$$

where $\mathbf{k}_0$ and $\mathbf{k}$ are the incident and scattering wave vectors with $|\mathbf{k}_0| = |\mathbf{k}| = 2\pi/\lambda$, λ denoting the wavelength, A_0 and $A_0 b/r$ are the scattering amplitudes of the two waves, and $\mathbf{r}$ is the vector which

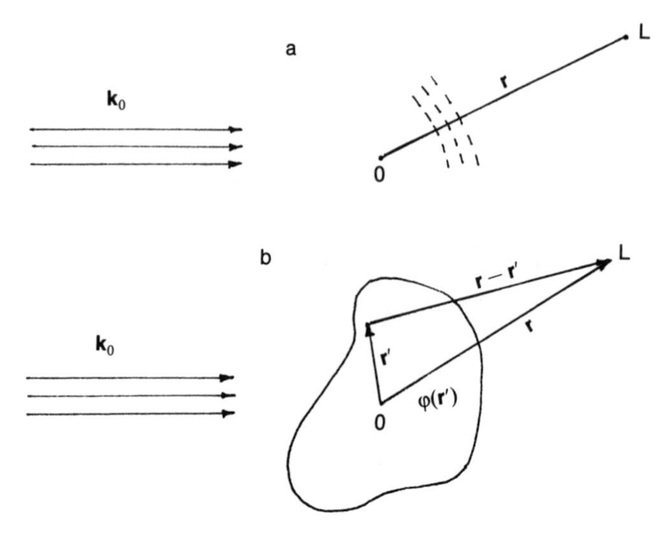

Figure 1.1. Scattering of a plane wave by a point scatterer (a) and a bounded potential field $\varphi(\mathbf{r})$ (b).

determines an observation point L corresponding to the scattering center O. The stronger the interaction between the incident wave and the point center O, the greater the constant b. This quantity has the dimension of length and is called the "scattering length," or scattering amplitude, of the center.

We consider the scattering of a plane monochromatic wave of X rays or neutrons by a real object consisting of an accumulation of nuclei and electrons — a group of point scatterers. In this case the scattering ability of the object may be characterized by the scattering density function $\varphi(\mathbf{r})$. This function, the potential field, is determined within a finite range of space; in some cases $\varphi(\mathbf{r})$ may depend on time. In the case of X-ray scattering $\varphi(\mathbf{r})$ represents the distribution of the electric charge density (see Section 1.4), while in neutron scattering this function represents the nuclear and spin density distribution (see Section 1.5).

Now we examine the interaction of the plane wave with matter. The wave interacts with all the nuclei and electrons, which become the sources of secondary waves. The superposition of these waves gives the first approximation to the scattering wave. In turn, the secondary waves are scattered by all the centers, and superpositioning of all these waves gives the second approximation, and so on. Successive approximations of this type converge into the resultant wave when interaction between the incident wave and separate centers is not great. Calculation of these approximations is very difficult and, as a rule, is performed only for the first approximation. The use of the first approximation is appropriate for weakly scattering centers or fields only (Landau and Lifshitz, 1985a). The calculation of a scattering wave as a manifold superposition of spherical waves is based on the disturbance theory. The wave $\psi(\mathbf{r})$ scattered by field $\varphi(\mathbf{r})$ is sought as a solution of the wave equation

$$[\Delta + k_0^2 + v\varphi(\mathbf{r})]\psi(\mathbf{r}) = 0$$

where $k_0 = |\mathbf{k}_0|$ represents the wave number for the incident wave in free space, Δ is the Laplace operator, and v is a parameter specifying the strength of the interaction with a potential field.

The solution of this equation is sought as a power series in v. The free term of this series represents an incident wave, and the leading term is accordingly the first approximation to the scattering wave. When we restrict ourselves to the first approximation, the scattering amplitude at a point L is equal to the sum of two components,

$$\psi^{(0)}(\mathbf{r}) + \psi^{(1)}(\mathbf{r}) = A_0 \exp(i\mathbf{k}_0\mathbf{r}) + \frac{A_0 v}{4\pi}\int \frac{\exp[i\mathbf{k}(\mathbf{r}-\mathbf{r}')]}{|\mathbf{r}-\mathbf{r}'|}\varphi(\mathbf{r}')\exp(i\mathbf{k}_0\mathbf{r}')d\mathbf{r}' \tag{1.2}$$

The observation point L is usually placed in such a manner (Figure 1.1b) that $|\mathbf{r}-\mathbf{r}'|$ is very large compared to the dimensions of the scattering area (see Figure 1.1). This is the case for Fraunhofer diffraction and we obtain the asymptotic form for the scattering wave:

$$\psi^{(0)}(\mathbf{r}) + \psi^{(1)}(\mathbf{r}) = A_0 \exp(i\mathbf{k}_0\mathbf{r}) + \frac{A_0 v}{4\pi}\frac{\exp(i\mathbf{k}_0\mathbf{r})}{r}\int \varphi(\mathbf{r}')\exp(i\mathbf{s}\mathbf{r}')d\mathbf{r}' \tag{1.3}$$

where $\mathbf{s} = \mathbf{k} - \mathbf{k}_0$ represents the scattering vector and $|\mathbf{s}| = (4\pi \sin\theta)/\lambda$, 2θ being the scattering angle.

Comparison of equations (1.3) and (1.1) shows that, for the potential field $\varphi(\mathbf{r})$, the function

$$f(\mathbf{s}) = \frac{v}{4\pi}\int \varphi(\mathbf{r})\exp(i\mathbf{s}\mathbf{r})d\mathbf{r} \tag{1.4}$$

plays the same role as that of factor b in the case of the point center. Thus $f(\mathbf{s})$ is the amplitude of elastic scattering by the field $\varphi(\mathbf{r})$. This expression is a first Born approximation, in other words, a single-scattering approximation.

Expression (1.4) represents a Fourier transform. Solution of the inverse problem — calculation of $\varphi(\mathbf{r})$ using the known function $f(\mathbf{s})$ — is given by the inverse Fourier transform

$$\varphi(\mathbf{r}) = \frac{1}{2\pi^2 v} \int f(\mathbf{r}) \exp(-i\mathbf{sr}) d\mathbf{s} \tag{1.5}$$

Thus application of Fourier transforms permits one to calculate both the scattering amplitude of the known system of scattering centers and the potential field for the known scattering amplitude if the conditions of the first Born approximation are fulfilled. For atomic and molecular structure determinations, one usually tries to perform an experiment in a way which permits this approximation to be satisfied. Hence we may use inverse integral Fourier transforms to calculate the scattering amplitude and the potential field. The case of strong scattering, when successive terms of the expansion must be taken into account, requires separate analysis.

In an experiment we cannot measure the scattering amplitude, but only the flow of scattering energy or the number of scattered particles (photons, neutrons), proportional to the square of the scattering amplitude:

$$\frac{dW}{d\Omega} = \frac{A_0^2}{r^2} |f(\mathbf{s})|^2 = \frac{A_0^2}{r^2} I(\mathbf{s}) \tag{1.6}$$

where Ω represents a solid angle. The function $I(\mathbf{s})$ is known as the scattering intensity (a term used in X-ray structure analysis), or the differential cross section (for neutron scattering). It is evident that the dimension of this function is length squared.

One significant feature of an object is the total scattering intensity (or total scattering cross section), which is obtained by integrating equation (1.6) over all the angles:

$$\sigma = \int_{\Omega} \frac{I(\mathbf{s})}{r^2} d\Omega \tag{1.7}$$

The main problem in the structure analysis of matter is the reconstruction of the scattering density distribution according to the measured function $I(\mathbf{s})$. The theory of Fourier transforms provides the basic mathematical apparatus for structural determinations. The major features of this concept are now examined.

1.2. Fourier Transforms. Convolutions

Fourier transforms establish the interrelation between functions in real space (**r**-space, object space) and in reciprocal space (**s**-space, image space). The pair of Fourier transforms is described by (Sneddon, 1951)

$$f(\mathbf{q}) = \mathscr{F}[\varphi(\mathbf{r})] = \int \varphi(\mathbf{r})\exp(2\pi i\,\mathbf{qr})d\mathbf{r} \tag{1.8}$$

and

$$\varphi(\mathbf{r}) = \mathscr{F}^{-1}[f(\mathbf{q})] = \int f(\mathbf{q})\exp(-2\pi i\,\mathbf{qr})d\mathbf{q} \tag{1.9}$$

For a higher symmetry of our expressions we have introduced a variable $\mathbf{q} = \mathbf{s}/2\pi$. Components of vector $\mathbf{r}$ are denoted by (x, y, z) and components of $\mathbf{q}$ by (X, Y, Z). Expressions (1.8) and (1.9) for one-dimensional functions may be rewritten in the form

$$f(X) = \mathscr{F}[\varphi(x)] = \int_{-\infty}^{\infty} \varphi(x)\exp(2\pi ixX)dx \tag{1.10}$$

and

$$\varphi(x) = \mathscr{F}^{-1}[f(X)] = \int_{-\infty}^{\infty} f(X)\exp(-2\pi ixX)dX \tag{1.11}$$

The main properties of the Fourier transforms are listed in Table 1.1 for a one-dimensional case, but the same expressions are valid for three dimensions. It should be emphasized that subdivision into two spaces — real and reciprocal — is purely provisional. Both spaces are fully equivalent; in particular, all the relationships in Table 1.1 are valid from left to right, and vice versa.

Particular cases of sine and cosine Fourier transforms are used for practical purposes. When $\varphi(x)$ is an even (real) function we have

$$f_c(X) = \mathscr{F}_c[\varphi(x)] = 2\int_0^{\infty} \varphi(x)\cos(2\pi xX)dx$$

Table 1.1. Basic Properties of Fourier Transforms

Function	Fourier image
$\varphi(x)$	$f(X)$
$\varphi^*(x)$	$f(-X)$
$\varphi(ax)$	$f(X/a)/a$
$\alpha\varphi_1(x) + \beta\varphi_2(x)$	$\alpha f_1(X) + \beta f_2(X)$
$\varphi(x-a)$	$\exp(2\pi iaX)\,f(X)$
$d[\varphi(x)]/dx$	$-2\pi iXf(X)$

which is the cosine Fourier image of function $\varphi(x)$. If $\varphi(x)$ is an odd function, then

$$f_s(X) = \mathscr{F}_s[\varphi(x)] = 2\int_0^\infty \varphi(x)\sin(2\pi xX)dx$$

is its sine Fourier image. Tables of Fourier integrals (see, for example, Erdélyi, 1954) usually contain sets of various even and odd functions when any arbitrary real function may be represented as a sum of even and odd functions.

The convolution integral of the two functions $\varphi(r)$ and $\psi(r)$ is frequently used when employing Fourier transforms. This integral is defined by

$$\varphi(\mathbf{r})*\psi(\mathbf{r}) = \int \varphi(\mathbf{u})\psi(\mathbf{r}-\mathbf{u})d\mathbf{u} \tag{1.12}$$

In one dimension

$$\varphi(x)*\psi(x) = \int_{-\infty}^{\infty} \varphi(y)\psi(x-y)dy \tag{1.13}$$

We shall now consider (without derivation) two major relationships, namely, the multiplication theorem

$$\mathscr{F}[\varphi(\mathbf{r})\psi(\mathbf{r})] = \mathscr{F}[\varphi(\mathbf{r})]*\mathscr{F}[\psi(\mathbf{r})] = [f(\mathbf{q})*g(\mathbf{q})] \tag{1.14a}$$

which states that the Fourier transform of a product of two functions is the convolution of their Fourier transforms, and the convolution theorem

$$\mathscr{F}[\varphi(\mathbf{r})*\psi(\mathbf{r})] = \mathscr{F}[\varphi(\mathbf{r})]\mathscr{F}[\psi(\mathbf{r})] = f(\mathbf{q})g(\mathbf{q}) \tag{1.14b}$$

i.e., the Fourier transform of the convolution of two functions is the product of their Fourier transforms.

Having defined the convolution integral, it is appropriate to introduce the Dirac delta function $\delta(x)$, which satisfies the expression

$$\int_{-\infty}^{\infty} \varphi(x)\delta(x-a)dx = \varphi(a) \tag{1.15}$$

for a bounded function $\varphi(x)$ and any value of a. The basic properties of the δ-function are

$$\int_{-\infty}^{\infty} \delta(x)dx = 1 \quad \text{and} \quad \delta(x-a) = 0 \qquad \text{when } x \neq a \tag{1.16}$$

The convolution integral and δ-function are frequently applied in the analysis of many theoretical and experimental problems involved in elastic wave scattering. For instance, the scattering intensity of the system comprising N point centers with scattering lengths b_i located at points $\mathbf{r}_i$ is given by

$$\varphi(\mathbf{r}) = \sum_{i=1}^{N} b_i \delta(\mathbf{r} - \mathbf{r}_i) \tag{1.17}$$

Another purely experimental case shows the possibility of using the convolution integral. If $I(X)$ is the intensity of scattering by an object and there is a narrow slit in front of the detector with propagation function $G(X)$, then the energy registered by the detector at a point X_0 is

$$J(X_0) = \int_{-\infty}^{\infty} I(X)G(X - X_0)dX$$

i.e., the convolution integral of functions $I(X)$ and $G(X)$, where $G(X)$ is inverted with respect to the origin of coordinates.

The scattering intensity may be expressed with the aid of the convolution integral. The Fourier transform of function $I(\mathbf{s})$ in equation (1.6) is (see Table 1.1)

$$P(\mathbf{r}) = \mathscr{F}^{-1}[|f(\mathbf{q})|^2] = \mathscr{F}^{-1}[f(\mathbf{q})f^*(\mathbf{q})]$$
$$= \varphi(\mathbf{r}) * \varphi(-\mathbf{r}) = \int \varphi(\mathbf{u})\varphi(\mathbf{u} + \mathbf{r})d\mathbf{u}$$

i.e., the Fourier-transformed scattering intensity represents the convolution of function $\varphi(\mathbf{r})$ with itself, being inverted at the origin $\mathbf{r} = 0$. This convolution product $P(\mathbf{r})$ is known as a self-convolution, or the Patterson function in structure analysis problems (Patterson derived his equation in 1935). Hence we may write

$$P(\mathbf{r}) = \int I(\mathbf{q})\exp(-2\pi i\mathbf{qr})d\mathbf{q}, \quad I(\mathbf{q}) = \int P(\mathbf{r})\exp(2\pi i\mathbf{qr})d\mathbf{r} \tag{1.18}$$

i.e., the scattering intensity and the Patterson function form a reciprocal pair of transforms exactly as in the case of the scattering amplitude and the scattering density.

The use of Fourier transforms permits one to derive the expression for the total scattering intensity:

$$\int I(\mathbf{q})d\mathbf{q} = \int \varphi^2(\mathbf{r})d\mathbf{r} \tag{1.19}$$

i.e., this value is determined only by the total amount of scattering matter. Expression (1.19) is one of the basic invariants of potential scattering at a small amplitude.

1.3. Scattering by Simple Objects

Now we deal with some characteristic types of scattering objects and calculate their scattering amplitudes and corresponding scattering intensities. The scattering properties of three-dimensional objects will be calculated but we shall analyze scattering by one- and two-dimensional objects in three-dimensional space as a limiting case. For details, see Hosemann and Bagchi (1962) and Vainshtein (1966).

The relationships for Fourier integrals (Erdélyi, 1954) are very useful when calculating scattering amplitudes involving objects of simple shape. Table 1.2 presents some practically important pairs of one-dimensional Fourier transforms.

1.3.1. Rectangular Parallelepiped

We consider a homogeneous parallelepiped (scattering density φ_0) with edges a, b, and c. In this case we can write

$$\varphi(\mathbf{r}) = \varphi(x, y, z) = \varphi_0 \Pi(x, a)\Pi(y, b)\Pi(z, c) \tag{1.20}$$

where function Π is defined in Table 1.2. The scattering amplitude is then given by

$$f(\mathbf{q}) = \varphi_0 V \delta(a, X)\delta(b, X)\delta(c, Z) \tag{1.21}$$

where $V = abc$ represents the parallelepiped volume and $\delta(u, v) = \sin(\pi uv)/\pi uv$. Figure 1.2 shows the general appearance of this function.

Table 1.2. Fourier Images of Some Functions

$\varphi(x)$	$f(X)$
$\exp(-\pi x^2)$	$\exp(-\pi X^2)$
1	$\delta(X)$
$\cos(2\pi ax)$	$[\delta(X+a)] + [\delta(X-a)]/2$
$\sin(2\pi ax)$	$[\delta(X+a)] - [\delta(X-a)]/2$
$\Pi(x, a) = \begin{cases} 1, & \lvert x \rvert \leqslant a/2 \\ 0, & \lvert x \rvert > a/2 \end{cases}$	$a \sin(\pi aX)/\pi aX$
$\Lambda(x, a) = \begin{cases} 1 - \lvert x \rvert/a, & \lvert x \rvert \leqslant a \\ 0, & \lvert x \rvert > a \end{cases}$	$a[\sin(\pi aX)/\pi aX]^2$

The longer one or another edge of the parallelepiped, the narrower the first and all subsidiary maxima. This is a consequence of the general metrical properties of Fourier transforms: the distances in real and reciprocal space are mutually inverse. One can see (Figure 1.2) that the main part of the scattering energy for a parallelepiped with edge a is concentrated in the range $0 \leqslant X \leqslant 1/a$.

Our information may be generalized as follows: to obtain data on inhomogeneities with linear dimension D one must measure the scattering intensity in the range of scattering angles

$$|\mathbf{q}| \leqslant 1/D \qquad (\text{or } |\mathbf{s}| \leqslant 2\pi/D) \tag{1.22}$$

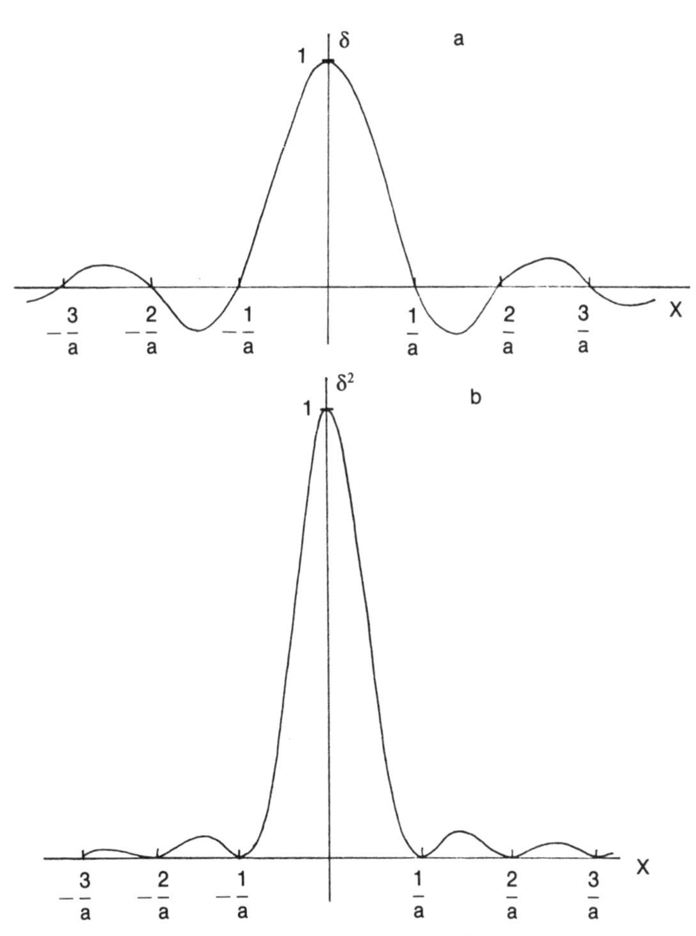

Figure 1.2. Functions $\delta(a, X)$ (a) and $\delta^2(a, X)$ (b).

This general relationship relates the linear characteristics of real and reciprocal spaces and represents the general definition of the resolution limits of any diffraction experiment.

If $a, b, c \to \infty$ we obtain

$$f_\infty(\mathbf{q}) = \varphi_0 \delta(X)\delta(Y)\delta(Z)$$

i.e., the whole scattering is concentrated at point $\mathbf{q} = 0$.

Thus the scattering intensity of a parallelepiped is given by the relationship (Figure 1.2)

$$I(\mathbf{q}) = \varphi_0^2 V^2 \delta^2(a, X)\delta^2(b, Y)\delta^2(c, Z)$$

1.3.2. Homogeneous Thin Plate

This case is obtained from the previous one (Section 1.3.1) when $c \to 0$. Here

$$\varphi(\mathbf{r}) = \varphi_0 \Pi(x, a)\Pi(y, b)\delta(z)$$

and

$$f(\mathbf{q}) = \varphi_0 S\delta(a, C)\delta(b, Y)$$

where $S = ab$ is the area of the plate. If $a, b \to \infty$ (infinitely thin plate), one obtains

$$f(\mathbf{q}) \to \varphi_0 \delta(X)\delta(Y)$$

i.e., the scattering amplitude differs from zero only along the straight line $X = 0$, $Y = 0$.

1.3.3. Homogeneous Thin Rod

This object is also a particular case of a homogeneous parallelepiped when $a, b \to 0$, so

$$\varphi(\mathbf{r}) = \varphi_0 \delta(x)\delta(y)\Pi(z, c), \qquad f(\mathbf{q}) = \varphi_0 c\delta(c, Z)$$

If $c \to \infty$ (infinitely long rod)

$$f(\mathbf{q}) = \varphi_0 \delta(Z)$$

i.e., the whole scattering is concentrated in the plane $Z = 0$.

1.3.4. Inhomogeneous Parallelepiped

We now consider the case when the distribution of matter can be represented by function $\varphi_\infty(\mathbf{r})$ but we are interested only in scattering by a body which is a part of the object. Hence one can write for the body

$$\varphi(\mathbf{r}) = \varphi_\infty(\mathbf{r})\phi_{\mathrm{p}}(\mathbf{r})$$

where

$$\phi_{\mathrm{p}}(\mathbf{r}) = \begin{cases} 1, & \mathbf{r} \text{ inside the body} \\ 0, & \mathbf{r} \text{ outside the body} \end{cases}$$

and represents the form function.

For a parallelepiped this function is given by equation (1.20) up to a constant factor. Then

$$f(\mathbf{q}) = \mathscr{F}[\varphi_\infty(\mathbf{r})] * \mathscr{F}[\phi(\mathbf{r})] = f_\infty(\mathbf{q}) * \psi(\mathbf{q})$$

i.e., the scattering amplitude is given by the convolution of the Fourier transform of the density distribution with the Fourier transform of the form factor $\psi(\mathbf{q})$ [for a parallelepiped given by equation (1.21)].

1.3.5. Periodic Set of Centers

For a one-dimensional unlimited set of identical equidistant point scattering centers with unit scattering amplitude,

$$\varphi_\infty(x) = \sum_{n=-\infty}^{\infty} \delta(x - na)$$

where a is the distance between centers, one may obtain

$$f_\infty(X) = \sum_{n=-\infty}^{\infty} \exp(2\pi i X n a)$$

This may be rewritten (see Cowley, 1975) in terms of the δ-function as

$$f_\infty(X) = \frac{1}{a} \sum_{n=-\infty}^{\infty} \delta\left(X - \frac{n}{a}\right) \tag{1.23}$$

i.e., the scattering amplitude represents an infinite set of diffraction δ-like peaks. The distance between the peaks is equal to $1/a$. We consider the finite set of centers which comprises $N + 1$ centers

$$\varphi_L(x) = \varphi_\infty(x) L(x)$$

where $L(x)$ differs from zero only for $x = na$ ($n = 0, 1, \cdots, N$). Then

$$f_N(X) = \frac{1}{a} \sum_{n=-\infty}^{\infty} \delta\left(X - \frac{n}{a}\right) * [Na\delta(Na, X)] \tag{1.24}$$

and the distances between peaks do not change but spread out (half-width equal to $1/Na$).

1.3.6. Spherically Symmetric Body

When $\varphi(\mathbf{r}) = \varphi(r)$ ($r = |\mathbf{r}|$), then using spherical coordinates in equation (1.8) one may write

$$f(s) = 4\pi \int_0^\infty \varphi(r) \frac{\sin(sr)}{sr} r^2 dr \tag{1.25}$$

where we recall that $s = 2\pi q = (4\pi \sin \theta)/\lambda$. This variable s is convenient for writing relationships in the case of isotropic scattering and most small-angle scattering problems. It is clear that for spherically symmetric bodies the scattering patterns also have spherical symmetry. For scattering by a homogeneous solid sphere

$$\varphi(r) = \Pi(r - R) = \begin{cases} 1, & r \leqslant R \\ 0, & r > R \end{cases} \tag{1.26}$$

and one may obtain the following important relationship by direct calculation:

$$f(s) = 3(sR)^{-3}[\sin(sR) - sR \cos(sR)] = \phi(sR) \tag{1.27}$$

where $f(0) = 1$ (normalized scattering amplitude).

These examples illustrate the application of Fourier transforms and convolution integrals to the solution of certain scattering problems. The relationships given above will be used later when analyzing certain general problems of small-angle scattering.

1.4. Scattering of X Rays by Atoms

The scattering of X rays by matter is determined almost entirely by the interaction of incident radiation with electrons. A part of nuclear scattering is negligible in that the mass of the nuclei is more than 10^3 times greater than the electron mass and the nuclear scattering energy is accordingly 10^6 times less than the electron scattering energy. We now consider scattering by the electron shell of an atom, but the case of a single free electron will be studied first.

For a plane monochromatic incident wave

$$\mathbf{E} = \mathbf{E}_0 \exp[i(\mathbf{k}_0\mathbf{r} - \omega t)]$$

where $\mathbf{E}$ represents the electric field intensity, classical wave theory (Landau and Lifshitz, 1985b; Bassani and Altarelli, 1983) gives the following expression for the amplitude of a scattered wave:

$$E_s = -E_0 \frac{e^2}{mc^2} \frac{1}{r} \sin \psi$$

where e and m are the charge and mass of the electron, c is the velocity of light, r the distance from the electron to the point of observation, and ψ the angle between the scattering beam and the direction of acceleration of the electron. The phase of scattered radiation relative to the phase of the incident beam changes by 180°.

Thus according to expression (1.1), the X-ray scattering length of a single electron is

$$b_x = \frac{e^2}{mc^2} \sin \psi = r_0 \sin \psi \tag{1.28}$$

where $r_0 = e^2/mc^2 = 2.82 \times 10^{-13}$ cm is the classical Thomson electron radius. Accordingly the scattering intensity, or differential cross section, by the electron in a given direction is

$$i_x = b_x^2 = r_0^2 \sin^2 \psi$$

For small-angle scattering $\psi = 90°$ and, moreover, the angular factors caused by polarization of incident and scattered beams are unimportant. Hence we write $i_x \approx r_0^2 = 7.95 \times 10^{-26}$ cm^2.

Calculation of the total scattering energy scattered by a single electron (the integral cross section), taking account of the angular factors, gives the following value:

$$\sigma_x = \frac{8\pi}{3} r_0^2 = 6.66 \times 10^{-25} \text{ cm}^2 \tag{1.29}$$

In calculating the scattering amplitude by an atom containing Z electrons, we introduce the density of the electron shell $\rho_a(\mathbf{r})$. This function represents the time-averaged probability of the electron distribution in an atom and may be accepted as a spherically symmetric function for free atoms. The scattering amplitude (or length) may then be written as

$$f_a(s) = r_0 L(\theta) \int_0^\infty \rho_a(r) \frac{\sin(sr)}{sr} 4\pi r^2 dr$$

where $L(\theta)$ is the polarization factor. For small scattering angles $L(\theta) \approx 1$ and

$$\int_0^\infty \rho_a(r) 4\pi r^2 dr = Z$$

in accordance with the definition of function $\rho_a(r)$. Thus, for small-angle scattering studies, the scattering length of an atom is just $f_a = r_0 Z$, or $f = f_a / r_0 = Z$ in electron units.

These relationships for the atomic scattering length are valid when the frequency of incident radiation is much greater than the frequency corresponding to the energy excitation of *K*-, *L*-, and other shells of the atom. The dispersion corrections play an important role only when the atom has an absorption edge close to the incident X-ray frequency. In general, the scattering amplitude depends on the wavelength of the X rays:

$$f(\lambda) = f_0 + f'(\lambda) + i f''(\lambda) \qquad (1.30)$$

i.e., $f(\lambda)$ is a complex function. The corrections f' and f'' are significant near the absorption edge (anomalous scattering, *Handbuch der Physik*, 1957) and may cause a relative alteration of $f(\lambda)$ up to 50%. Figure 1.3 indicates schematically the variations of f' and f'' near λ_k (the wavelength of the *K* absorption edge for an atom). These functions are connected by the Kramers–Kronig relationship

$$f'(\omega) = \frac{2}{\pi} \int_0^\infty \frac{\omega' f''(\omega') d\omega'}{\omega^2 - \omega'^2}, \qquad \omega = 2\pi c / \lambda$$

One may use the simplified expression

$$f'(\omega) = -\frac{1}{\pi} \int_{-\delta}^{\delta} f''(\omega + \varepsilon) \varepsilon^{-1} d\varepsilon$$

where the order of magnitude of δ is 10^{-2}. Values of f'' near the absorption edge are usually calculated using the experimentally mea-

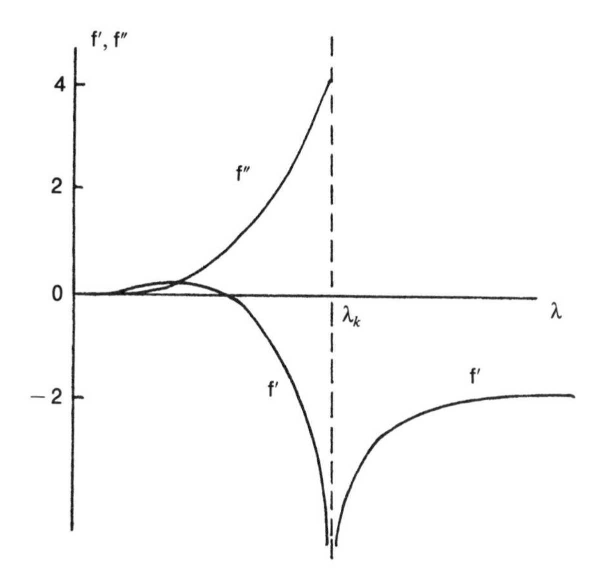

Figure 1.3. Real and imaginary parts of X-ray anomalous scattering factors f' and f'' as functions of wavelength λ_k (corresponding to the K_α absorption edge).

sured mass absorption factor (see below Section 1.6)

$$\mu_m = \frac{2N_A}{A}\lambda r_0 f'' \approx 337\,A^{-1}\lambda f'' \tag{1.31}$$

where μ_m is given in mm^{-1} per unit density, N_A is Avogadro's number, and A is the atomic mass of the resonant atom. The anomalous scattering phenomena are used to search scattering amplitude phases in structure analysis of crystals and, in a certain way, also in small-angle scattering studies (see Section 4.4.2).

1.5. Scattering of Thermal Neutrons by Nuclei

Contrary to X rays, neutrons are not scattered appreciably by electrons. They interact mainly with the nucleus. The wavelength of neutrons is determined by the well-known expression $\lambda = h/mv$, where m and v are the mass and velocity of a neutron and h is the Planck constant. When the neutron beam passes through the moderator at temperature T, the mean-square velocity of the neutrons $\langle v^2 \rangle$ satisfies the relationship

$$m\langle v^2\rangle/2 = 3k_B T/2$$

where k_B is the Boltzmann constant. The maximum intensity of the thermal neutron spectrum is at wavelength

$$\lambda_{max} = (h^2/3mk_B T)^{1/2} \tag{1.32}$$

Hence to obtain $\lambda_{max} = 1$, 3, or 10 Å, it is necessary to have the neutron moderator at 660, 73, and 6.6 K respectively.

Let us now study the main features of neutron scattering by the nucleus. We consider a nucleus with zero spin that scatters a monochromatic beam of neutrons. Then the scattering length is determined by the general expression (1.1), but the value $b = b_n$ depends on the character of the interaction of a neutron with the nucleus. Leaving aside the details of this interaction we shall now consider the most important conclusions and write some expressions, which are necessary for calculating the intensity of scattering by real substances (see, for example, Bacon, 1975; Nozik *et al*., 1986).

Owing to the fact that the dimensions of the nucleus are much smaller than the neutron wavelength, it follows that the scattering length for thermal neutrons will be isotropic, independent of the scattering angle, and given by a constant value b_n. The neturon scattering may be divided into two parts: "potential scattering" and "resonance scattering." In both cases we speak of elastic scattering, i.e., there is no change in the energy of the scattered neutrons.

The neutron scattering length b_n for the unique resonance level is determined by the Breit–Wigner relation

$$b_n = b_0 + \frac{\Gamma_n^{(r)}(2k)^{-1}}{(E - E_r) - (i/2)(\Gamma_n^{(r)} + \Gamma_a^{(r)})} \tag{1.33}$$

where b_0 corresponds to potential scattering, E is the energy of the incident neutron, E_r is the energy which would give resonance, $\Gamma_n^{(r)}$ is the width of the resonance for re-emission of a neutron with its original energy, $\Gamma_a^{(r)}$ is the width of the resonance for absorption, and $k = 2\pi/\lambda$ is the wave number of the neutron.

Expression (1.33) simplifies when the energy of the incident neutrons is far from the resonance level:

$$b_n = b_0 - \Gamma_n/2k_0 E_r$$

The scattering length of a neutron by the nucleus is a constant real value over a wide energy range. The value b_0 is close to the nucleus radius (of the order of magnitude 10^{-13} cm) and should increase as the cube root of A, where A is the mass number of the nucleus. Contrary to the X-ray scattering length f, the neutron scattering length b_n may have a positive or negative sign, depending on the contribution of the two terms in expression (1.33) (the imaginary part vanishes for the majority of nuclei). For many isotopes the second term in equation (1.33) has a significant value and for some isotopes it is greater than the first term. In the latter case, the scattering amplitude reverses its sign for nuclei such as ^{1}H, ^{7}Li, ^{48}Ti, ^{51}V, ^{53}Mn, and ^{62}Ni. The experimentally determined values of b_n for some elements and individual isotopes, as well as the X-ray scattering lengths f, are given in Table 1.3 in the same units.

The integral cross section of neutron scattering is expressed by a simpler formula than in X-ray scattering, namely

$$\sigma_n = 4\pi b_n^2$$

because, as we discussed above, the elastic neutron scattering is isotropic and does not depend on the scattering angle.

The previous relationships were derived for an unmovable nucleus. This situation is close to a solid state, while in gases the nuclei may be treated as free particles. The cross section for a free nucleus is slightly smaller than for a tightly bound one:

$$\sigma_f = \left(\frac{A}{A+1}\right)^2 \sigma_n$$

This decrease in σ is most striking for hydrogen atoms: in this case the integral cross section decreases by a factor of four. When the atomic mass A increases, the decrease gradually becomes negligibly small.

A simple equation for calculating the integral cross section becomes more complicated for nuclei having nonzero spin. The neutron itself possesses a spin of $\frac{1}{2}$, so if the scattering nucleus has spin I the two possible compound nuclei can be formed, having spins $I + \frac{1}{2}$ and $I - \frac{1}{2}$ with relative probabilities $\omega_+ = (I+1)/(2I+1)$ and $\omega_- = I/(2I+1)$. These two states have different scattering lengths b_+ and b_- respectively.

The total cross section is determined as

$$\sigma_n = 4\pi(\omega_+ b_+^2 + \omega_- b_-^2)$$

and a coherent part, which contributes to scattering and has the form

$$\sigma_{cn} = 4\pi(\omega_+ b_+ + \omega_- b_-)^2$$

The difference between the two expressions permits one to calculate the incoherent scattering $\sigma_{in} = \sigma_n - \sigma_{cn}$. Incoherent scattering by accumulation of atoms gives a continuous background over all scattering angles, not connected with a definite atomic or molecular configuration.

Table 1.3. Atomic Scattering Amplitudes b and Mass Absorption Coefficients μ/d of Some Elements and Isotopes for X-Ray and Thermal Neutron Scattering

Element or isotope	X rays				Thermal neutrons	
		μ/d (cm^2/g)				μ/d (cm^2/g)
	b (10^{-12} cm)	$\lambda = 1$ Å	$\lambda = 1.5$ Å	$\lambda = 2$ Å	b (10^{-12} cm)	$\lambda = 1.08$ Å
(1)	(2)	(3)	(4)	(5)	(6)	(7)
H	0.282	0.395	0.431	0.493	−0.374	0.11
D	0.282	0.395	0.431	0.493	0.667	0.0001
^{6}Li	0.846	0.316	0.672	1.36	0.18 + 0.025i	49
^{7}Li	0.846	0.316	0.672	1.36	−0.233	—
^{10}B	1.41	0.782	2.21	4.98	0.14 + 0.11i	128
B	1.41	0.782	2.21	4.98	0.54 + 0.02i	24
^{12}C	1.69	1.4	4.25	9.76	0.665	0.00015
^{14}N	1.97	2.21	6.95	16.1	0.94	0.048
O	2.26	3.31	10.6	24.6	0.580	0.00001
Na	3.10	8.52	27.8	64.3	0.36	0.007
Mg	3.38	11.0	35.6	82.0	0.52	0.001
Al	3.67	13.8	44.9	103	0.35	0.003
Si	3.94	17.3	56.0	128	0.42	0.002
P	4.23	21.2	68.5	156	0.51	0.002
S	4.51	25.6	82.4	186	0.28	0.0055
Cl	4.79	30.6	97.6	219	0.96	0.33
K	5.36	42.0	132	290	0.37	0.018
Ca	5.64	48.4	150	326	0.47	0.0037
Ti	6.20	63.5	193	409	−0.34	0.044
V	6.49	71.9	217	454	−0.038	0.033
Cr	6.77	80.9	242	500	0.363	0.021
Mn	7.05	89.9	265	62.7	−0.37	0.083
Fe	7.33	99.1	288	72.9	0.95	0.015
Co	7.61	108	310	84.3	0.28	0.21
Ni	7.90	118	42.2	97.1	1.03	0.028
Cu	8.18	128	48.9	112	0.76	0.021
Ge	9.02	157	69.8	160	0.819	0.011
As	9.31	166	77.0	176	0.64	0.02
Ag	13.25	67.3	203	423	0.60	0.20
Pt	22.0	—	187	380	0.95	0.015
Au	22.28	—	194	393	0.76	0.17
Hg	22.56	—	201	406	1.27	0.63
Pb	23.12	74.2	216	432	0.94	0.0003

For most nuclei the coherent scattering is stronger than the incoherent one. But in some cases the situation is the opposite. The hydrogen atom, having $I = \frac{1}{2}$, is a classical and important example. The scattering amplitude of a triplet state (having spin $\frac{3}{2}$) is equal to $+1.04 \times 10^{-12}$ cm and of a singlet state (having spin $\frac{1}{2}$) is -4.7×10^{-12} cm. Then $\sigma_n = 81 \times 10^{-24}$ cm^2, $\sigma_{cn} = 2 \times 10^{-24}$ cm^2, and $\sigma_{in} = 79 \times 10^{-24}$ cm^2.

The scattering length also depends on the isotopic composition of a sample, determined by the natural abundance of isotopes. The scattering length of the isotopic mixture is

$$b_{ni} = \sum_k \omega_k b_k$$

where ω_k is the mass part of the k type of isotope and b_k its scattering length. Let us consider an example of the influence of the isotopic composition. Natural metal Ni consists of three isotopes (each with zero spin), the scattering length is equal to $+1.44$ (^{58}Ni), $+0.3$ (^{60}Ni), and -0.87×10^{-12} cm (^{62}Ni), and the averaged value b_n for a natural element is 1.03×10^{-12} cm.

Finally, we mention magnetic scattering, which is usually less interesting from the small-angle viewpoint but has many applications when studying magnetic materials. Magnetic scattering is a result of interaction of the magnetic moment of a neutron with the moment carried by atoms in a magnetic substance. Each magnetic atom has a magnetic scattering length p, which may be calculated in terms of the expression

$$p = \frac{e^2\gamma}{2mc^2} gJf = 0.27gJf\ 10^{-12}\ \text{cm}$$

The numerical factor is determined by e and m, i.e., the charge and mass of the electron, and γ is the magnetic moment of the neutron expressed in nuclear magnetons. The other quantities are the total quantum number J, the splitting factor g, and the atomic form-factor f, which depends on the scattering angle like the atomic factor for X-ray scattering by atoms.

In many cases the orbital contribution to the magnetic moment of an atom is negligible if compared to the spin contribution. Then, the expression for p simplifies to

$$p = 0.54\ Sf \times 10^{-12}\ \text{cm}$$

where S is the spin quantum number. If the incident beam of neutrons is unpolarized, there is no interference between nuclear and magnetic scattering, i.e., the corresponding intensities are additive. As a rule, magnetic scattering is normally essentially weaker than nuclear scattering and is used to determine the spin density in paramagnetic, ferro-, ferri, and antiferromagnetic materials.

1.6. Absorption of X Rays and Neutrons

When an incident wave (X rays, neutrons) passes through matter, the intensity of the beam decreases. This weakening is a result of two effects: a change in the direction of the scattering particle, i.e., a scattering phenomenon (see Section 1.1), and the "true" absorption, which is associated with the "disappearance" of the photons or neutrons. True absorption of X rays results from ionization of an atom owing to the intrinsic photoeffect. Then the atom returns to its normal state by a set of various transitions (Blokhin, 1957) in which we are not interested here. Analogous effects occur at true absorption of neutron beams and their weakening on penetration into matter.

We shall consider in a general way a decrease in the intensity of the incident beam in terms of its interaction with a single atom. The integral scattering cross section of an atom is a part subtracted from the incident beam. An analogous cross section may also be determined to calculate the true absorption by atoms. Then, if scattering and true absorption effects are independent, the absorption cross section of an atom may be written as the sum of two cross sections,

$$\mu_a = \sigma_a + \tau_a$$

where σ_a and τ_a are partial cross sections of the scattering and true absorption, respectively.

The decrease of the incident beam in a macrosample may then be represented by the following expression:

$$\left(\frac{dW}{d\Omega}\right)_x = \left(\frac{dW}{d\Omega}\right)_0 \exp\left(-\mu_a \frac{dN_A}{A} x\right) = \left(\frac{dW}{d\Omega}\right)_0 \exp(-\mu x)$$

where $(dW/d\Omega)_0$ is the incident beam flow density, d the substance density, x the distance from the sample surface, and $\mu = \mu_a dN_A/A$ the linear absorption factor. The mass absorption factor μ_m is introduced in terms of the equation

$$\left(\frac{dW}{d\Omega}\right)_m = \left(\frac{dW}{d\Omega}\right)_0 \exp(-\mu_m m)$$

where $m = xd$ is the respective substance mass up to point x. The mass absorption factor $\mu_m = \mu N_a/A$ is determined by the atomic properties of the sample and does not depend on the aggregate state of a sample. It is very useful for many practical calculations. Certainly, linear and mass absorption factors may be represented as the sum of two parts determined by scattering and true absorption:

$$\mu = \sigma_l + \tau_l, \qquad \mu_m = \sigma_m + \tau_m$$

We now consider the case of X-ray propagation and write expressions for both parts of the absorption. When relation (1.29) is applied, the scattering absorption factors are derived as follows:

$$\sigma_{lx} = \sigma_{ax} \frac{dN_A}{A} = \frac{8\pi}{3} r_0^2 Z \frac{dN_A}{A}$$

and

$$\sigma_{mx} = \frac{8\pi}{3} r_0^2 Z \frac{N_A}{A}$$

A very interesting result follows: as Z/A is approximately equal to 0.5 for almost all the atoms, $\sigma_{mx} \approx 0.2\ \text{cm}^2/\text{g}$, i.e., the latter value is a constant that is virtually independent of both the composition of the scattered object and the wavelength of the X rays.

This is not the case for true absorption for which the approximate expression may be written as (Blokhin, 1957)

$$\tau_{ax} \approx CZ\lambda^3 \qquad (1.34)$$

where C varies slightly, depending on the wavelength. It follows that the absorption of X rays increases rapidly with an increase in the atomic number of the element and the wavelength. These very simple expressions determine in practice the potentialities of using X rays for small-angle research (see Section 8.3). It is noteworthy that to use X rays with a wavelength greater than 2 Å in practice is rather difficult owing to the great absorption of this radiation. In small-angle studies, the absorption of incident and scattering beams is determined mainly by true absorption since τ_{mx} is significantly greater than σ_{mx}. This is not so for neutron diffraction, because for most elements and wavelengths the true absorption is of the same order as, or less than, the apparent absorption due to scattering. It should be noted that, with some exceptions, the absorption cross sections for neutrons are usually much smaller than the respective values for X rays. The values of the mass absorption coefficients for X rays and neutrons are given in Table 1.3.

1.7. Conclusion

In this chapter the reader can find the minimum information concerning the general theory of elastic scattering and the interaction of X rays and neutrons with matter necessary for practical application of the small-angle scattering technique for structure investigations. The general problems of X-ray and neutron scattering are discussed on the basis of a single approach. At the same time, some specific features pertaining to the interaction of X rays and neutrons with matter are given to stress the difference between these two main sources for small-angle structure studies. The material given here should by no means be regarded as comprising a reference book on diffraction theory. The books of Hosemann and Bagchi (1962), Vainshtein (1966, 1981), and Cowley (1975) are recommended for those interested in more detailed information on the mathematical apparatus of scattering theory. The *International Tables of X-ray Crystallography*, edited by McGillavary and Rieck (1983), is a most comprehensive reference book on X-ray and neutron scattering.

2

General Principles of Small-Angle Diffraction

X-ray and neutron scattering can be used to investigate a wide range of substances, such as poly- and single crystals, metals and alloys, amorphous solids and liquids, synthetic polymers, and biological macromolecules, among others. The properties of the diffraction patterns for objects with different ordering are considered in Section 2.1. When analyzing the structure of the majority of highly disperse systems, it is sufficient to take into account the very-low-angle diffraction region — "small-angle scattering." Section 2.2 presents the principal relationships for the case of small-angle diffraction.

The model in which a disperse system is regarded as a matrix (such as a homogeneous solvent) with particles embedded in it (regions occupied by the other phase) is of general value in small-angle scattering studies. Scattering by such a model is described in Section 2.3. It appears that if all the particles in "solution" are identical, the total scattering intensity is proportional to the scattering intensity of a particle, averaged over all orientations. Particle scattering is examined in Section 2.4, while scattering by systems for which the "matrix-particle" model does not hold is dealt with in Section 2.5. The connection between the structure of an object and its scattering intensity in small-angle investigations of disperse systems is then described schematically.

2.1. Scattering by Objects with Different Ordering

The degree of ordering in the structure of an object strongly affects the diffraction pattern as well as the possibilities of extracting structural information from it. Generally speaking, the less ordered an object, the less informative the scattering pattern. Since various types of structure can be investigated by small-angle scattering, we shall examine aspects of the diffraction on different crystalline and noncrystalline objects before deriving the basic equations of small-angle diffraction. The problems arising in the course of diffraction studies of different objects will be dealt with explicitly during this examination.

It was seen in Chapter 1 that the amplitude of elastic scattering by any ensemble containing N atoms can be written as

$$A(\mathbf{q}) = \sum_{i=1}^{N} b_i(\mathbf{q})\exp(2\pi i\mathbf{q}\mathbf{r}_i) \tag{2.1}$$

where $b_i(\mathbf{q})$ and $\mathbf{r}_i$ are the scattering lengths and coordinates of the atoms. The main problem of structure studies is to restore the spatial arrangement of atoms from the diffraction pattern, namely, the scattering intensity distribution $I(\mathbf{q}) = A(\mathbf{q})A^*(\mathbf{q})$. However, this task is of practical importance only when the range of summation in equation (2.1) is sufficiently small. Actually, in an experiment the number of atoms in the sample is about 10^{20} or more, so one can hope to restore the atomic coordinates only for highly ordered samples. Owing to the lack of ordering, only some general features of the structure can be determined.

2.1.1. Single Crystals

The atomic positions in the unit cell suffice to describe the structure of a single crystal. The whole structure can be constructed by periodic repetitions of the cell in space, the cell parameters a, b, and c determining corresponding periodicities along the x, y, and z axes. Therefore, the periodicity of function $\rho(\mathbf{r})$ should be taken into account in the general equation for the scattering amplitude

$$A(\mathbf{q}) = \int \rho(\mathbf{r})\exp(2\pi i\mathbf{q}\mathbf{r})d\mathbf{r} \tag{2.2}$$

where ρ denotes the scattering density (scattering length density) distribution, as in X-ray structure analysis. It was shown in Section 1.3.5 that this approach allows $A(\mathbf{q})$ to assume nonzero values only for definite values of $\mathbf{q}$, namely

$$\mathbf{q} = \mathbf{H}_{hkl} = h\mathbf{a}^* + k\mathbf{b}^* + l\mathbf{c}^*$$

where $\mathbf{a}^* = [\mathbf{bc}]/\Omega$, $\mathbf{b}^* = [\mathbf{ca}]/\Omega$, and $\mathbf{c}^* = [\mathbf{ab}]/\Omega$; h, k, and l are integers and $\Omega = (\mathbf{a}[\mathbf{bc}])$ is the volume of the unit cell.

Thus the scattering amplitude in this case is represented by the discrete set

$$A(\mathbf{q}) = \sum_{hkl} F_{hkl}\delta(\mathbf{q} - \mathbf{H}_{hkl})$$

where

$$F_{hkl} = \Omega^{-1} \int \rho(\mathbf{r})\exp(2\pi i \mathbf{H}_{hkl}\mathbf{r})d\mathbf{r}$$

here, integration is carried out within a cell.

The number of terms in sum (2.1) for single crystals is therefore determined by the number of atoms in the independent part of the unit cell. The latter is of order 10^1–10^2 for inorganic crystals and 10^3–10^4 for biological crystals. Thus the problem of the atomic-structure restoration can be formulated and solved in this case.

This is the problem encountered in single-crystal structure analysis. The main difficulty in the restoration of $\rho(r)$ is to determine the phases φ_{hkl} of the complex values $F_{hkl} = |F_{hkl}|\exp(\varphi_{hkl})$ while their moduli are measured experimentally. There are several approaches to overcome this difficulty [direct methods, Patterson functions, isomorphous replacements, etc.; see, for example, Vainshtein (1981)]. Resolution up to 0.5 Å can be achieved for inorganic crystals, and up to 1.0 Å for biological crystals.

2.1.2. One-Dimensional Periodic Structures

We consider the system, periodic in one dimension, say z, so that $\rho(x, y, z) = \rho(x, y, z + c)$. Then the diffraction pattern will be represented by a set of layer lines spaced $1/c$ apart, the scattering amplitude along these lines being a continuous function due to lack of periodicity along the x and y axes:

$$A(\mathbf{q}) = A_l(X, Y) = \int_{-\infty}^{\infty}\int_{-\infty}^{\infty}\int_{0}^{c} \rho(x, y, z)\exp[2\pi i(xX + yY + lz/c)]dz\,dy\,dx \quad (2.3)$$

Hence the soution of the phase problem for the meridional amplitudes $A_l(0, 0)$ enables one to construct the structure projection on the z axis,

$$\sigma(z) = c^{-1} \sum_{l=-\infty}^{\infty} A_l(0, 0)\exp(-2\pi i l z/c)$$

The projection on the plane $z = 0$ can be calculated from the scattering amplitude along the zero-layer line,

$$\sigma(x, y) = \int_{-\infty}^{\infty} \int_{-\infty}^{\infty} A_0(X, Y)\exp[-2\pi i(xX + yY)]dXdY$$

We therefore see that the restoration of the atomic structure is inapplicable in this case (integration over x and y is performed over the whole sample). Moreover, the methods of phase determination developed in single-crystal structure analysis cannot be applied to a continuous function $A_l(X, Y)$, essentially hampering restoration of the projections of function $\rho(x, y, z)$.

2.1.3. Cylindrically Symmetric Objects

Samples with a preferred direction (z) are frequent in research. Different polymeric and liquid crystalline systems, where molecules are packed in a parallel order with arbitrary angular (azimuthal) orientation, represent such objects. In this case we have function $\rho(\mathbf{r})$, and consequently $I(\mathbf{s})$, with statistically cylindrical symmetry.

Cylindrical coordinates r, ψ, z ($x = r\cos\psi$, $y = r\sin\psi$, $z = z$) are convenient to describe the diffraction in this case. In terms of these coordinates, equation (2.2) assumes the form

$$A(R, \Psi, Z) = \int_{-\infty}^{\infty} \int_{0}^{2\pi} \int_{0}^{\infty} \rho(x, y, z)\exp\{2\pi i[Rr\cos(\psi - \Psi) + zZ]\}r\,dr\,d\psi\,dz \quad (2.4)$$

Objects with a preferred direction usually possess a periodic arrangement of the molecules along this direction. As we already know, this fact forces the scattering intensity to be distributed at the layer lines with $Z = l/c$, where c is the periodicity on the z axis. Taking into account the well-known expression for the Bessel functions,

$$2\pi i^n J_n(u) = \int_0^{2\pi} \exp(iu\cos\varphi + in\varphi)d\varphi$$

where $\varphi = 2\pi rR$, one can reduce equation (2.4) to

$$A_1(R, \Psi, Z) = \sum_{n=-\infty}^{\infty} A_{nl}(R)\exp(in\varphi)$$

where

$$A_{nl}(R) = \exp(in\pi/2)\int_0^{\infty} \rho_{nl}(r)J_n(2\pi rR)2\pi r\,dr$$

with

$$\rho_{nl}(r)=\frac{1}{2\pi}\int_0^{2\pi}\int_0^c \rho(r,\psi,z)\exp[-i(n\psi-2\pi lz/c)]d\psi dz$$

These relations are used, for example, in the analysis of chain molecules. The helical molecular structure specifies selection rules for the Bessel functions. Their diffraction patterns have an appearance of "inclined crosses" owing to displacements of the corresponding Bessel-function maxima (Vainshtein, 1966).

For cylindrically symmetric objects the ψ dependence vanishes, thus substantially simplifying the structure interpretation. Hence the inverse transformation of the zero-layer line amplitude

$$\rho_0(r)=\int_0^\infty A_0(R)J_0(2\pi rR)dR$$

gives the radial scattering density distribution in the plane $z=0$. The cylindrical Patterson function

$$P(r,z)=2\int_0^\infty\int_0^\infty I(R,Z)J_0(2\pi rR)\cos(2\pi zZ)2\pi R\,dR\,dZ$$

may also provide some information on the structure. It is noteworthy, however, that determination of the molecular structure of statistically, cylindrically symmetric systems is rather difficult, unless additional information on the molecules is available. As a rule, the main problem while investigating such systems is to analyze the features of the molecular packing (shear or rotational displacement; see Deas, 1952), rather than to determine the structure of the molecules themselves.

2.1.4. *Isotropic Systems*

In the absence of long-range order in the sample (isotropic objects, such as amorphous solids, powders, liquids, and particle solutions), all the atoms should be taken into account in equation (2.1) for the scattering amplitude. There are two considerations as to why the reverse task of obtaining the atomic structure is pointless in this case. First, the structure with such a great number of independently placed atoms can by no means be restored using available experimental setups and data-treatment procedures. Second, there is no need for such a detailed description of these systems: only their general features (e.g., the dimensions of inhomogeneities or the magnitude of statistical density fluctuations) are to be determined for a clear understanding of their properties. It is the isotropy of the samples that makes the scattering intensity also isotropic (i.e., the intensity depends only on the magnitude of the scattering vector **s**). Using spherical coordinates, one can express the radial Patterson function with the aid of the Fourier transform of the scattering intensity

$$P(r) = \frac{1}{2\pi^2} \int_0^\infty I(s) \frac{\sin(sr)}{sr} s^2 ds$$

This function yields certain conclusions as to typical distances between the atoms in liquids or gases, the dimensions of clusters or pores in amorphous solids, or the particle dimensions in solutions. It is evident that the possibilities of structural studies of such systems are rather limited. However, we shall see below that in a number of cases (such as two-phase monodisperse systems) more detailed analysis of their constitution is possible.

2.2. Small-Angle Scattering by Disperse Systems

It is clear from the results in the previous section that the possibilities for structural interpretation of the diffraction pattern diminishes with disordering of the object under investigation. By studying low-ordered samples, the spatial resolution that can be reached is much worse than atomic sizes and interatomic distances. In crystal-structure analysis, the measured diffraction region should be as wide as possible while for low-ordered systems it is sufficient to take into account a central part of the diffraction pattern (the region of small scattering vectors). This leads to the "small-angle scattering" method, which serves to study inhomogeneities of large sizes (in comparison with interatomic distances) in various disperse systems.

2.2.1. Scattering Intensity by a Disordered Object

Study of disordered samples by diffraction methods involves measuring the scattering intensity, which is a function of the microscopic state of a system. Therefore, the system should be discribed using some general statistical function of atomic positions as well.

We let the system under investigation consist of N identical atoms (or, more generally, N identical atomic sets: molecules, cells, or particles) with scattering length (amplitude) $f(\mathbf{s})$. Thus, from equation (2.1), the intensity of scattering by the whole ensemble is given by

$$I(\mathbf{s}) = \sum_{i=1}^{N} \sum_{j=1}^{N} f_i(\mathbf{s}) f_j(\mathbf{s}) \exp[i\mathbf{s}(\mathbf{r}_i - \mathbf{r}_j)] \quad (2.5)$$

where $\mathbf{r}_i$ and $\mathbf{r}_j$ are the coordinates of atoms or vectors describing the positions of the sets. Separating the terms with $i = j$ and averaging over the ensemble, one has

$$\langle I(\mathbf{s})\rangle = \left\langle \sum_{i=1}^{N} f_i^2(\mathbf{s}) \right\rangle + \left\langle \sum_{i \neq j}\sum f_i(\mathbf{s}) f_j(\mathbf{s}) \cos(\mathbf{s}\mathbf{r}_{ij}) \right\rangle$$

$$= \langle N \rangle \langle f^2(\mathbf{s})\rangle + \langle f(\mathbf{s})\rangle^2 \left\langle \sum_{i \neq j}\sum \cos(\mathbf{s}\mathbf{r}_{ij}) \right\rangle \tag{2.6}$$

where $\mathbf{r}_{ij} = \mathbf{r}_i - \mathbf{r}_j$. The average value $\langle N \rangle$ is written here instead of N, because the number of atoms (particles) can also fluctuate in the general case. We suppose the system to be isotropic, i.e., the atomic arrangement does not depend on the direction. In order to calculate the mean value of the double sum in equation (2.6), a probability function p_{ij} is convenient. The function gives the probability of finding a particle in volume dv_i pointing along vector $\mathbf{r}_i$, and simultaneously another particle in volume dv_j pointing along $\mathbf{r}_j$. Averaging $\cos(\mathbf{sr})$ over all orientations yields

$$\langle \cos(\mathbf{sr})\rangle = \int_0^{\pi} \cos[sr\cos(\varphi)] \frac{\sin(\varphi)}{2}\, d\varphi = \frac{\sin(sr)}{sr} \tag{2.7}$$

where φ is the angle between vectors $\mathbf{s}$ and $\mathbf{r}$, while $\sin(\varphi)d\varphi/2$ represents the probability of the angle lying inside the interval $(\varphi, \varphi + d\varphi)$. Thus we obtain

$$\left\langle \sum_{i \neq j}\sum \cos(\mathbf{s}\mathbf{r}_{ij}) \right\rangle = \iint_{V_0} \frac{\sin(sr_{ij})}{sr_{ij}}\, p_{ij}\, dv_i\, dv_j$$

where V_0 is the irradiated volume of a sample.

It is more convenient to use the normalized function $P_{ij} = p_{ij}/v_1^2$ instead of p_{ij}, where $v_1 = V_0/N$ is the mean volume per particle. On substituting $P(r) = 1 - [1 - P(r)]$, one can write

$$\langle I(\mathbf{s})\rangle = \langle N \rangle \langle f^2(\mathbf{s})\rangle + \langle f(\mathbf{s})\rangle^2 \iint_{V_0} \frac{\sin(sr_{ij})}{sr_{ij}} \frac{dv_i}{v_1} \frac{dv_j}{v_1}$$

$$+ \langle f(\mathbf{s})\rangle^2 \iint_{V_0} [1 - P(r_{ij})] \frac{\sin(sr_{ij})}{sr_{ij}} \frac{dv_i}{v_1} \frac{dv_j}{v_1} \tag{2.8}$$

Let us examine equation (2.8). It can be easily seen that the first integral on the right-hand side corresponds to scattering by the particle with volume V_0 and constant density. This term is negligible for scattering vectors with $s > 2\pi/D_0$, where $D_0 \approx V^{1/3}$ (see Section 1.3.1), so it cannot be measured experimentally.

As regards the second integral, one should bear in mind that, when viewed from the standpoint of the normalization of $P(r_{ij})$, this function tends to unity as r_{ij} tends to infinity (the positions of atoms are practically independent for $v_1^{1/3} \ll r_{ij} \ll V_0^{1/3}$ due to the absence of long-range order). Thus one can write

$$\int_{V_0} \frac{dv_j}{v_1} \int_{V_0} [1 - P(r_{ij})] \frac{\sin(sr_{ij})}{sr_{ij}} \frac{dv_i}{v_1} \approx \langle N \rangle \int_0^\infty [1 - P(r)] \frac{\sin(sr)}{sr} \frac{4\pi r^2}{v_1} dr$$

where the first integral equals V_0/v_1 and spherical coordinates have been used in the second one. Therefore, for all $s > 2\pi/D_0$,

$$I(\mathbf{s}) = \langle N \rangle \left\{ \langle f^2(\mathbf{s}) \rangle - \frac{\langle f^2(\mathbf{s}) \rangle}{v_1} \int_0^\infty [1 - P(r)] \frac{\sin(sr)}{sr} 4\pi r^2 dr \right\}$$

If the atomic (the particle's) scattering amplitude is assumed to be spherically symmetric (or nearly so), then $\langle f^2(\mathbf{s}) \rangle = \langle f(\mathbf{s}) \rangle^2$; hence we can finally write

$$I(s) = \langle N \rangle F^2(s) \left\{ 1 - \frac{1}{v_1} \int_0^\infty [1 - P(r)] \frac{\sin(sr)}{sr} 4\pi r^2 dr \right\} \tag{2.9}$$

where $I(s)$ and $F^2(s)$ represent averaged scattering intensities of the system and an atom (particle), respectively. This is a classical equation, derived by Zernike and Prins (1927). It is quite important and expresses the scattering intensity as a function of the form factor of an atom (particle) as well as of the statistical function governing their arrangement. Thus, for example, one can search for $P(r)$ at given $F^2(s)$ (or vice versa) instead of making pointless attempts to restore the atomic coordinates in the system.

We have now examined a particular case of a disordered system, namely, an isotropic ensemble of identical atoms (particles). Similar equations can be derived for nonisotropic systems and nonidentical particles, although they would depend on the special features of the given system. It should be noted, however, that equation (2.9) is very important, being widely applicable in practice. In particular, for the case of nonidentical particles, function $F^2(s)$ can be regarded as obtained by averaging contributions of the scattering intensities of individual atoms (particles).

2.2.2. Scattering at Low Angles

It was shown in the previous section that, while studying disordered objects, there is a peak corresponding to finite dimensions of the sample at the beginning of a scattering curve. This is also the case for ordered samples: in single-crystal structure analysis, every reflection widening, including the zero reflex, is determined by the crystal size; see Section 1.3.5. Such peaks are, however, unobservable in experiments using wavelength about 1 Å, because they are obscured by the primary bean (the samples have macroscopic dimensions).

On the other hand, it is seen from equation (2.9) that scattering by a disordered system depends strongly on the form factors of the scattering inhomogeneities (atoms, cells, and particles). We consider the case when the sizes of the inhomogeneities substantially exceed the wavelength, but are sufficiently small on a macroscopic scale. As noted in Section 1.3.1, to obtain enough information on the inhomogeneities with typical size D, one should measure the scattering curve up to approximately $s = 2\pi/D$. It is easily seen that for "colloidal" inhomogeneities (size 10^4–10^1 Å) to be investigated, the scattering region up to $s = 0.0006$–0.6 Å^{-1} should be measured, that is, at wavelength $\lambda = 1.54$ Å (the X-ray Cu K_α line) the angular region up to $2\theta = 0.008$–8^0. A variety of highly disperse systems possesses such inhomogeneities, so that measuring the central part of the diffraction pattern ("small-angle scattering") is an appropriate method for studying their inner structure. Here we list the objects investigated successfully by the method:

1. *Biologically active substances*. Biological macromolecules and their complexes (proteins, nucleic acids, viruses, membranes, etc.) are studied by small-angle scattering, in water–salt solutions, i.e., in their native state.

2. *Polymeric materials*. The properties of chain packing and some general molecular parameters of natural and synthetic polymers can be investigated by the method both in the solid state and in solution.

3. *Amorphous solids and liquids*. Small-angle scattering enables one to analyze thermodynamic parameters and the cluster structure of liquids, as well as fluctuations and phase decompositions in glasses and other amorphous materials.

4. *Metals, alloys, and powders*. Different features of the disperse structure of solids, such as defects in single crystals, miscibility gap limits, and phase separation processes in alloys, grain or pore dimensions in powders and porous materials, magnetic domains in ferromagnetics, and many others, may be examined using this method.

Various characteristics of disperse systems can thus be studied with small-angle scattering techniques. Two important points are to be noted. First, since atomic-scale inhomogeneities do not affect the small-angle part of the scattering curve, the method allows determination of the disperse structure of an object irrespective of the substance it consists of (e.g., both the cluster structure of liquids and the porosity of graphites require the same methods of data analysis). Second, in most cases an object can be regarded as a matrix (solution, ground solid phase, or gas surrounding) with intrinsic inhomogeneities (particles, pores, or defects). It is worth noting, however, that small-angle scattering can be applied to nonparticulate systems as well. All systems possessing inhomogeneities of colloidal size (from 10^1 to 10^4 Å) can in principle be investigated by the method.

2.3. Particle Solutions

The representation of an object as a compound consisting of particles and ground phase (matrix) proves very helpful in small-angle scattering. Below we shall deal frequently with such a representation. Therefore, while deriving the basic relationships, the investigated inhomogeneities will be denoted as "particles" and the ground phase as "solvent." The meaning of these terms depends on the substance under investigation (see Section 2.2.2).

2.3.1. Monodisperse and Polydisperse Systems

In deriving equation (2.9) we assumed that all the particles are identical. There was the monodisperse system, for which the explicit connection between particle form factor $F^2(s)$ and scattering intensity $I(s)$ could be obtained. For polydisperse systems, where particles may have different shapes, sizes, and compositions, such an expression does not hold. It has already been noted that the averaging caused by the fact that the particle form factors may differ from each other would be included in the overall averaging over the ensemble in equation (2.9).

We now consider the polydisperse system. It is appropriate to describe the particles by a certain effective size R. A function $D_N(R)$ can then be introduced in such a way that $D_N(R)dR$ corresponds to the number of particles whose sizes fall within the interval $(R, R + dR)$. The contribution of these particles to the averaged form factor is thus given by the product $\langle F_0^2(s, R)\rangle m^2(R)D_N(R)dR$, where $\langle F_0^2(s, R)\rangle$ is the averaged, normalized form factor of the particles of size R $[F_0(0, R) = 1]$, while $m(R)$ connects the chosen effective size with the full scattering length of a particle.

The total averaged form factor may therefore be written as

$$\langle F^2(s)\rangle = \int_0^\infty \langle F_0^2(s, R)\rangle m^2(R) D_N(R) dR$$

In the particular case of an isotropic polydisperse system, when all the particles are of different size but identical shape, one obtains

$$F^2(s) = \int_0^\infty i_0(s, R) m^2(R) D_N(R) dR \qquad (2.10)$$

where $i_0(sR)$ is averaged over all orientations and the normalized form factor of a particle.

One can easily see from here that while for monodisperse systems it is possible to search for the structure of a particle, for polydisperse systems this problem can be stated only when function $D_N(R)$ is known. There is another task that is more important in this research, namely to search for $D_N(R)$ assuming $i_0(sR)$ given. It is evident that monodisperse systems allow much more detailed analyses of the structure of an object than polydisperse ones. On the other hand, polydispersed objects are frequently dealt with in small-angle investigations. (Almost the only, though very important, practical case of a perfectly mondisperse system is represented by the solutions of biological macromolecules.)

2.3.2. Conception of Contrast

In our discussion until now solvent scattering has been neglected (the particles under investigation were supposed to be in vacuum). As a matter of fact, the latter are placed into a solvent with a nonzero mean value of the scattering density ρ_s. By adopting the scattering density of a particle equal to $\rho(\mathbf{r})$ (the case of a monodisperse system is examined for the sake of simplicity), an object can be regarded as a superposition of the irradiated volume V_0, filled with pure solvent, and N particles with scattering density approximately equal to

$$g(\mathbf{r}) = \rho(\mathbf{r}) - \rho_s \qquad (2.11)$$

(the "excess scattering density"). Here, we believe that the structure inhomogeneities of the solvent are essentially smaller in size than the particle structure inhomogeneities, so that $g(\mathbf{r})$ can be obtained from $\rho(\mathbf{r})$ by simple subtraction of ρ_s. Since the sample volume V_0 is much greater than the particle volume V, one can omit the cross term and express the measured scattering density as a sum of two terms, namely, scattering by an ensemble of N particles with excess density $g(\mathbf{r})$, and scattering by pure solvent. Thus, to separate the scattering from particles, measurements on the solution and pure solvent are performed followed by subtraction of the latter from the former. The above-derived expressions hold for the difference scattering curve, substituting $\rho(\mathbf{r})$ by $g(\mathbf{r})$.

It is obvious, however, that the difference curve may alter substantially depending on the relation between $\rho(\mathbf{r})$ and ρ_s. The contrast $\Delta\rho$, namely, the difference between the mean particle density and the solvent density

$$\Delta\rho = \langle g(\mathbf{r})\rangle = V^{-1}\int_V [\rho(\mathbf{r}) - \rho_s]d\mathbf{r} = \bar{\rho} - \rho_s \tag{2.12}$$

is a very convenient parameter to describe this relation. One can easily see that at high values of $\Delta\rho$ ($\Delta\rho \gg \rho_s$ or $\Delta\rho \ll \rho_s$) ("strong contrast") the particles (their shapes) are seen clearly, "contrastively," on the solvent background. On the other hand, at low contrast ($\bar{\rho} \approx \rho_s$, $\Delta\rho \approx 0$) the difference curve is determined by particle inhomogeneities rather than shapes, as if the particles are "smeared." Contrast is one of the most important sample parameters in small-angle studies. The possibility of choosing the contrast (e.g., by changing the solvent density) may provide additional structural information, as described in Chapter 4.

2.3.3. Concentrated and Dilute Systems

General expression (2.9) shows that the scattering intensity consists of two terms: the first depends only on particle structure, while the second is also a function of their statistical distribution $P(r)$ and the volume v_1 per particle. Moreover, function $P(r)$ itself should also depend on v_1: $P(r) = P(r, v_1)$. If equation (2.9) is expressed in the form

$$I(s) = \langle N\rangle\langle F^2(\mathbf{s})\rangle - \langle F(\mathbf{s})\rangle^2 \frac{\langle N\rangle^2}{V}\int_0^\infty [1 - P(r, v_1)]\frac{\sin(sr)}{sr}4\pi r^2 dr \tag{2.13}$$

one can see that the first term corresponds to independent scattering from $\langle N\rangle$ particles, while the second term, proportional to $\langle N\rangle^2$, is caused by interparticle interference.

The interference effects are unlikely to be treated rigorously, the main difficulty being just the dependence of P on v_1. It was stressed by Guinier and Fournet (1955) that contradiction is inevitable in those approaches where P is assumed to be independent of v_1, because when studying $I(s)$ as a function of concentration one should not believe that one of the most important functions describing $I(s)$ is independent of this parameter.

For single-phase liquids and gases, it is possible to describe interparticle interference with the help of an equation of state (see Section 2.5.1). Debye (1927) calculated $I(s)$ for the simplest model of nonpenetrating solid spheres. Taking $P(r, v_1)$ in the form

$$P(r, v_1) = P(r) = \begin{cases} 0, & 0 \leqslant r \leqslant 2R \\ 1, & r > 2R \end{cases}$$

where R is the radius of a sphere, one obtains

$$I(s) = \langle N \rangle \phi^2(sR) \left[1 - \frac{8V}{v_1} \phi(2sR) \right]$$

where $V = 4\pi R^3/3$ is the sphere volume and $\phi(sR)$ is its normalized scattering amplitude [equation (1.27)]. This relation describes interference effects satisfactorily at low concentrations. At high concentrations, however, the already-mentioned contradiction manifests itself clearly, so that for $V/v_1 > \frac{1}{8}$ pointless negative values of $I(0)$ are obtained. It should be noted that no correct treatment of interference effects based on equation (2.13) has yet been presented. There are also other approaches to this problem [in particular, the "surrounding function" of Porod (1972)], but they only give qualitative results.

The case when the mean sample volume per particle is much greater than its own volume ($v_1 \gg V$) is of great importance and is very favorable for studying the structure of particles in solution. Here, the second term on the right-hand side of equation (2.9) can be neglected leading to

$$I(s) = \langle N \rangle \langle F^2(\mathbf{s}) \rangle \tag{2.14}$$

This classical result states that for a dilute system of identical particles the small-angle scattering intensity is proportional to the scattering of a particle averaged over all orientations. Hence the shape as well as the inner structure of the particle in question can be investigated in this case. Preparation of dilute monodisperse solutions followed by small-angle scattering experiments is therefore one of the most significant ways of studying the structure of noncrystalline objects. Structure analysis of biological macromolecules and their complexes is the branch where this technique has been applied most effectively.

If the particles are of the same shape but different sizes, then equation (2.10) yields

$$I(s) = \langle N \rangle \int_0^\infty i_0(sR) \cdot D_N(R) \cdot m^2(R) dR \tag{2.15}$$

Calculation of the size distribution $D_N(R)$ for the known shape of the particles is of the most practical value in this case. The problem arose when investigating, for example, porous materials, and the cluster structure of liquids and glasses (see Section 7.2). In addition determination of some general parameters of multiphase objects is also possible (see Section 2.5.2).

As the concentration increases, the scattering curve begins to alter owing to interparticle interference. The interference effects may be described qualitatively as follows:

1. They increase with concentration.
2. They are significant only at the very beginning of the curve, because the function $\sin(sr)$ at larger values of s starts oscillating rapidly compared with $P(r, v_1)$, and the integral in equation (2.13) tends to zero very quickly.
3. They decrease the observed scattering intensity, since $P(r, v_1) \leq 1$ for all r.

For solutions of intermediate concentration ($V/v_1 \sim 0.1$) it is possible to reduce these effects by performing measurements on the samples with different concentration (see Section 3.3.1).

For highly concentrated solutions, interference effects become quite significant. As a matter of fact, they are paracrystalline structures rather than disordered systems (see Hosemann and Bagchi, 1962). Fibrous structures, liquid crystals, and solid polymers are examples of such objects. To study such samples other methods, different from those employed to investigate isotropic solutions, are required. Some of them will be considered in Chapter 6.

2.4. Isolated Particle

We saw in Section 2.3 that the scattering intensity by a dilute monodisperse solution is equal, within a constant factor, to the scattering by a particle averaged over all orientations (we remind the reader that the solvent scattering is omitted). In this section we shall derive basic relations dealing with small-angle scattering by an isolated particle.

2.4.1. Debye Equation

It is quite easy to write an explicit equation for the averaged scattering intensity, the coordinates of atoms within a particle being given. If there are n atoms with scattering lengths $f_i(\mathbf{s})$ [which may be regarded as constants in the small-angle region: $f_i(\mathbf{s}) = f_i(0) = f_i$] and coordinates $\mathbf{r}_i$ within a particle, then, similar to equations (2.5)–(2.7), one can write for the averaged intensity

$$I(s) = \sum_{i=1}^{n} \sum_{j=1}^{n} f_i f_j \frac{\sin(sr_{ij})}{sr_{ij}} \tag{2.16}$$

where $r_{ij} = |\mathbf{r}_i - \mathbf{r}_j|$ while the sum is taken over all the atoms in a particle. This equation was derived by Debye (1915) and bears his name. It provides an exact expression for the small-angle scattering intensity by a particle, but there is no question of solving the inverse problem because of the large number of terms in the double sum. It is noteworthy, however, that small-angle scattering does not deal with restoration of the atomic structure of a particle principally, owing to the restricted angular range taken into account. It will be seen later that one can use the Debye equation to calculate the scattering intensity by model bodies (normally not atomic models, but more approximate ones; see Section 3.5.3).

If a density distribution $\rho(\mathbf{r})$ and not an atomic set is used to describe a particle, the Debye equation takes the form

$$I(s) = \int\int_V \rho(\mathbf{r}_1)\rho(\mathbf{r}_2)\frac{\sin(sr_{12})}{sr_{12}}\,d\mathbf{r}_1 d\mathbf{r}_2 \tag{2.17}$$

We shall see below (Sections 4.1 and 5.3) that this form is quite appropriate for applications in many cases.

2.4.2. Correlation Function

Let us consider another approach to the calculation of the small-angle scattering intensity. Equation (1.18) for the scattering intensity by a fixed particle can be expressed in the form

$$I(\mathbf{s}) = \int_V P(\mathbf{r})\cos(\mathbf{rs})\,d\mathbf{r} \tag{2.18}$$

where the fact that even function $P(\mathbf{r})$ corresponds to real function $\rho(\mathbf{r})$ was used and the cosine Fourier transform was applied (see Section 1.2). The small-angle scattering intensity is obtained by averaging $I(\mathbf{s})$ over the solid angle in reciprocal space. Having passed to polar coordinates in real space (r, θ, φ; $d\omega = \sin\theta \cdot d\theta\, d\varphi$ is an element of solid angle), we obtain

$$\begin{aligned} I(s) &= (4\pi)^{-1}\int_\Omega I(\mathbf{s})\,d\Omega \\ &= \frac{1}{(4\pi)}\int_0^\infty\int_0^{4\pi} r^2 dr\,d\omega \int_0^{4\pi} d\Omega P(\mathbf{r})\cos(\mathbf{sr}) \\ &= \int_0^\infty r^2 dr \int_0^{4\pi} P(\mathbf{r})\,d\omega \frac{1}{4\pi}\int_0^{4\pi}\cos(\mathbf{sr})\,d\Omega \end{aligned} \tag{2.19}$$

If equation (2.7) is taken into account, one has

$$I(s) = 4\pi \int_0^\infty \gamma(r) \frac{\sin(sr)}{sr} r^2 dr \tag{2.20}$$

where

$$\gamma(r) = \frac{1}{4\pi} \int_0^{4\pi} P(\mathbf{r}) d\omega = \langle \rho(\mathbf{r}) * \rho(-\mathbf{r}) \rangle \tag{2.21}$$

(the averaged self-convolution of the density distribution) is the correlation function of a particle, first introduced by Debye and Bueche (1949). The inverse transform can be written as

$$\gamma(r) = \frac{1}{2\pi^2} \int_0^\infty I(s) \frac{\sin(sr)}{sr} s^2 ds \tag{2.22}$$

Equations (2.20)–(2.22) are of great importance in small-angle scattering data analysis. The correlation function, which can be calculated directly from the small-angle intensity curve, allows one to immediately reach a number of conclusions on the particle structure. In particular, it is clear from equation (2.21) that $\gamma(r) \equiv 0$ for $r > D$, where D is the largest dimension in a particle. Thus, equation (2.20) may be rewritten as

$$I(s) = 4\pi \int_0^D \gamma(r) \frac{\sin(sr)}{sr} r^2 dr \tag{2.23}$$

Such a form proves very appropriate for undertaking further calculations (see Section 2.4.4).

Function $\gamma(r)$ as well as the function

$$p(r) = r^2 \gamma(r) \tag{2.24}$$

(the distance distribution function) are widely used in small-angle scattering. These functions depend both on a particle's geometry, expressing numerically the set of distances joining the volume elements within a particle, and on a particle's inner inhomogeneity distribution. They are especially illustrative for the case of uniform particles, which will be considered next.

2.4.3. Uniform Particles

The density distribution function of a uniform particle can be expressed in the form

$$\rho_u(\mathbf{r}) = \begin{cases} \rho & \text{if } \mathbf{r} \text{ lies inside the particle} \\ 0 & \text{if } \mathbf{r} \text{ lies outside the particle} \end{cases}$$

One can easily see that in this case $\gamma(0) = \rho^2 V$, where V is the volume of a particle. We consider the function

$$\gamma_0(r) = \frac{\gamma(r)}{\gamma(0)} = \frac{1}{4\pi} \int d\omega \left[\frac{1}{V} \int \frac{\rho_u(\mathbf{u})}{\rho} \frac{\rho_u(\mathbf{u}+\mathbf{r})}{\rho} d\mathbf{u} \right] \qquad (2.25)$$

It is evident that for any $\mathbf{u}$ belonging to a particle the expression after the integral sign will equal 1 if $\mathbf{u} + \mathbf{r}$ also belongs to the particle and 0 if $\mathbf{u} + \mathbf{r}$ lies outside it. Therefore, function $\gamma_0(r)$ [the characteristic function; Porod (1951)] depends only on the shape of a particle and represents the probability of finding a point within a particle at distance r from the given point and equal to unity at $r = 0$ and to zero at $r > D$. Wilson (1949) gave a simple geometrical construction for $\gamma_0(r)$: if one shifts by a vector $\mathbf{r}$ the homogeneous body of volume V (Figure 2.1) and takes the mean value of the remaining common volumes $V_c(\mathbf{r})$ (with respect to all orientations), then $\gamma_0(r) = \langle V_c(\mathbf{r}) \rangle / V$.

On the other hand, in the majority of cases one should integrate equation (2.21) numerically to calculate function $\gamma_0(r)$ for the given particle shape, since analytical expressions are very seldom available. The case of a solid sphere is one of the few cases when this is possible. Here, the expression for the common volume does not depend on the

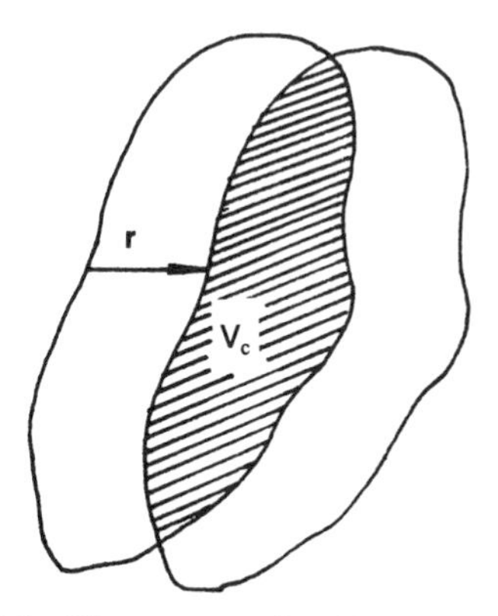

Figure 2.1. The concept of "shifted volume" $V_c(\mathbf{r})$.

direction of vector **r**, so a simple calculation gives

$$\gamma_0(r) = 1 - \frac{3}{4}\frac{r}{R} + \frac{1}{16}\left(\frac{r}{R}\right)^3 \tag{2.26}$$

where R is the sphere radius. Figure 2.2 shows functions $\gamma_0(r)$ and $p(r)$ for this case.

Function $p(r)$ can also be illustrated geometrically. We consider a volume element v_i within a particle. The probability of another volume element v_j being situated at distance r from v_i also within the particle is thus given by $\gamma_0(r)$. The number of elements v_i is proportional to the particle volume V, while the number of elements v_j is determined by the surface of the sphere with radius r, namely $4\pi r^2$. The product $4\pi r^2 V\gamma_0(r) = 4\pi p(r)/\rho$ therefore represents the number of distances r joining two arbitrary volume elements within a particle.

Hence functions $\gamma_0(r)$ and $p(r)$ for uniform particles allow obvious geometrical interpretation. It follows from the definition of the Patterson function itself that they are nonnegative everywhere. The same functions for inhomogeneous particles do not have a clear meaning, because each distance is weighted by the product of corresponding excess scattering-length densities in the joined volume elements. Thus if there are regions with opposite density signs in a particle, functions $\gamma(r)$ may exhibit negative values.

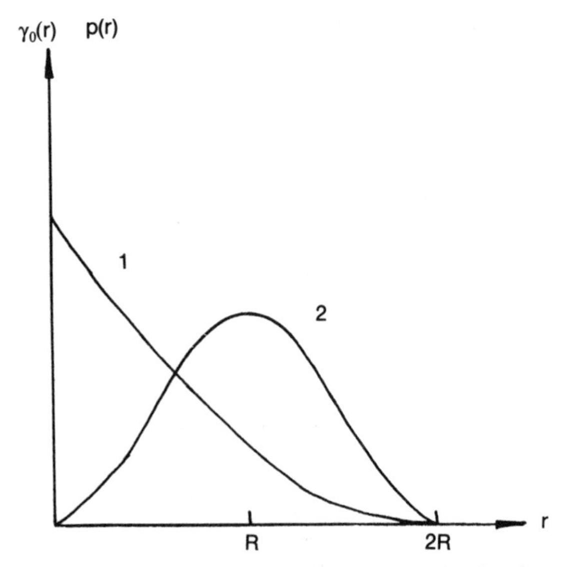

Figure 2.2. Characteristic function $\gamma_0(r)$ and distance distribution function $p(r)$ for a solid sphere of radius R.

We have already seen that the value $\gamma(0)$ is related to the particle volume. We shall also examine the derivatives of $\gamma(r)$ at the point $r = 0$. Here

$$\frac{d\gamma(r)}{dr} = \frac{1}{4\pi}\int_0^{4\pi} d\omega \int_V \frac{d}{dr}[\rho(\mathbf{u})\rho(\mathbf{r}+\mathbf{u})]d\mathbf{u}$$

$$= \frac{1}{4\pi}\int_0^{4\pi} d\omega \int_V \rho(\mathbf{u})\nabla\rho(\mathbf{r}+\mathbf{u})\mathbf{e}d\mathbf{u}$$

where $\mathbf{e} = \mathbf{r}/r$. The particle is assumed to be compact in shape and to possess a smooth closed boundary. Thus the volume of integration is $V = V_- + \Gamma$, where V is the open area and Γ is its boundary. Therefore, integration over V_- and round $\Gamma(\mathbf{r})$ separately yields

$$\left.\frac{d\gamma(r)}{dr}\right|_{r=0} = \frac{1}{4\pi}\int_0^{4\pi} \mathbf{e}d\omega \int_{V_-} \rho(\mathbf{u})\nabla\rho(\mathbf{r}+\mathbf{u})d\mathbf{u}\bigg|_{r=0}$$

$$+ \frac{1}{4\pi}\int_0^{4\pi} d\omega \int_{\Gamma(\mathbf{r})} \rho(\mathbf{u})\rho(\mathbf{r}+\mathbf{u})(-\mathbf{n}\mathbf{e})dS_{\Gamma(\mathbf{r})}\bigg|_{r=0}$$

where $\mathbf{n}$ is the outer normal vector. It is evident that the first integral equals zero because $\rho(\mathbf{r})$ is constant within V_-. In the second integral only the directions of $\mathbf{r}$ obeying the condition $(\mathbf{nr}) < 0$ should be taken into account; since r tends to zero, the value of $\rho(\mathbf{r}+\mathbf{u})$ can differ from zero only for the directions "inside" the particle. So one can write

$$\left.\frac{d\gamma(r)}{dr}\right|_{r=0} = \frac{1}{4\pi}\int_{\omega_-} d\omega \int_{\Gamma(\mathbf{r})} \rho(\mathbf{r})(-\mathbf{n}\mathbf{e})\rho(\mathbf{r}+\mathbf{u})dS_{\Gamma(\mathbf{r})}\bigg|_{r=0}$$

$$= +\frac{1}{4\pi}\int_{\Gamma(0)} \rho^2(\mathbf{u})dS_\Gamma \int_{\omega_-} (\mathbf{ne})d\omega$$

where the value $\mathbf{r} = 0$ has been substituted and the order of integration changed; ω_- represents the area where $(\mathbf{ne}) \leqslant 0$. Since $\mathbf{n}$ and $\mathbf{e}$ are unit vectors, we have

$$\int_{\omega_-} (\mathbf{ne})d\omega = \int_0^{\pi} d\theta \int_{\pi/2}^{3\pi/2} \sin^2\theta\cos\theta d\theta d\varphi = -\pi$$

The surface $\Gamma(0)$ is just the particle surface. Taking into consideration that the density inside the particle is constant, we finally obtain

$$\left.\frac{d\gamma(r)}{dr}\right|_{r=0} = \frac{1}{4\pi}(-\pi)\int_S \rho^2 dS = -\frac{\rho^2 S}{4} \tag{2.27}$$

Therefore, for uniform particles the first derivative of the correlation function at $r = 0$ is proportional to the particle surface S.

Certain structural meaning can be attributed also to the value of $\gamma''(0)$. For a better understanding we consider another function describing the particle geometry — the "chord distribution" (Porod, 1967). This function is defined in such a way that $G(l)dl$ gives the probability that a chord (a line connecting two surface points) chosen at random is of length between l and $l + dl$. We shall derive the relation between $\gamma_0(r)$ and $G(l)$. Any distance r between two intraparticle points can be extended to give a chord of length $l \geqslant r$. The number of the distances that can be measured on the chord is proportional to the difference $l - r$. Bearing in mind that $\gamma_0(r)$ provides the probability of finding the given distance (r) within a particle, one has

$$\gamma_0(r) = \frac{1}{\bar{l}} \int_r^D (l - r)G(l)dl$$

where

$$\bar{l} = \int_0^D lG(l)dl$$

is a normalizing constant ("mean chord" value). On differentiating twice, one has

$$d^2\gamma_0(r)/dr^2 = G(r)\bar{l} \tag{2.28}$$

i.e., the chord distribution function is proportional to the second derivative of the characteristic function. In particular, $\gamma''(r)$ at $r = 0$ equals $G(0)/\bar{l}$. It is evident that for smooth surfaces $G(0) = 0$ (the probability of finding the chord of length l tends to zero as $l \to 0$ owing to the smoothness of the surface). Thus a zero value of $\gamma''(0)$ [or $G(0)$] means that there are no angular or flat regions on a particle surface.

Several attempts have been made to develop the concept of $G(l)$ further (Schmidt, 1967; Wu and Schmidt, 1971; Luzzati *et al*., 1976), but it still has no wide practical application. The main reason is that it is very difficult to calculate $G(l)$ [being the second derivative of function $\gamma_0(r)$, determined from experimental data] with sufficient accuracy. This same reason renders the question of the structural meaning of higher derivatives of $\gamma_0(r)$ at $r = 0$ of purely theoretical interest, since their values cannot be reliably determined experimentally.

Thus we see that functions $\gamma(r)$ and $p(r)$ for homogeneous particles allow for intuitive interpretation and have obvious geometrical meaning. The above considerations hold to some extent also for inhomogeneous particles when the inner inhomogeneities are not too large (see Section 3.3.3 for a more detailed description).

2.4.4. Asymptotic Behavior of Intensity. The Porod Invariant

We now consider the behavior of $I(s)$ as s tends to infinity. Allowing for $\gamma(D) = 0$ one has

$$I(s) = -\frac{8\pi}{s^4}\gamma'(0) + \frac{4\pi}{s^3}D\gamma'(D)\sin(sD) + \frac{4\pi}{s^4}[2\gamma'(D) + D\gamma''(D)]\cos(sD) - \frac{4\pi}{s^4}\int_0^D [r\gamma(r)]^{(3)}\cos(sr)dr \quad (2.29)$$

We assume that function $r\gamma(r)$ has at least continuous first and second derivatives but that $[r\gamma(r)]^{(3)}$ may have discontinuities. This assumption is valid for compact particles (Filipovich, 1956), where function $\gamma(r)$, resulting from spherical averaging of the Patterson function (2.21), is smooth. According to the general theory of Fourier integrals (Bochner, 1959), the integral on the right-hand side of equation (2.29) decreases as $\Sigma\, c_i \cos(sR_i) + O(1/s)$, where c_i are constants, and R_i are the discontinuities of the function $[r\gamma(r)]^{(3)}$ over the interval $[0, D]$. Thus equation (2.29) can be rewritten as

$$I(s) = -\frac{8\pi}{s^4}\gamma'(0) + \frac{A\sin(sD)}{s^3} + \frac{B\cos(sD)}{s^4} + \frac{1}{s^4}\sum_i c_i\cos(sR_i) + O(s^{-5})$$

where $A = 4\pi D\gamma'(D)$ and $B = 4\pi[2\gamma'(D) + D\gamma''(D)]$. All the terms proportional to s^{-3} and s^{-4}, except the first, are oscillating, so that the main asymptotic trend of $I(s)$ as s tends to infinity is given by

$$I(s) \approx -\frac{8\pi}{s^4}\gamma'(0) \quad (2.30)$$

This general law of intensity decrease at large scattering angles has wide application.

It is worth stressing, however, that the above considerations hold only in the case of particles for which the scattering density distribution $\rho(\mathbf{r})$ is a continuous smooth function. If this is not so function $\gamma(r)$ may not meet the smoothness requirements. Bearing in mind that the second derivative of $\gamma(r)$ corresponds to the chord distribution $G(l)$ within a particle, one cannot assume $\gamma''(r)$ to be a smooth function for noncompact particles. For the latter, the asymptotic behavior of $I(s)$ will be mainly determined by the discontinuities in the corresponding derivatives, and equation (2.30) is not applicable. In particular, for a particle structure at atomic resolution, where $\rho(\mathbf{r})$ is regarded as a sum of δ-functions according to equation (1.17), an additional constant term appears in the expression for $I(s)$ (see Section 3.3.3).

For homogeneous particles the derivative $\gamma'(0)$ has a simple geometrical meaning. Substitution of equation (2.27) into equation (2.30) yields

$$I(s) \underset{s\to\infty}{\approx} \frac{2\pi}{s^4}\rho^2 S \tag{2.31}$$

Hence the asymptotic trend of the scattering intensity for s tending to infinity is readily related to the particle surface. This equation, the "Porod law," was derived independently by Debye and Bueche (1949) and Porod (1951).

The Porod invariant (Porod, 1952) is an important integral characteristic of the scattering intensity:

$$Q = \int_0^\infty s^2 I(s)ds \tag{2.32}$$

Comparison with equation (1.19) clearly shows that Q is proportional to the total scattered energy (the factor s^2 is due to the use of polar coordinates). Parseval's equality implies that Q describes the total scattering length of the object. For particle scattering, equations (2.21) and (2.22) give

$$Q = 2\pi^2\gamma(0) = 2\pi^2 \int_V \rho^2(\mathbf{r})d\mathbf{r} \tag{2.33}$$

that is, Q is proportional to the mean-square density fluctuation caused by a particle [for a solution, one has to replace $\rho(\mathbf{r})$ by the excess density $g(\mathbf{r})$; see Section 2.3.2]. For scattering by an arbitrary system, Q provides the mean-square density fluctuation over the whole system (see Section 2.5.2). We shall see below that this invariant is widely used for calculating a number of structural parameters. The volume of a homogeneous particle can be readily obtained from the value of the invariant, since in this case

$$Q = 2\pi^2\rho^2 V \tag{2.34}$$

2.4.5. Special Types of Particle

We now examine cases where it is possible to simplify the basic equations of small-angle scattering by using the specific features of the shape or structural organization of the particles under investigation.

2.4.5.1. Spherically Symmetric Particle

In this case the particle density $\rho(\mathbf{r}) = \rho(r)$, and therefore the scattering amplitude of a particle in a fixed orientation, depend on the modulus of the vector $\mathbf{s}$ [see equation (1.25)]. Averaging does not alter the diffraction pattern, and one can write

$$I(s) = |A(s)|^2 = \left[4\pi \int_0^\infty \rho(r) \frac{\sin(sr)}{sr} r^2 dr\right]^2 \tag{2.35}$$

2.4.5.2. Rodlike Particle

Scattering by particles with dimensions considerably larger along one (z) axis than along the other two also has specific features. We shall consider a columnar structure, i.e., a structure where the density distribution in a cross section perpendicular to the z axis does not depend on z. If

$$\rho(x, y, z) = \rho(x, y) \cdot \Pi(z, L)$$

then the three-dimensional intensity is

$$I(\mathbf{s}) = A^2(\mathbf{s}) = A_0(X, Y) \cdot L^2 \cdot \delta^2(L, Z)$$

where

$$A_0(X, Y) = \int\int_{S_c} \rho(x, y) \exp[i(xX + yY)] dx dy$$

while S_c is the cross section of the particle in the xy plane. [Functions $\Pi(z, L)$ and $\delta(L, Z)$ were defined in Section 1.3.] Since we assumed L to be sufficiently large, function $\delta(L, Z)$ is nearly equal to $\delta(Z)$ and most of the scattering intensity is concentrated near the XY plane.

The small-angle scattering intensity of such an object can be calculated by two-dimensional averaging of the intensity $A_0^2(X, Y)$ followed by rotation of the plane $Z = 0$. That is to say, functions $A_0^2(X, Y)$ and $\delta^2(L, Z)$ may be averaged independently. We write $\delta(Z, L)$ as

$$\delta(Z, L) = \frac{\sin(\pi L |\mathbf{q}| t)}{\pi L |\mathbf{q}| t} = \frac{\sin(Lst/2)}{Lst/2}$$

where $t = \cos\alpha$, α being the angle between the scattering vector $\mathbf{s}$ and the Z axis in reciprocal space. The averaging is carried out by integrating $\delta^2(Z, L)$ over t. In principle, the upper limit of integration should be unity. If, however, the condition $s \geqslant 2\pi L$ is satisfied, $\delta^2(Z, L)$ is negligibly small at $t > 1$, so that integration up to infinity can be performed. Thus one obtains

$$\langle L^2\delta^2(L, Z)\rangle = \langle F_L^2\rangle = L^2 \int_0^\infty \frac{\sin(sLt/2)}{sLt/2}\, dt = L\frac{\pi}{s}$$

and finally

$$I(s) = \langle F_L^2\rangle\langle A_0^2(X, Y)\rangle = L\frac{\pi}{s} I_c(s) \tag{2.36}$$

It is therefore possible for columnar structures to split the scattering intensity into two factors: the first ("length factor" $\langle F_L^2\rangle$) proportional to s^{-1}, while the second [intensity $I_c(s)$] corresponds to the structure of the particle in the cross section $\rho(x, y)$. This approach can also be used when function $\rho(x, y)$ varies little along the z axis: in this case $I_c(s)$ corresponds to some average with respect to a function of z [$\langle\rho(x, y)\rangle$].

A cross-section characteristic function can be introduced, as for the whole particle:

$$\gamma_c(r) = \frac{1}{2\pi}\int_0^{2\pi} d\omega \int_{S_c} \rho(\mathbf{u})\rho(\mathbf{u} + \mathbf{r})dxdy$$

where vectors $\mathbf{u}$ and $\mathbf{r}$ correspond to the xy plane and φ is the angle between these vectors. On performing calculations as in equations (2.19)–(2.21) and taking into account that two-dimensional averaging of $\exp[i(xX + yY)]$ results in the function $J_0(sr)$, where J_0 is the Bessel function of zero order and $r = (x^2 + y^2)^{1/2}$, one gets

$$I_c(s) = 2\pi \int_0^d \gamma_c(r)J_0(sr)r\, dr \tag{2.37}$$

where d is the diameter of the cross section S_c. The inverse equation can be obtained by the inverse Hankel transform:

$$\gamma_c(r) = \frac{1}{2\pi}\int_0^\infty I_c(s)J_0(sr)s\, ds \tag{2.38}$$

Functions $I_c(s)$ and $\gamma_c(s)$ are of great value in investigations of rodlike particles, as we shall see below. One should bear in mind, however, that basic relation (2.36) is valid only for $s > 2\pi/L$.

2.4.5.3. Lamellar Particle

It is also possible to separate two components in the scattering intensity of particles such that two dimensions are much greater than the third. If one writes the density in the form

$$\rho(x, y, z) = \rho(x) \cdot \Pi(y, L_y) \cdot \Pi(z, L_z)$$

then, having performed the calculations as in the previous section, one obtains

$$I(s) = (2\pi S_l/s^2) I_t(s) \tag{2.39}$$

where S_l is the cross section of the lamella. The characteristic function of the thickness is

$$\gamma_t(r) = \int_0^T \rho(u)\rho(u+r)du$$

where r is a coordinate in the x direction and T is the lamella thickness ($T \ll L_y$, $T \ll L_z$). Function $\gamma_t(r)$ is related to $I_t(s)$ by the transforms

$$I_t(s) = \pi \int_0^T \gamma_t(r)\cos(sr)dr \tag{2.40}$$

and

$$\gamma_t(r) = \frac{1}{\pi} \int_0^\infty I_t(s)\cos(sr)ds \tag{2.41}$$

As for rodlike particles, these representations are not valid for very small angles since the condition $s > 2\pi/S_l^{1/2}$ must be fulfilled.

The basic relations between the particle structure and its small-angle scattering curve were derived in this section. It is clear from the results that only the direct problem [i.e., the calculation of function $I(s)$ using a given function $\rho(\mathbf{r})$] permits a unique solution. The reverse restoration cannot be unique, because of the loss of information on the directions of the vectors joining volume elements within a particle owing to spherical averaging. Hence, only information about the moduli of the vectors remains [cf. the Patterson function and the characteristic function; see equation (2.21)]. Therefore, determination of the parameters of the particle structure from small-angle scattering data requires special methods, which will be dealt with in the subsequent chapters.

2.5. Nonparticulate Systems

A number of disperse systems (such as liquids, glasses, amorphous polymers, metals, and alloys) cannot be treated as a superposition of particles and a matrix (solution). One distinguishes a single-phase system, where scattering is caused by statistical fluctuations, and two- (many-) phase systems, where scattering is due to differences in the scattering densities of the phases. The two cases will be outlined here.

2.5.1. Scattering Due to Statistical Fluctuations

We consider scattering by a statistically homogeneous sample of volume V_0. It was shown in Section 2.2.1 that, ignoring statistical density fluctuations within an object, one would obtain only a nonmeasurable zero scattering peak caused by the sample shape. Density fluctuations give rise to statistical inhomogeneities and, obviously, the larger their amplitude, the larger their contribution to the scattering intensity.

The zero-angle scattering caused by density fluctuations will be estimated quantitatively. Equation (2.9) yields

$$I(0) = F^2(0)\langle N\rangle \left\{1 - \frac{1}{v_1}\int_0^\infty [1 - P(r)]4\pi r^2 dr\right\} \tag{2.42}$$

where we recall that $\langle N\rangle$ is a random value fluctuating around the mean value of V_0/v_1. In order to calculate the integral on the right-hand side of equation (2.42) one can first evaluate the mean number of atomic pairs $\langle N_p\rangle$ in the volume V_0. Bearing in mind the definition of function $P(r)$ (see Section 2.2.1) one can write

$$\langle N_p\rangle = \langle N\rangle \int_V P(r)\frac{dv}{v_1} = \frac{\langle N\rangle}{v_1}\int_0^\infty P(r)4\pi r^2 dr$$

On the other hand, it is obvious that

$$\langle N_p\rangle = \langle N(N-1)\rangle = \langle N^2\rangle - \langle N\rangle$$

so that one obtains

$$I(0) = F^2(0)\langle (N - \langle N\rangle)^2\rangle$$

It is known from the kinetic theory of gases [see, for example, Leontovitch (1983)] that the isothermal compressibility coefficient is

$$\beta = -\frac{1}{V_0}\left(\frac{\partial V_0}{\partial p}\right)_T = \frac{v_1}{k_B T\langle N\rangle}\langle (N - \langle N\rangle)^2\rangle$$

where p is the pressure and T the absolute temperature. Hence

$$I(0) = F^2(0)\langle N \rangle \frac{k_B T \beta}{v_1} \tag{2.43}$$

In the particular case of an ideal gas $\beta = 1/p$, $pv_1 = k_B T$, and $I(0) = F^2(0)\langle N \rangle$, corresponding to independent scattering of $\langle N \rangle$ atoms (molecules) with form factor $F^2(s)$.

The zero-scattering intensity caused by statistical density fluctuations is thus closely related to the thermodynamic properties of the system. It is clear that the larger the compressibility of a system (i.e., the closer it is to the gaseous state), the larger the value of $I(0)$ (for a fixed number of molecules). This can be explained by the fact that atoms (molecules) in liquids are much closer to each other than in gases. This leads to considerable interference effects, and a reduction in scattering at very low angles takes place.

Figure 2.3 illustrates schematically the scattering curves of gaseous and liquid samples. It is worth noting that in the small-angle region in which we are interested, scattering by statistical fluctuations is practically constant. Wide-angle scattering studies enable one to investigate the molecular structure of an object; for example, the structure of gas molecules can be examined by means of the electron-diffraction method; see Vilkov *et al.* (1978). Some applications of small-angle scattering to the study of single-phase amorphous and liquid substances will be discussed in Section 7.3.

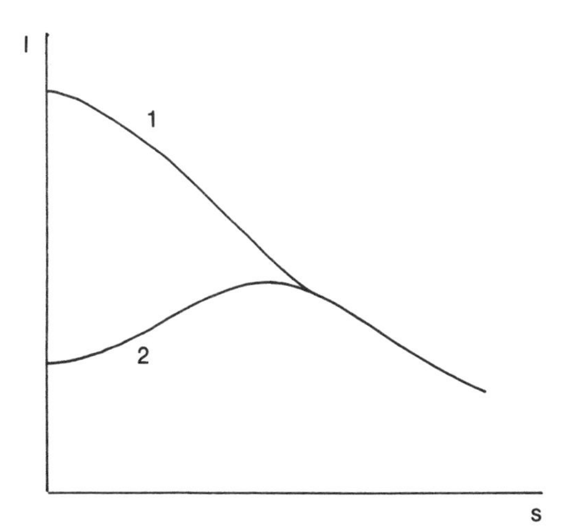

Figure 2.3. Gas-type (1) and liquid-type (2) scattering curves.

2.5.2. Two-Phase and Multiphase Systems

We shall now consider another type of object for which the "particle–solution" model is inapplicable. Such an object consists of two (or several) phases with different scattering densities. Here, small-angle scattering is caused by the presence of regions of colloidal size, occupied by different phases.

A two-phase (binary) system is illustrated schematically in Figure 2.4 (the scattering densities ρ_1 and ρ_2 within the phases are assumed constant). It is obvious that the "particle" concept is pointless for such systems and only information on their general features can be contained in diffraction data. In particular, the contrast (see Section 2.3.2) will be represented by a mean-square fluctuation

$$\langle \Delta\rho^2 \rangle = \langle [\rho(\mathbf{r}) - \bar{\rho}]^2 \rangle = (\rho_1 - \rho_2)^2 \varphi_1 \varphi_2 \tag{2.44}$$

rather than by the difference between the mean-particle and solvent densities. Here $\bar{\rho} = \varphi_1\rho_1 + \varphi_2\rho_2$ is an average density of a system, φ_1 and φ_2 are volume fractions of the phases, and $\varphi_1 + \varphi_2 = 1$. Correspondingly, the invariant Q in equation (2.33) is given by

$$Q = 2\pi^2 V_0 \langle \Delta\rho^2 \rangle = 2\pi^2 \varphi_1 \varphi_2 (\rho_1 - \rho_2)^2 V_0 \tag{2.45}$$

It is worth noting that, according to Babinet's principle, addition of a constant term to the scattering density of a system does not alter the scattering curve (it has already been indicated that scattering by the large volume V_0 produces only an unobservable central peak). The system may

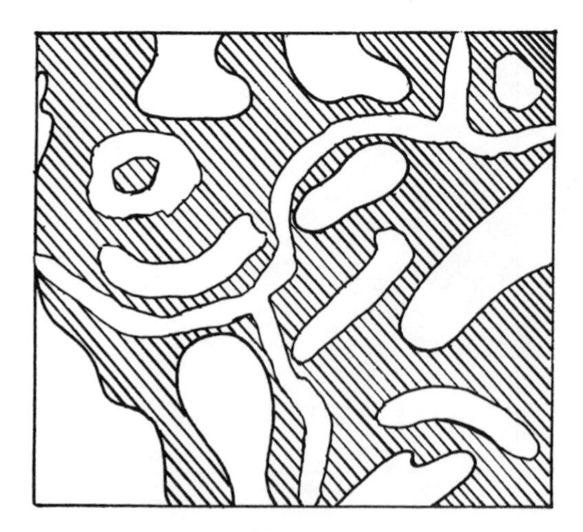

Figure 2.4. Two-phase nonparticulate system.

therefore be treated, for example, as a porous body with density $\rho_1 - \rho_2$. Moreover, it does not matter which (first or second) phase is regarded as being filled with the substance. This means that all relationships for the determination of structure parameters of binary systems should be symmetric with respect to ρ_1 and ρ_2, φ_1 and φ_2.

Basic expressions for the integral parameters of a binary system can be obtained using equations similar to those derived above for particle solutions. The meaning of the structural parameters is, however, not the same. Thus the characteristic function $\gamma_0(r)$ (see Section 2.4.3), as generalized for a two-phase system, also has a probabilistic meaning, but its geometrical interpretation is more complicated (Porod, 1951). If one chooses at random a line of length r within the sample, then

$$\gamma_0(r) = \tfrac{1}{2}[P_{11}(r) + P_{22}(r)] - \tfrac{1}{2}[P_{12}(r) + P_{21}(r)]$$

where P_{ij} gives the probability for the segment to connect points belonging to the ith and jth phases. At $r = 0$, $P = 1$ for $i = j$ and $P = 0$ for $i \neq j$, so that $\gamma_0(0) = 1$. If the value of r substantially exceeds the characteristic dimensions of inhomogeneities within the system, the probabilities of joining the same or different phases become practically equal, and $\gamma_0(r) \approx 0$. There is no clear concept of "maximal dimension" here, and $\gamma_0(r)$ approaches zero asymptotically as r tends to infinity.

Thus the characteristic function for a binary system does not possess an intuitive geometrical sense, as for a particle, although it contains an estimate of average dimensions and shapes of inhomogeneities. Similarly, the "volume" in equation (2.34) can be regarded only as an approximate estimate of an average volume of the inhomogeneities (phase clusters).

General expression (2.27) for the derivative of the characteristic function at $r = 0$ is still valid, assuming S to be the full surface of phase separation within a sample, while ρ should be replaced by $\rho_1 - \rho_2$. The asymptotic trend (2.29), describing the behavior of the intensity at large angles, also holds, except that the oscillating terms proportional to s^{-3} and s^{-4} vanish due to the fact that no largest dimension exists. If equation (2.45) for the invariant Q is used, one obtains the specific surface of phase separation in the form

$$\frac{S}{V_0} = \frac{\pi\varphi_1\varphi_2}{Q} \lim_{s\to\infty} [s^4 I(s)] \tag{2.46}$$

Similar relations can be derived also for multiphase systems. Thus for the ternary system

$$\bar{\rho} = \rho_1\varphi_1 + \rho_2\varphi_2 + \rho_3\varphi_3 \tag{2.47}$$

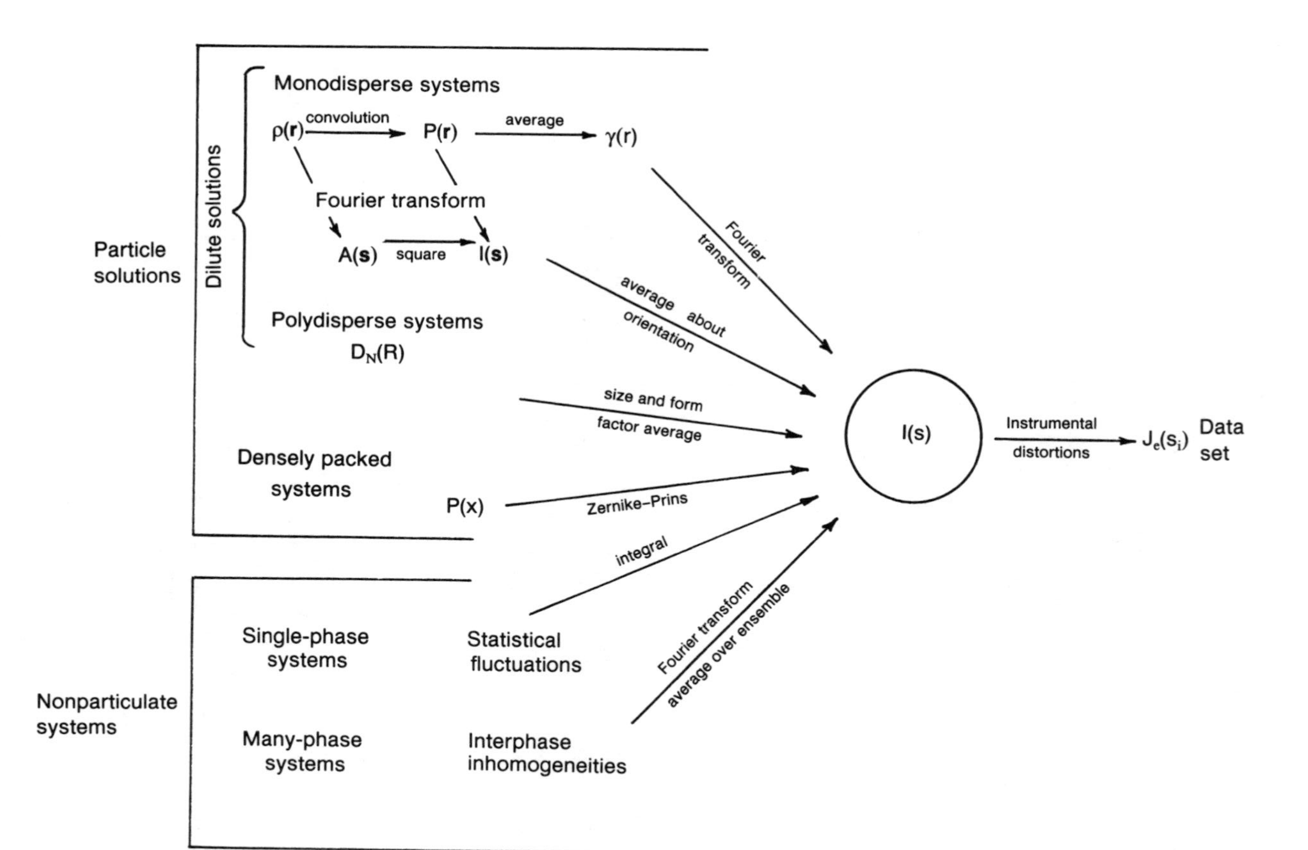

Figure 2.5. Scheme for small-angle structure investigations of isotropic systems.

and

$$\langle \Delta\rho^2 \rangle = (\rho_3 - \rho_2)^2 \varphi_3 \varphi_2 + (\rho_3 - \rho_1)^2 \varphi_3 \varphi_1 + (\rho_2 - \rho_1)^2 \varphi_2 \varphi_1 \qquad (2.48)$$

where ρ_3 and φ_3 correspond to the density and volume fraction of the third phase. Quantitative determination of general structure parameters for three- and more-phase systems is, however, less accurate than for binary ones.

Hence small-angle scattering can provide information on heterogeneous systems of rather different phase composition. It should be noted, however, that the above estimates are only rough and break down when the characteristic dimensions of inhomogeneities are comparable with the interatomic distances. Moreover, the presence of colloidal inhomogeneities within one and the same phase may also distort the results. This makes the quantitative analysis of heterogeneous systems a difficult task. Examples of small-angle scattering investigations of such systems are presented in Sections 6.3.1, 7.1.2, and 7.3.2.

2.6. Conclusion

The outstanding feature of the mathematical tools for analyzing scattering by disordered systems, as opposed to crystal structure analysis, is that neither the structure of the object nor the scattering intensity is described in terms of the atomic coordinates. It is convenient for such systems to introduce some general distribution function [such as the statistical function $P(r)$ or characteristic function $\gamma(r)$], which can be related to the scattering intensity by simple transformations. Figure 2.5 presents a general diagram for small-angle diffraction investigations of disperse isotropic systems on the basis of the results obtained in this chapter. Transition from experimental values $J_e(s_i)$, which are affected by different instrumental distortions, to the ideal scattering curve $I(s)$ (discussed in the present chapter) requires processing of experimental data using methods described in Chapter 9.

In the present chapter, the main types of disperse system have been considered, and relations between the small-angle scattering intensity and structure derived. In particular, it was shown how the scattering intensity could be obtained directly for a given structure. Further consideration will be given to the more important inverse problem, namely, determination of the structure parameters of an object using small-angle scattering data.

II

Monodisperse Systems

3

Determination of the Integral Parameters of Particles

Spherical averaging of the intensity of the scattering curve caused by random orientation of particles in solution (matrix) leads to a considerable loss of information contained in the scattering data. Therefore, even if all the particles are identical and the interference effects can be neglected, usually only some integral parameters can be evaluated from the isotropic scattering curve without *a priori* information. Traditional methods of analyzing small-angle scattering curves from isotropic monodisperse systems will be considered is this chapter.

Basic relations for the invariants of a scattering curve are derived in Section 3.1, while Section 3.2 deals with the problem of the number and accuracy of the structural parameters that can be determined from the given scattering curve. Section 3.3 treats practical approaches to the calculation of the invariants.

One of the most widely used approaches to the interpretation of small-angle scattering data is a "trial-and-error" modeling method. The experimental scattering intensity is compared with the scattering curves of a number of model bodies chosen on the basis of *a priori* information on the particle. Scattering by simple geometrical bodies is considered in Section 3.4, while Section 3.5 is devoted to approximation techniques for calculating model intensity curves. Some applications of the described methods are given in Section 3.6.

3.1. Geometrical and Weight Invariants

In this section equations for the invariants of small-angle scattering curves will be obtained. The term "invariants" denotes the structural parameters of a particle that can be directly related to the intensity curve $I(s)$.

3.1.1. Total Scattering Length and Radius of Gyration

First, the behavior of $I(s)$ at very low angles will be examined. To this end one can substitute the Mclaurin series

$$\sin(sr)/sr = 1 - s^2r^2/6 + s^4r^4/120 - \cdots \tag{3.1}$$

into equation (2.23). If we restrict ourselves to the first two terms, then in the vicinity of $s = 0$

$$I(s) = I(0)(1 - s^2R_g^2/3) \tag{3.2}$$

where

$$I(0) = 4\pi \int_0^D \gamma(r)r^2dr \tag{3.3}$$

and

$$R_g^2 = \frac{1}{2}\int_0^D \gamma(r)r^4dr \bigg/ \int_0^D \gamma(r)r^2dr \tag{3.4}$$

The expression on the right-hand side of equation (3.2) can be regarded as the first two terms of the Mclaurin series of function $\exp(-s^2R_g^2/3)$. Thus, to an accuracy of terms proportional to s^4, one can write for the beginning of the scattering curve

$$I(s) = I(0)\exp(-s^2R_g^2/3) \tag{3.5}$$

This is the Guinier equation, derived nearly 50 years ago (Guinier, 1939). In order to relate parameters $I(0)$ and R_g to the structure of a particle, we substitute equation (3.1) into the Debye equation (2.17) and obtain

$$I(s) = \int\int \rho(\mathbf{r}_1)\rho(\mathbf{r}_2)d\mathbf{r}_1d\mathbf{r}_2 - \frac{s^2}{6}\int\int \rho(\mathbf{r}_1)\rho(\mathbf{r}_2)|\mathbf{r}_1 - \mathbf{r}_2|^2d\mathbf{r}_1d\mathbf{r}_2 \tag{3.6}$$

Comparison of equations (3.6) and (3.2) enables one to deduce from the leading terms on the right-hand side that

$$I(0) = \left|\int \rho(\mathbf{r})d\mathbf{r}\right|^2 \tag{3.7}$$

This is just the square of the total particle scattering length.

On setting equal the coefficients of the s^2 terms one has

$$R_g^2 = \frac{\int\int \rho(\mathbf{r}_1)\rho(\mathbf{r}_2)|\mathbf{r}_1 - \mathbf{r}_2|^2 d\mathbf{r}_1 d\mathbf{r}_2}{2\int\int \rho(\mathbf{r}_1)\rho(\mathbf{r}_2) d\mathbf{r}_1 d\mathbf{r}_2} \tag{3.8}$$

If we place the origin of coordinates at a point $\mathbf{r}_0$, then simple calculations give

$$R_g^2 = \frac{\int \rho(\mathbf{r} - \mathbf{r}_0)|\mathbf{r} - \mathbf{r}_0|^2 d\mathbf{r}}{\int \rho(\mathbf{r}) d\mathbf{r}} - \left[\frac{\int \rho(\mathbf{r} - \mathbf{r}_0)(\mathbf{r} - \mathbf{r}_0) d\mathbf{r}}{\int \rho(\mathbf{r}) d\mathbf{r}}\right]^2 \tag{3.9}$$

It should be noted that the integrals in equation (3.8) do not depend on the actual position of the origin of coordinates, since both integrals are taken over the whole space. Hence one can match point $\mathbf{r}_0$ with the center of mass of a particle. The second term in equation (3.9) thus becomes equal to zero and

$$R_g^2 = \int_V \rho(\mathbf{r}) r^2 d\mathbf{r} \Big/ \int_V \rho(\mathbf{r}) d\mathbf{r} \tag{3.10}$$

which is the well-known expression for the radius of gyration of a particle about its center of mass.

Therefore, regardless of the structure of a particle, the beginning of the scattering curve can be described with the help of two parameters, namely $I(0)$, characterizing the total amount of scattering matter, and R_g, bearing information on its distribution with respect to the particle center of mass.

3.1.2. Volume and Surface

For homogeneous particles equation (3.7) yields $I(0) = (\rho V)^2$. By taking into account equation (2.34) it is possible to write

$$V = 2\pi^2 I(0)/Q = 2\pi^2/Q_0 \tag{3.11}$$

where

$$Q_0 = \int_0^\infty i(s) s^2 ds \quad \text{and} \quad i(s) = I(s)/I(0)$$

are the normalized Porod invariant and normalized scattering intensity. Consequently, knowledge of the scattering intensity in relative units is sufficient to determine the volume of homogeneous particles. This fact reflects properties of the invariant Q, which according to Section 2.4.4 can be used to eliminate the absolute scattering density from routine equations. Thus, having substituted equation (2.34) into equation (2.31), the specific surface of a particle can be written in the form

$$\frac{S}{V} = \frac{\pi}{Q_0} \lim_{s \to \infty} [s^4 i(s)] = \frac{\pi}{Q_0} c_4 \tag{3.12}$$

Up to some accuracy, equations (3.11) and (3.12) can also be applied for inhomogeneous particles (see Section 3.3.3).

3.1.3. Largest Dimension and Correlation Length

It was noted in Section 2.4.2 that the largest dimension (maximum diameter) D of a particle can be estimated directly from the correlation function satisfying the condition $\gamma(r) \equiv 0$ for $r > D$. The "correlation length" is another important linear parameter and is given by

$$l_{\mathrm{m}} = \frac{2}{\gamma(0)} \int_0^D \gamma(r) dr \tag{3.13}$$

It provides an estimate of the width of the distribution $\gamma(r)$. For homogeneous particles l_{m} allows a simple geometrical explanation. The relationship (2.28) between $\gamma_0(r)$ and $G(l)$ enables one to readily see that

$$l_{\mathrm{m}} = 2 \int_0^D \gamma_0(r) dr = \frac{1}{\bar{l}} \int_0^D G(r) r^2 dr$$

i.e., l_{m} is a weight-averaged chord in a particle.

Substitution of equation (3.13) into equation (2.22) with allowance for equation (2.33) yields

$$\begin{aligned} l_{\mathrm{m}} &= \frac{2}{\gamma(0)} \int_0^\infty dr \frac{1}{2\pi^2} \int_0^\infty I(s) \frac{\sin(sr)}{sr} s^2 ds \\ &= \frac{2}{Q} \int_0^\infty I(s) s\, ds \int_0^\infty \frac{\sin(sr)}{sr} dr \\ &= \frac{2\pi}{Q_0} \int_0^\infty i(s) s\, ds \end{aligned}$$

Therefore, this parameter of a particle can also be evaluated from the scattering curve expressed in relative units with the help of the invariant Q.

3.1.4. Anisometric Particles

We have already seen that for rodlike and lamellar particles one can separate in the scattering intensity the contribution of their cross section and thickness, respectively (Section 2.4.5). Thus, in the former case, the expansion of the Bessel function $J_0(x)$ in a Mclaurin series

$$J_0(x) = 1 - x^2/4 + \cdots$$

gives, from equation (2.37) for small values of s,

$$I_c(s) = I_c(0)\exp(-s^2R_c^2/2) \tag{3.14}$$

where $I_c(0)$ denotes the amount of scattering matter in the cross section $[I_c(0) = \bar{\rho}_c^2 S_c^2$, where $\bar{\rho}_c$ is an average scattering density in the cross section and S_c is the area thereof], while R_c is its radius of gyration. Asymptote (2.30) still holds for large scattering angles, consequently $I_c(s)$ decreases as s^{-3} in this region [see equation (2.36)]. Equation (3.12) for the specific surface turns out to be the ratio of the cross-section perimeter l_c to its area S_c [actually, for rodlike particles of length L, one has $S = L_cL$ and $V \approx S_cL$, so quantity L cancels in equation (3.12)].

For lamellar particles equation (2.40) gives

$$I_t(s) = I_t(0)\exp(-s^2R_t^2) \tag{3.15}$$

where $I_t(0) = \bar{\rho}_t T^2$ ($\bar{\rho}_t$ is an average thickness density) and R_t is the thickness radius of gyration. Function $I_t(s)$ decreases as s^{-2} when s tends to infinity, and the left-hand side of equation (3.12) simply reduces to $1/T$.

3.2. Information Content in Small-Angle Scattering Data

The invariants of small-angle scattering curves were defined by means of the calculations in Section 3.1, i.e., the structural parameters which can be evaluated from the scattering curves, if minimum *a priori* information on an object is available. In this section we estimate the general information content in small-angle scattering data, namely, the number of independent structural parameters of a particle that can be evaluated from a given scattering curve. The question as to how their accuracy can be related to the accuracy of experimental data will be examined as well.

3.2.1. General Approach

We state the problem in general form. A set of experimental data is given, namely vector $\mathbf{Y} = (Y_i)$, $i = 1, 2, \ldots, N$, $Y_i = Y(s_i)$, measured within a finite region of scattering angles ($s_1 = s_{\min}$, $s_N = s_{\max}$). The accuracy of the data is defined by the standard deviations $\sigma_i = \sigma(s_i)$. Furthermore, we describe the particle by a set of n parameters:

$$\rho(\mathbf{r}) \sim \mathbf{P}(\mathbf{X}) \tag{3.16}$$

where $\mathbf{X} = (X_j)$, $j = 1, \ldots, n$, while an operator $\mathbf{A}$ solves the direct problem for the chosen set of parameters, i.e., the vector

$$\mathbf{Y}^{\mathrm{t}} = \mathbf{A}[\mathbf{X}] \tag{3.17}$$

with $\mathbf{Y}^{\mathrm{t}} = (Y_i^{\mathrm{t}})$, $i = 1, 2, \ldots, N$, provides the theoretical scattering intensity by model (3.16) in the s_i. In this case optimum values of the parameters (vector $\mathbf{X}_0 = \{X_j^0\}$) can be determined, for instance, by minimization of the functional

$$\phi = \sum_{i=1}^{N} \frac{1}{\sigma_i^2} (Y_i - Y_i^{\mathrm{t}})^2 \tag{3.18}$$

Hence, the description of $\rho(\mathbf{r})$ with $\mathbf{P}(\mathbf{X}_0)$ will provide the best representation of a particle within the framework of the chosen parametrization.

Such an approach (it can be termed the "indirect method" of data treatment) requires that the following questions be answered:

1. Based on what reasoning do we select the parameters X_j and how can they be related to the structural features of a particle (i.e., what is the form of operator $\mathbf{P}$)?
2. What is the connection between the number of parameters n and the accuracy of their determination when the number of experimental points is N, the interval of measurements is ($s_{\min}$, $s_{\max}$), and the experimental errors are σ_i?
3. How can one determine optimum values of X_j^0 (in other words, which form of operator $\mathbf{A}$ allows the determination)?

It is certainly impossible to answer the questions in the general case, therefore specific properties of small-angle scattering data must be used to select and determine an appropriate set of independent structural parameters.

3.2.2. Number of Independent Parameters

It follows from equation (2.23) that function $sI(s)$ is the sine-Fourier image of function $r\gamma(r)$, which differs from zero within the finite range $0 \leqslant r \leqslant D$. Thus, the well-known sampling theorem (Shannon and Weaver, 1949; Kotelnikov and Nikolaev, 1950) is valid for function $sI(s)$,

$$sI(s) = \sum_{k=1}^{\infty} s_k I(s_k) \left\{ \frac{\sin[D(s-s_k)]}{D(s-s_k)} - \frac{\sin[D(s+s_k)]}{D(s+s_k)} \right\}$$
$$= \sum_{k=1}^{\infty} s_k I(s_k) \phi(s, k, D) \qquad (3.19)$$

where $s_k = k\pi/D$. This means that the scattering intensity can be fully defined by the values of $I(s)$ at the set of discrete points s_k, $k = 1, 2, \ldots$. Figure 3.1 illustrates schematically several terms of such a series.

Equation (3.19) itself does not define the number of independent parameters describing function $I(s)$, since the series is infinite. However, measurements using sampling interval $\Delta s > \pi/D$ lead to a loss of information (Damaschun *et al*., 1968; Damaschun and Pürschel, 1969). On the other hand, it does not follow from equation (3.19) that it would suffice to measure $I(s)$ only at the points s_k. In practice, a smaller sampling interval is necessary because of experimental errors and termination effects (a more detailed consideration is given in Section 9.3).

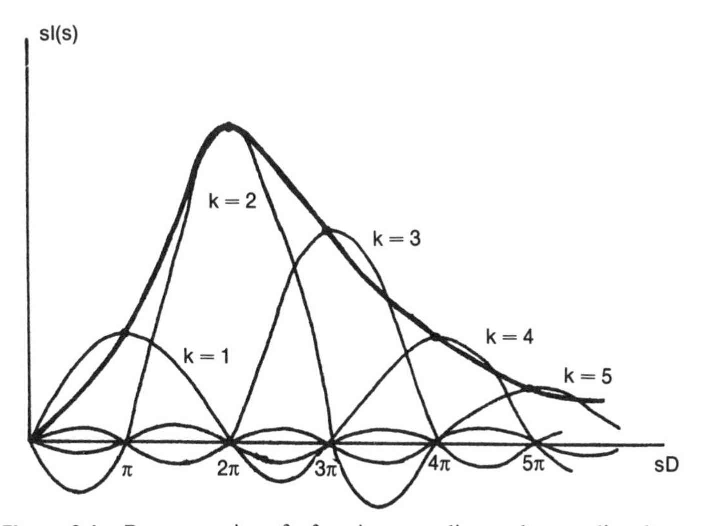

Figure 3.1. Representation of a function according to the sampling theorem.

Figure 3.1 shows that the behavior of function $sI(s)$ in the vicinity of s_k is determined mainly by the kth term of series (3.19), as the remaining terms are close to zero in this region. Furthermore, the maximum contribution of each term is less than the corresponding vaue of $s_kI(s_k)$. Since $sI(s)$ tends to zero when s tends to zero or infinity, one can assume that if exact values of $I(s_k)$ are known for $s_{\min} \leqslant s_k \leqslant s_{\max}$, then the function

$$s\tilde{I}(s) = \sum_{k=k_{\min}}^{k_{\max}} s_kI(s_k)\phi(s, k, D) \tag{3.20}$$

where

$$k_{\min} = [s_{\min}D/\pi] + 1, \qquad k_{\max} = [s_{\max}D/\pi]$$

is a good approximation of $sI(s)$ over this interval. The number of independent parameters sufficient to describe $I(s)$ in the interval $[s_{\min}, s_{\max}]$ to the accuracy of the termination effect in series (3.19) is equal to the number of "basic points" s_k belonging to the interval.

It seems natural to conclude that this number is equal to the number of independent structural parameters of a particle that can be evaluated from the scattering curve. Unfortunately, this statement is of no practical use. The unique derivation of $n = k_{\max} - k_{\min}$ real structural parameters of a particle from n terms of series (3.20) is, to say the least, a very indeterminate problem, while a parametrization allowing an approach like (3.16)–(3.18) to be applied has not yet been proposed.

Nevertheless, the sampling theorem enables parametrization to be introduced in a number of cases. Actually, one is not obliged to regard the components of vector **X** as real structural parameters of a particle. For example, Taupin and Luzzati (1982) chose the actual values of $s_kI(s_k)$ as the components of **X**. The asymptotic trend of $I(s)$ at high scattering angles was used to reduce termination effects (see Sections 2.4.4 and 3.3.3). The termination for $s < s_{\min}$ is much more critical and, for approximation (3.30) to be useful, it is necessary that $k_{\min} = 1$, i.e., $s_{\min} < \pi/D$, which sets one more condition for carrying out the experimental determination of $I(s)$. Taupin and Luzzati (1982) proposed a method of data treatment on the basis of an approach which allows one to estimate the accuracy of the results, as explained in more detail in Section 9.7.3.

The parametrization of function $\gamma(r)$:

$$r\gamma(r) = \sum_{i=1}^{n} c_j\varphi_j(r), \qquad \mathbf{X} = \{c_j\} \tag{3.21}$$

is the most important in practice. Here c_i are constants, and $\{\varphi_j(r)\}$ is some set of orthogonal functions. Moore's (1980) approach employing a trigonometric series is quite interesting:

$$r\gamma(r) = \frac{2}{\pi} \sum_{i=1}^{n} c_j \sin \frac{\pi r j}{D} \tag{3.22}$$

The operator **A** is then given by an integral transform (2.23) to the reciprocal space and vector (3.17) is linear with respect to coefficients c_j. Thus, the task of minimizing functional (3.18) is reduced to the system of normal linear equations

$$\partial\phi/\partial c_j = 0, \qquad j = 1, 2, \ldots, n \tag{3.23}$$

the solution of which provides the values of c_j. These methods will be examined in detail in Section 9.7; here, we note only the features of Moore's approach that are of interest from the standpoint of information content.

Substitution of equation (3.22) into equation (2.23) and comparison of the resulting expression with equation (3.20) shows that the c_j are nothing but the values of $s_k I(s_k)$. In this sense the methods of Taupin and Luzzati (1982) and Moore (1980) are equivalent. Thus, the requirements $k_{min} = 1$ and $n = k_{max} - 1$ also stand for the latter. The main feature of Moore's approach is that it does not consider the number of structural parameters to be evaluated, but is concentrated on the accuracy which can be achieved when calculating the invariants (Section 3.1) from the given set of experimental data. Explicit relations are given for the determination of the invariants and their accuracy from the set of c_j values. The accuracy can be calculated from the experimental values of σ_i using the propagating error law when solving system (3.23). Distortions caused by the termination of series (3.20) at $k = k_{max}$ can also be estimated. Thus, in this approach, the accuracy of the invariants is related to the number of basic points rather than the number of independent structural parameters.

The samping theorem as applied to small-angle scattering therefore leads to the following conclusions:

1. The experimental value of s_{min} should not exceed π/D, and the sampling increment Δs must be smaller than this value.
2. The number of independent parameters ("degrees of freedom," according to Luzzati) sufficient to describe function $I(s)$ in the region (s_{min}, s_{max}) is approximately $n = [(s_{max} - s_{min})D/\pi]$.
3. For a given experimental set of data, the number of degrees of freedom is connected explicitly to the accuracy of the determination of the small-angle invariants.

It is difficult to relate the number of structural parameters which can be evaluated to the number of degrees of freedom. As a rule, the invariants can be calculated reliably provided the requirements listed above for carrying out the experiment are fulfilled. It will be shown below that the number of independent parameters used to describe the structure of a particle is, to a large extent, determined by the *a priori* information available.

3.3. Evaluation of the Invariants

The equations for the invariants derived in Section 3.1 deal with the exact curve $I(s)$ known in the whole region $0 \leqslant s < \infty$. Information available in research always contains distortions; in particular, $I(s)$ can be measured only in a finite interval $s_{\min} \leqslant s \leqslant s_{\max}$. Moreover, the assumption used to derive the equations in question (e.g., leaving out terms of order s^4 in the Guinier law, or the hypothesis that a particle is homogeneous while obtaining expressions for the volume and specific surface) may be fulfilled only approximately. Thus, in a number of cases, the calculation of invariants from experimental data appears to be a complicated task.

An approach to the evaluation of invariants has already been examined in Section 3.2, where the parametrization of functions $\gamma(r)$ and $I(s)$ allows the invariants to be calculated with an estimation of the accuracy. The approach, however, has a limited range of application, requiring the scattering curve to be measured over a wide range of angles in order to achieve reliable accuracy of the invariants (the number of degrees of freedom should be large enough). Meanwhile, when calculating, say, the radius of gyration by the Guinier method, the very beginning of the scattering curve is only to be measured. In the present section, practical approaches to calculating invariants will be considered as well as the conditions necessary for the validity of the derived equations.

3.3.1. Accuracy of Calculation of the Radius of Gyration

First, we shall consider the possibilities of evaluating the best known small-angle invariant: the radius of gyration R_g. We already saw in Section 3.1 that this value determines the behavior of the scattering intensity in the vicinity of $s = 0$. By taking a natural logarithm of equation (3.5) one has

$$\ln I(s) \approx \ln I(0) - s^2 R_g^2/3 \tag{3.24}$$

therefore R_g can be determined from the slope of the linear part of the ln $I(s)$–s^2 plot (the Guinier plot). Expressions for the radius of gyration via the dimensions of simple uniform geometrical bodies are presented in Table 3.1.

The accuracy of calculation of R_g depends on a number of factors. To begin with, as noted in Section 2.3.3, interparticle interference may somewhat influence the initial part of the scattering curve. To eliminate the influence one usually extrapolates to infinite dilution, namely, having measured several samples with different (small) concentrations one reduces the data to the "zero" concentration situation assuming linear dependence of interference effects on concentration. Figure 3.2 shows an example of such a procedure.

Possible aggregation of particles also leads to distortions of scattering curves at very small values of s, i.e., to a sharp increase in scattering intensity. This effect can hardly be estimated quantitatively. Sometimes it is possible to detect visually the presence of aggregates from the Guinier plot (if several leading points deviate sharply from the straight line) and to determine R_g omitting these points.

However, when estimating R_g from the Guinier plot the main difficulties are caused by a limiting value of s_{min}. The value is mainly determined by the minimum scattering angle at which it is possible to detect scattering radiation, i.e., s_{min} depends on the geometry of the experimental setup (see Section 8.2). Thus, while calculating R_g, one

Table 3.1. Radii of Gyration of Some Homogeneous Bodies

Sphere of radius R	$R_g^2 = \frac{3}{5} R^2$
Spherical shell with radii $R_1 > R_2$	$R_g^2 = \frac{3}{5} \frac{R_1^5 - R_2^5}{R_1^3 - R_2^3}$
Ellipse with semiaxes a and b	$R_g^2 = \frac{a^2 + b^2}{4}$
Ellipsoid with semiaxes a, b, and c	$R_g^2 = \frac{a^2 + b^2 + c^2}{5}$
Prism with edges A, B, and C	$R_g^2 = \frac{A^2 + B^2 + C^2}{12}$
Elliptical cylinder with semiaxes a and b and height h	$R_g^2 = \frac{a^2 + b^2}{4} + \frac{h^2}{12}$
Hollow circular cylinder with radii $R_1 > R_2$ and height h	$R_g^2 = \frac{R_1^2 + R_2^2}{2} + \frac{h^2}{12}$

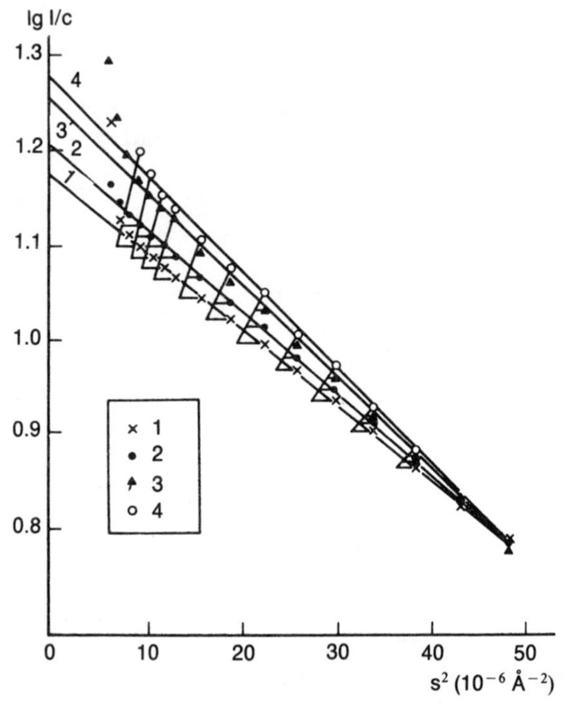

Figure 3.2. Extrapolation to zero concentration of the scattering curve for bacteriophage T7 (Rolbin *et al.*, 1980): (1)–(4) denote concentrations $c = 24$ mg/ml, 18 mg/ml, 6 mg/ml, and "infinite dilution" $c = 0$, respectively.

should always check whether the Guinier approximation is valid for the interval $(s_{\min}, s_1)$ in which the linear dependence (3.24) is sought. Hence the coefficients of the dependence can evaluated more precisely with increasing s_1 (on account of the statistics), but on the other hand this leads to bigger systematic errors, since decomposition (3.1) is only valid for very small values of s.

We now estimate the accuracy of the Guinier approximation quantitatively. Taking into account terms proportional to s^4 in equations (3.5) and (3.1) and denoting

$$\int_0^D r^k \gamma(r) dr = D^{k+1} \int_0^1 x^{k+1} \gamma(xD) dx = \mu^{k+1} R_g^{k+1} M_k$$

where $\mu = D/R_g$ and M_k is the normalized kth moment of function $\gamma(t)$, one has from equation (3.1)

$$I(s) \approx I(0) \left[1 - \frac{1}{6} \frac{M_4}{M_2} \mu^2 (sR_g)^2 + \frac{M_6}{120 M_2} \mu^4 (sR_g)^4 \right]$$

while from equation (3.5)

$$I_G(s) \approx I(0)\left[1 - \frac{1}{6}\frac{M_4}{M_2}\mu^2(sR_g)^2 + \frac{M_4^2}{72M_2^2}\mu^4(sR_g)^4\right]$$

Thus, to an accuracy of terms proportional to s^6, the following estimate is obtained for deviations from the Guinier law:

$$\Delta(s) = \frac{I(s) - I_G(s)}{I(0)} = \frac{3M_6M_2 - 5M_4^2}{360M_2^2}\mu^4(sR_g)^4 = \Delta M\mu^4(sR_g)^4$$

It is evident that the values of ΔM and μ may differ, essentially depending on the density distribution $\rho(\mathbf{r})$. By considering only nearly homogeneous particles one may note that functions $\gamma(r)$ exhibit a specific (decreasing with r) behavior. Since ΔM depends only on the relations between the moments of $\gamma(r)$, its estimate can be calculated. For the majority of shapes $\Delta M = (1\text{–}2) \times 10^{-4}$ (e.g., for a solid sphere $\Delta M = 1.7 \times 10^{-4}$, and for an infinitely thin rod $\Delta M = 1.9 \times 10^{-4}$). If $\Delta M = 2 \times 10^{-4}$, the deviation (in %) from the Guinier equation for a given value of s is given by

$$n \approx (\mu/2.7)^4(sR_g)^4 \tag{3.25}$$

The value of μ does not vary very much for different shapes (for example, it equals 2.46 for a solid sphere, 2.82 for an infinitely thin disk, 3.46 for an infinitely long rod and an arbitrary prism). Moreover, one cannot directly relate quantity μ and particle anisometry; thus $\mu = 3.56$ for the elipsoid of rotation with $c/a = 2$. Therefore, the conventional statement that the more anisometric a particle, the greater the deviation from the Guinier law, is far from true. Figure 3.3 shows the Guinier approximation together with scattering curves for a solid sphere, an infinitely thin disk, and an infinitely long rod, the latter two being extremely anisometric bodies. Estimates of deviations according to equation (3.25) with $\mu = 4$ are also given.

It should be stressed that since approximations neglecting higher orders of sR_g have been used [the Guinier law is valid to an accuracy of $(sR_g)^4$, while the estimate is valid to $(sR_g)^6$], all the above calculations are valid mainly in the interval $s_1R_g < 1$. Here, as can be seen from equation (3.25), systematic deviations do not exceed several percent and hence allow reliable determination of R_g. When $s_1R_g = 1.5$, deviations may reach 20–30% while for $s_1R_g > 2$, the approximations employed break down. However, it is difficult to perform similar calculations for essentially inhomogeneous particles, in which case one may assert that the Guinier approximation is valid at least for $s_1R_g < 1$ because of rapid convergence of power series in the interval.

Thus R_g can be determined from the Guinier plot by proceeding as follows:

1. Estimate this value roughly.
2. Choose the interval $s_1 < 1/R_g$.
3. Refine R_g.

In a good experiment using this procedure R_g may be calculated to an accuracy better than 1%.

Therefore it is seen that, generally speaking, it is undesirable to use points with $sR_g > 1$ for estimating values of R_g. In practice, however, especially when large particles are being investigated, the region $s_{min} < s < 1/R_g$ often contains very few experimental points (or s_{min} even appears to be greater than $1/R_g$). Hence the use of the Guinier approxima-

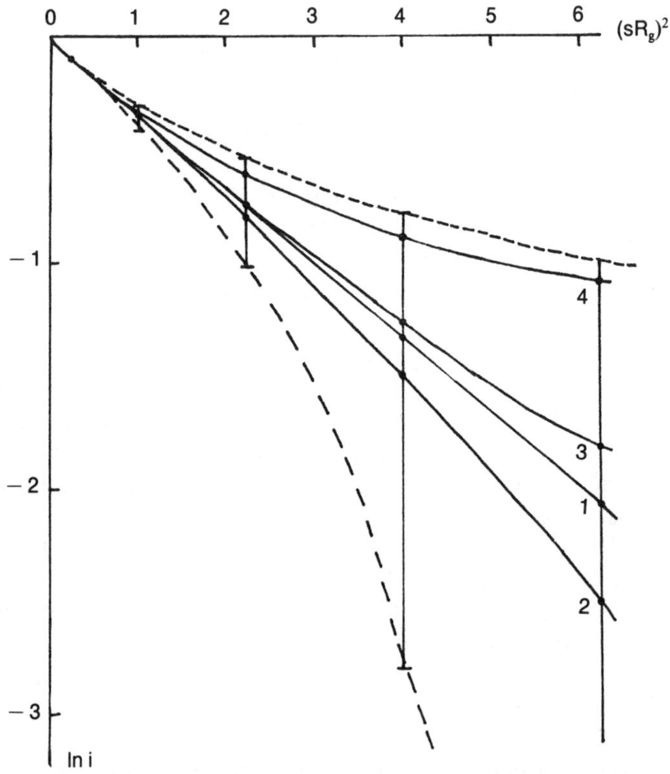

***Figure* 3.3.** Accuracy of the Guinier law: (1) Guinier approximation with estimate (3.25); (2)–(4) correspond to scattering by a solid sphere, an infinitely thin disk, and an infinitely long rod.

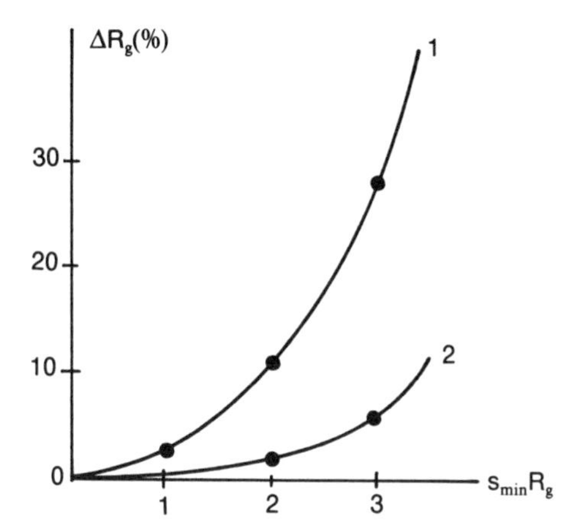

Figure 3.4. Accuracy of determination of the radius of gyration for a solid sphere: (1) according to the Guinier approximation; (2) by means of the indirect transformation.

tion inevitably leads to inherent systematic distortions. (Similar difficulties may arise if one wishes to eliminate concentration or aggregation effects by omitting leading experimental points.)

In these cases it is convenient to determine R_g from function $\gamma(r)$ using equation (3.4). Function $\gamma(r)$ itself can be calculated by one of the indirect transform methods (see Sections 3.2.2 and 9.7.2) that are not so sensitive to the termination of $I(s)$ at very small angles. The accuracies of determination of R_g by means of the Guinier plot and Glatter's technique (Section 9.7.2) are compared in Figure 3.4. Indirect methods can be employed to both calculate R_g more precisely and to some extent eliminate concentration and aggregation effects by omitting the very beginning of the scattering curve. One should bear in mind, however, that in using these methods the range of measurements should be wide enough and the condition $s_{min} < \pi/D$ fulfilled.

3.3.2. Absolute Measurements. Molecular-Mass Determination

In was shown in Section 3.1 that the value of $I(0)$ (which cannot be measured experimentally, so extrapolation to the zero value of s must be used) is determined by the total particle scattering length (3.7), namely, by the sum of the scattering lengths of all atoms inside the particle. Therefore, the chemical composition of a particle being known, the evaluation of $I(0)$ allows the molecular mass of a particle M to be determined in both X-ray and neutron experiments.

Actually the concept of the excess scattering length (Section 2.3.2) can be used to express equation (1.6) for the scattered energy into unit solid angle in the form

$$\frac{dW(0)}{d\Omega}=\frac{A_0^2}{a^2}I(0)=\frac{A_0^2}{a^2}(b_p-\rho_s V_p)^2$$

where b_p is the particle scattering length, ρ_s the scattering density of a solvent, V_p the volume of a particle inaccessible to the solvent, and a the distance between the sample and the recording plane.

For a dilute solution containing N particles one has, according to equation (2.14),

$$\frac{dW(0)}{d\Omega}=N\frac{A_0^2}{a^2}\left(z_p-\frac{\rho_s V_p}{M}\right)^2 M^2$$

where z_p is the scattering length per unit molecular mass of a particle. If $N=S_A dcN_A/M$, where S_A is the cross section of the beam, d the thickness of the sample, and c the weight concentration of particles, and $A_0^2=W_0/S_A$, where W_0 is the total energy of the primary beam irradiating the sample, one can write

$$M=\frac{W(0)}{W_0}\frac{a^2}{(z_p-\bar{v}\rho_s/N_A)^2 dcN_A} \tag{3.26}$$

where $\bar{v}=V_p N_A/M$ is the specific partial volume of a particle. The equation connects explicitly zero-angle scattering with the molecular mass of a particle.

The parameters entering the second factor are determined by the geometry of the instrument used as well as the constitution of the sample. Absolute measurements, namely determination of the ratio $W(0)/W_0$, is the main task while determining the value of M. It is impossible to perform direct measurements in the primary beam since the counting rate of the detectors in use is not high enough. Therefore, for absolute measurements in both X-ray and neutron scattering, either attenuation of the primary beam or calibrated samples must be used.

If, in X-ray scattering, z_{px} and ρ_{sx} are the number of mole-electrons per gram of particle substance and per cubic centimeter of the solvent, respectively, and the scattering intensity by an electron is taken into account (see Section 1.4), then one gets

$$M=\frac{W(0)}{W_0}\frac{21.0a^2}{(z_{px}-\bar{v}\rho_{sx})^2 dc} \tag{3.27}$$

where a and d are expressed in centimeters and c in grams per cubic centimeter (Kratky *et al.*, 1951).

Several techniques have been proposed with which to make absolute measurements in X-ray scattering. The first group involves attenuation of the primary beam. For this purpose several approaches have been developed: attenuation with absorptive filters (Luzzati, 1960); mechanical attenuation by a rotating sectorial diaphragm ["rotator," Kratky, (1960); Kratky and Wawra (1963)] or by moving slits (Stabinger and Kratky, 1978); a change in the operational conditions of the source (Sosfenov and Feigin, 1970). The second group deals with substances of known structure allowing theoretical calculation of the ratio of the primary to the scattering beam intensity ["primary standards" like gases (Shaffer and Beeman, 1970), silica gel (Patel and Schmidt, 1971)]. Secondary standard "Lupolen" [a platelet of polyethylene calibrated in Prof. Kratky's laboratory (Graz, Austria) with the help of the rotator method (Kratky *et al.*, 1966)] is now widely used. The absolute intensity of Lupolen (the ratio of the intensity of the primary to the scattered beam) had been determined for the geometry of the Kratky camera (Section 8.2.4); however, it can be used for other diffractometers as well. Several studies [e.g., Pilz (1969)] have shown Lupolen to be stable with respect to radiation influence and temperature changes. Lupolen is now in use in many small-angle laboratories owing to its high reliability and easy handling properties.

Absolute measurements in neutron scattering can also be performed with attenuators (cadmium, gold) or standard samples (Jacrot, 1976). In studying aqueous solutions it is found convenient to use incoherent scattering by water as a standard. It is possible because the incoherent scattering cross section by hydrogen atoms is much greater than its coherent scattering and absorption cross sections. Thus, if the absorption coefficient of a water sample equals A_H, one can consider $A_H W_0$ of neutrons to be scattered incoherently. Hence the intensity of incoherent scattering I_{inc} may serve to estimate W_0. By performing experiments for a water specimen under the same conditions as for the studied case one can write (Jacrot and Zaccai, 1981)

$$M = \frac{W(0)}{I_{inc}(0)} \frac{A_H}{4\pi(1-A_s) f} \frac{1}{cN_A d(z_p - \bar{v}\rho_s/N_A)^2}$$

where A_s is the absorption coefficient of the sample and $f \leqslant 1$ is a factor accounting for the anisotropy of incoherent scattering ($f = 0.46$ for $\lambda = 1$ Å, 0.75 for $\lambda = 5$ Å, and 1 for $\lambda \geqslant 10$ Å).

The accuracy of the determination of the molecular mass depends on a number of factors, including errors in computing the ratio $W(0)/W_0$ and in estimting the particle concentration in the solution. The greatest difficulties, however, are caused by errors in the value of the partial specific volume $\bar{v}$. We consider, for example, X-ray investigations of proteins in solution. The scattering density of proteins is $\rho_p = 0.441$ eÅ^{-3}, i.e., $z_{px} = 0.535$ mole-electron/g; for water $\rho_H = 0.334$ eÅ^{-3}, i.e., $\rho_{sx} = 0.558$ mole-electron/cm^3. It can easily be seen that a 5% error in $\bar{v}$ (the average value for proteins is 0.74 cm^3/g) leads to a 30% error in M. Neutron scattering is preferable in this case. The average value of z_p for proteins is 2.27×10^{-14} cm/dalton, while $\bar{v}\rho_H/N_A = -0.7 \times 10^{-14}$ cm (the scattering length density for water is $\rho_H = -0.56 \times 10^{10}$ cm^{-2}); therefore the same error in v leads only to a 1.5% error in M. However, there are specific problems in neutron scattering, too. First, incoherent scattering of hydrogen, though useful for absolute measurements, at the same time considerably decreases the accuracy of the difference curve and of $W(0)$. Using buffers containing D_2O one should take into account H–D exchange (for details see Section 4.2.3). In general, one can say that the accuracy of determination of the molecular mass is about 5%.

It is also noteworthy that when extrapolation to $s = 0$ by the Guinier plot is unreliable, it is possible to calculate $I(0)$ (and futhermore M) from equation (3.3) with higher accuracy using one of the indirect methods considered in Section 3.2.1.

3.3.3. Possibilities of Homogeneous Approximation

We have already seen that, assuming the particle to be homogeneous, several structural parameters (volume, surface area, correlation length) can be evaluated. However, straightforward application of the relations derived in Section 3.1 is hardly possible for two reasons: first, the angular range of data collection is limited [function $I(s)$, even extrapolated to $s = 0$, can be known only in a finite interval $0 \leqslant s \leqslant s_{max}$], and second, any real particle contains inhomogeneities which may drastically distort the results. We shall examine how these features can be taken into consideration when calculating values of V, S, and l_m.

Modern experimental small-angle setups afford accurate data collection over a wide range of angles. Therefore, the possibilities of calculating invariants are limited by the interval $0 \leqslant s \leqslant s_{un}$ (where scattering from the given particle is in good agreement with scattering from its shape) rather than by the interval $0 \leqslant s \leqslant s_{max}$. Here, "shape scattering" denotes scattering by a homogeneous body with the same shape as the given particle. The former interval will be regarded as the "homogeneity region" [one should bear in mind that the concept is of some use only when the contrast (Section 2.3.2) is high enough] and this is the region to be used for calculating V, S, and l_m. The contribution of the inner inhomogeneities can be estimated correctly by contrast-variation techniques (see Section 4.2.2). Here, we shall consider approaches which allow one to eliminate their influence (at least partly) for a given contrast.

Initially we note that, having described a particle with density distribution function $\rho(\mathbf{r})$, inhomogeneities due to its atomic structure have been ignored. They are really unimportant here because they convey no information at small-angle resolution; however, that does not mean that they do not influence small-angle curves. Actually, by separating terms with $i = j$ in the Debye equation (2.16) it is seen that the scattering intensity contains the term $\Sigma f_i^2(s)$. Atomic scattering amplitudes may be regarded as constants in small-angle regions; this means that a constant term is involved in the scattering intensity. This does not contradict asymptotic expression (2.31), which was derived for the continuous distribution $\rho(\mathbf{r})$ rather than the singular sum $\Sigma \rho_i \delta(\mathbf{r} - \mathbf{r}_i)$. It merely means that for real particles, the asymptotic expression should be modified to

$$I(s) \underset{s \to \infty}{\approx} \sum_{i=1}^{N} f_i^2 + (2\pi\bar{\rho}^2/s^4)S = c_0 + c_4/s^4 \tag{3.28}$$

Normally, in practical applications one plots function $s^4 I(s)$ versus s^4. For sufficiently large s such a dependence should be linear (more exactly, it should oscillate about the straight line $c_0 s^4 + c_4$). Provided an appropriate region can be separated in the plot, the values of c_0 and c_4 may be determined, e.g., by a least-squares method. For futher calculations c_0 is to be subtracted from function $I(s)$ (Luzzati *et al.*, 1961). The lower the molecular mass of a particle, the relatively greater c_0 becomes. Actually, the constant term in the Debye equation contains N summands, so $N(N - 1)$ terms remain. Thus, the contribution from the atomic structure is approximately $1/N$ times the zero-angle scattering.

It is impossible to take into account larger-size inhomogeneities by this simple procedure. If the dimensions of the inhomogeneities are comparable with the dimensions of the particle (e.g., if the particle consists of two components with approximately equal volumes and considerably different excess scattering densities), then the homogeneous approximation cannot be used at all. In such cases, special methods for studying the structure of inhomogeneous particles must be applied (see Chapter 4). If, however, these dimensions are not too large and the contrast is sufficiently high (such as the inhomogeneities of polypeptide chains when studying proteins in solution), then the homogeneity region can be sufficiently large. Kayushina *et al.* (1974a) carried out model calculations to estimate the limits of the homogeneity region at different sizes of inhomogeneities, l, and their magnitudes ρ_{f}, characterized by the relative mean-square fluctuation

$$\sigma^2 = \int [\rho_{\mathrm{f}}(\mathbf{r})]^2 d\mathbf{r}/\bar{\rho}^2 V$$

A rectangular parallelepiped with the edges satisfying the ratio a: b: c =

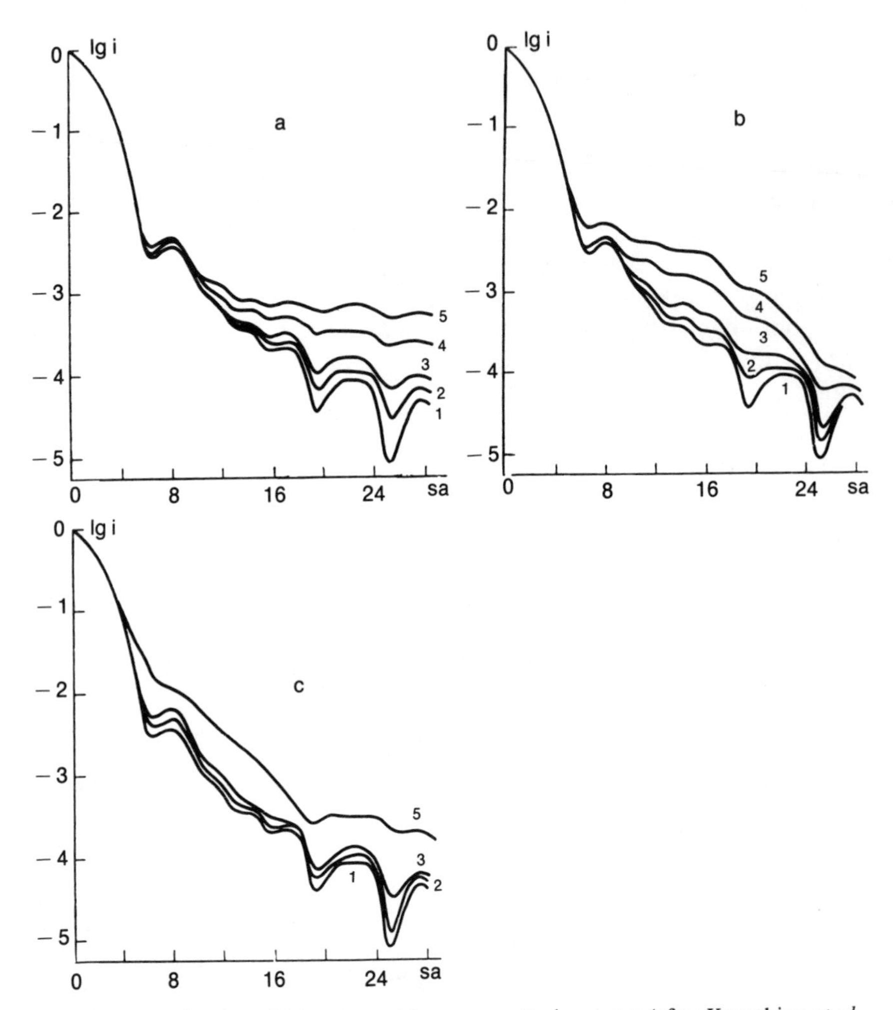

Figure 3.5. Influence of inhomogeneities on a scattering curve (after Kayushina *et al.*, 1974a) with $l = 0.1b$ (a), $0.2b$ (b), $0.3b$ (c). Curves (1)–(5) correspond to $\sigma = 0$ (homogeneous particle), 0.2, 0.33, 0.67, 1.0.

0.8:1:1.6 was chosen (merely to simplify the model calculations). Figure 3.5 shows the calculated model scattering curves for different values of l and σ with chaotic distribution of the inhomogeneities. This example, as well as other calculations presented in the paper of Kayushina *et al.* (1974a), allow one to reach the conclusion that the main parameter describing the homogeneity region $(0, s_{un})$ is the ratio $p = I(0)/I(s_{un})$. For "not very large" inhomogeneities (numerically, $l < 0.2b$ and $\sigma < 0.5$) the interval up to $\lg p = 2.0$–2.5 gives a good estimate of the homogeneity region [the curve $I(s)$ agrees well with the shape scattering in the interval]. Therefore, the region where $I(s)$ drops up to 2–2.5 orders of magnitude may normally be used to calculate the structural parameters of a particle (and, further, to fit its model) in the framework of the homogeneous approximation. We observe a similar situation in X-ray single-crystal structure analysis, where 15–20% of strong reflections contain 70–80% information about the structure, so allowing its preliminary model to be constructed (Vainshtein and Kayushina, 1966).

We now consider how to determine the structural parameters using a finite portion $(0, s_{un})$ of the scattering curve. It seems natural to apply the asymptotic trend (3.28) to this region. Having determined the constants c_0 and c_4 one obtains from equation (2.32), after subtracting c_0, that

$$\tilde{Q} = \int_0^{s_{un}} I(s)s^2 ds + \int_{s_{un}}^{\infty} (c_4/s^4)s^2 ds = Q(s_{un}) + c_4/s_{un} \tag{3.29}$$

Thus, for the invariants,

$$V = 2\pi I(0)/\tilde{Q} \tag{3.30}$$

and

$$S = 2\pi^3 c_4 I(0)/\tilde{Q}^2 \tag{3.31}$$

Similarly, the correlation length can be expressed as

$$l_m = \frac{2\pi}{\tilde{Q}}\left[\int_0^{s_{un}} I(s)s\, ds + 2c_4/s_{un}^2\right] \tag{3.32}$$

The asymptotic expression (3.28) can be employed to calculate the corresponding correlation function as well:

$$\tilde{\gamma}(r) = \frac{1}{2\pi^2}\left\{\int_0^{s_{un}} I(s)\frac{\sin(sr)}{sr}s^2 ds + \frac{c_4}{2s_{un}} \times\left[\cos(t) + \frac{\sin(t)}{t} + \frac{[\mathrm{Si}(t) - \pi/2]}{t}\right]\right\} \tag{3.33}$$

where $t = rs_{un}$ and $\mathrm{Si}(t)$ is the integral sine. Moreover, using $\gamma(r)$ one can calculate the invariants from equation (3.13) and the relations

$$V = 4\pi \int_0^D \tilde{\gamma}_0(r) r^2 dr \tag{3.34}$$

and

$$S = 4\pi \lim_{r\to 0}\{[1 - \tilde{\gamma}_0(r)]/r\} \tag{3.35}$$

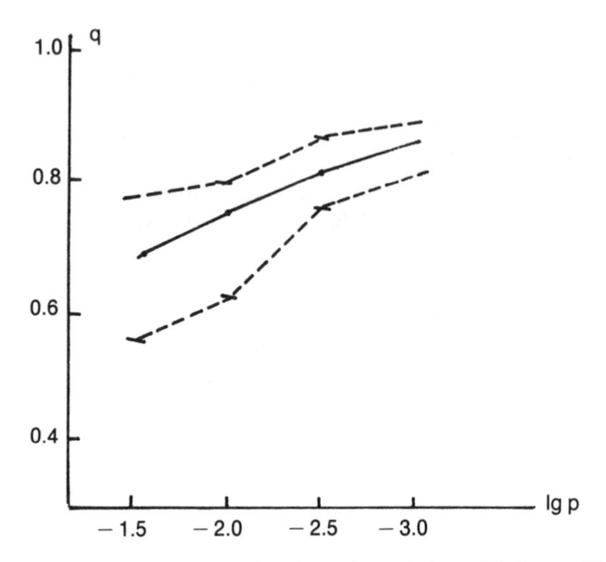

Figure 3.6. Values of q as a function of the intensity ratio ln p. All the considered models gave values of q within the area bounded by the dashed lines.

where $\tilde{\gamma}_0(r) = \tilde{\gamma}(r)/\gamma(0)$ [equations (3.34) and (3.35) are easily derived from equations (2.23), (2.25), and (2.31)]. If the values of the invariants calculated by the two approaches coincide, it confirms the validity of the homogeneous approximation and the correctness of the determined values of c_0 and c_4.

Normally, such an approach ensures good accuracy for calculating the invariants (approximately 5% for the volume and 10% for the surface). However, it is not always possible to reliably determine the parameters of the relationship (3.28). Therefore, the empirical procedure suggested by Kayushina *et al.* (1974b) to estimate the value of Q is of interest. It was shown by model examples that the error in the value of Q depends mainly on the value of p. The plot of $q = Q(s_{\text{max}})/Q$ versus lg p was obtained by averaging the results for as many as ten different homogeneous bodies, and is shown in Figure 3.6. The relationship can be fairly well approximated by the expression

$$q = -0.0217(\lg p)^2 + 0.2185 \lg p + 0.3993 \qquad (3.36)$$

(when lg $p = 2.5$, $q = 0.8$). Hence having chosen that part of the curve $(0, s_p)$ corresponding to a value of p, one can first determine the value of q from equation (3.36) and then calculate

$$\tilde{Q}_p = Q(s_p)/q(p) \qquad (3.37)$$

and

$$\tilde{c}_4 = s_p Q(s_p)[1 - q(p)]/q(p) \tag{3.38}$$

Experience shows that this approach also gives an accuracy of about 5–10%. A practical example of calculating the volume of bacteriophage T7 from the X-ray small-angle scattering curve is given in Figure 3.7 (Rolbin *et al*., 1980b). The calculated value $V = (1.24 \pm 0.05) \times 10^8$ Å^3 is the same for both considered techniques.

Therefore, to calculate the parameters V, S, and l_m for real particles one begins by choosing the homogeneous region. Thus, obtained values of the invariants will correspond to "the best homogeneous model," namely, to a uniform body, such that the scattering intensity coincides with the scattering by the given particle in this region. Apparently, more exact estimates of the invariants are impossible unless additional information about the dimensions and magnitude of the inhomogeneities is available.

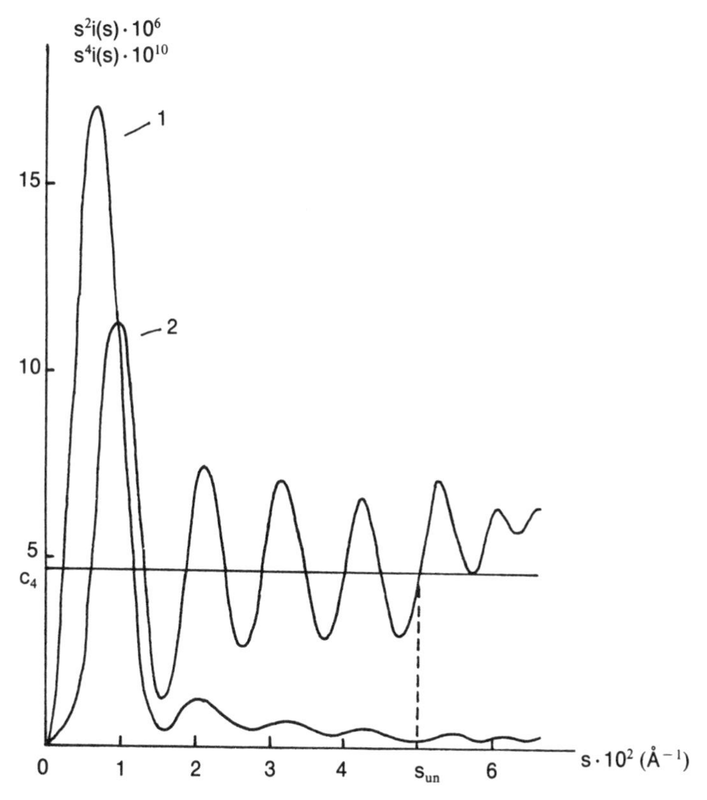

Figure 3.7. Calculation of the volume of bacteriophage T7: (1) curve $s^2 i(s)$, (2) curve $s^4 i(s)$; $c_0 = 0$, $c_4 = 4.7 \times 10^{-10}$, $s_{un} = 0.05$ Å^{-1}.

3.3.4. Estimate of the Largest Dimension

The largest dimension in a particle, D, besides its structural importance, plays a significant role in the treatment and interpretation of experimental data. It was shown in Section 3.2 that the value of D is critical for methods using the sampling theorem; several data-processing techniques also require this value to be specified (see Sections 9.3.3 and 9.7). Therefore in many cases treatment of data should start with the determination of D.

The condition $\gamma(r) \equiv 0$ for $r > D$ can be used to estimate easily the value of D, the correlation function itself having already been calculated (unless there are particle aggregates in the sample). We note only that the concentration effects manifest themselves by lowering the curve of function $\gamma(r)$ at large values of r. Normally, a negative peak appears in this region, the value of D corresponding to its minimum (Pilz *et al*., 1980).

In this way up to 5% accuracy can be achieved when estimating the value of D. However, in computing function $\gamma(r)$ with the aid of equation (2.22), a number of data-processing techniques should be included to restore function $I(s)$ from the set of experimental data (see Section 9.1.6) and to extrapolate to $s = 0$ and $s \to \infty$. Damaschun *et al*. (1974) and Müller *et al*. (1980) developed a method using sets of raw data. It was shown that the "frequency function"

$$F(\omega) = \int_0^\infty J(s)\cos(\omega s)ds \tag{3.39}$$

where $J(s)$ is a smeared curve (see Section 9.1.3), differs from zero in the same interval $0 \leqslant \omega \leqslant D$ as $\gamma(r)$ for arbitrary experimental conditions. Taking into consideration that $J(s)$ may contain statistical errors, which can be described by a function $\varepsilon(s)$, we obtain

$$F_\varepsilon(\omega) = F(\omega) + \xi(\omega)$$

where

$$\xi(\omega) = \int_0^\infty \varepsilon(s)\cos(\omega s)ds$$

The integral depends on the angular increment Δs and differs substantially from zero for frequencies ω equal to about $\pi/\Delta s$. Thus, for experiments with $\Delta s \ll \pi/D$ (the condition which should always be fulfilled; see Section 3.2.1) one has $F_\varepsilon(\omega) \approx F(\omega)$ for $\omega < D$. This means that the value of D can be estimated directly from the raw data, omitting smoothing and desmearing procedures. The accuracy decreases but, as shown by Müler *et al*. (1980) with model examples, it remains within the limits of 5–10%. A refinement of the value of D can be achieved during the course of further data treatment.

Nonetheless, one should bear in mind that such approaches allowing direct determination of D are very sensitive to the termination of $I(s)$ for $s < s_{min}$ and $s > s_{max}$. The latter is not of much importance when the scattering curve is measured up to sufficiently large angles [drops in $I(s)/I(0)$ of more than 2.5–3 orders of magnitude]. In this case, the methods described below in Section 9.6 and extrapolation to Porod's law [the trend $I(s) \sim s^{-4}$ is used for the point collimation and $I(s) \sim s^{-3}$ for the infinite slit; see Section 9.4.1] may be used to reduce the termination effect. Either the Guinier law or the sampling theorem should be used for extrapolation to $s = 0$ (Damaschun *et al*., 1971; Damaschun and Pürschel, 1971). The former technique requires the condition $s_{min} < 1/R_g$, and the latter $s_{min} < \pi/D$, to be fulfilled.

An application of the indirect methods (see Sections 3.2.2 and 9.7) provides another possibility for estimating the value of D. An approximate value $D \approx D_0$ is always known *a priori* with some degree of accuracy. (It is this fact that enables one to continue experiments in such a way that structural information can be further extracted.) Thus, by using indirect methods for different values of D near D_0, one may analyze the obtained deviations from the experimental data and structural parameters and at least select a stable region of values of D. A more accurate estimate of D can then be made by analyzing the profiles of functions $\gamma(r)$. As noted earlier, the condition $s_{min} < \pi/D$ is also necessary in order to apply indirect methods.

Hence the determination of this invariant, unlike others, normally involves visual data analysis. Nevertheless, the value of D can be estimated fairly accurately (up to 1%), provided high-quality experimentation [$(s_{min} < \pi/D$, $\Delta s \ll \pi/D$, $I(s_{max})/I(0) \sim 10^{-3})$] is available.

3.3.5. Evaluation of the Invariants of Anisometric Particles

In principle, one can calculate the invariants of rodlike and flat particles in the same manner as for globular ones. The only difference is that, in this case, it is always necessary to check the validity of the assumption about high anisometry of particles.

Thus equations (3.14) and (3.15), and plots of function $\ln[sI(s)]$ or $\ln[s^2I(s)]$ versus s^2, similar to the Guinier plot, can be employed to determine the radii of gyration of the cross section R_c for rodlike particles or of the thickness R_t for lamellar particles. However, these equations are invalid for both large and very small angles, since representations (2.26) and (2.39) break down as s tends to zero. Requiring conditions $s < 1/R_c$ and $s > 2\pi/L$ to be satisfied simultaneously (see Section 2.4.5), one

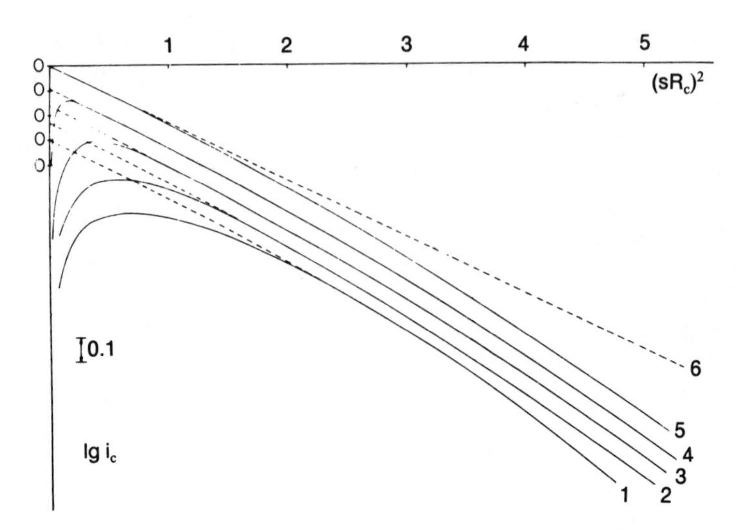

Figure 3.8. Theoretical cross-section factors for the prolate ellipsoid of revolution (after Mittelbach, 1964), where $i_c(s) = sI(s)/\lim[sI(s)]$ as $s \to 0$: curves (1)–(5) correspond to $c/a = 1.5, 1.8, 2.5, 5, \infty$; (6) represents the function $\exp(-s^2R_c^2/2)$ for an infinitely long ellipsoid.

obtains the estimate $L > 2\pi R_c$; for such particles R_c can be determined reliably. Suppose for prolate ellipsoids with ratio of semiaxes a: a: c the estimate gives $c/a > \pi/\sqrt{2} = 2.2$. This situation is illustrated in Figure 3.8, where the curves $\ln[sI(s)]$ of the ellipsoids 1: 1: c are presented for different values of c. It is evident that for $c > 2.5$ the linear region may be identified clearly. Similar calculations for flat particles show that the sizes of lamellae should be 2.5–3 times greater than their thickness in order to determine the value of R_t.

Mass characteristics of anisometric particles can be determined from the values of $I_c(0)$ and $I_t(0)$. By using equation (3.26) as well as equations (2.36) and (2.39) one obtains

$$M_L = \frac{M}{L} = \frac{\lim_{s\to 0}[sW(s)]}{\pi W_0} \frac{a^2}{(z_p - \bar{v}\rho_s/N_A)^2 dcN_A} \tag{3.40}$$

which is the molecular mass per unit length of a rod, while

$$M_S = \frac{M}{S_1} = \frac{\lim_{s\to 0}[s^2W(s)]}{2\pi W_0} \frac{a^2}{(z_p - \bar{v}\rho_s/N_A)^2 dcN_A} \tag{3.41}$$

is the molecular mass per unit area of a lamella. Absolute measurements are performed, as described in Section 3.3.2. Independent equations governing M and M_L (or M_s) allow one to estimate L (or S_l) without calculating the molecular masses themselves:

$$L = \pi I(0) \Big/ \lim_{s \to 0} [sI(s)] \tag{3.42}$$

and

$$S_l = 2\pi I(0) \Big/ \lim_{s \to 0} [s^2 I(s)] \tag{3.43}$$

Furthermore, by taking into account asymptotic expression (3.28) one obtains for rodlike particles

$$L_c/S_c = \pi c_4/Q \tag{3.44}$$

and for flat ones

$$1/T = \pi c_4/Q \tag{3.45}$$

Naturally, the requirement for particle homogeneity remains.

Certain conclusions about the anisometry of homogeneous particles can be reached on the basis of the profiles of function $p(r)$. Thus, for a solid sphere, this function attains a maximum at $r_m = 0.525D$. For deviations from spherical shape, the ratio r_m/D will decrease in the following sequence: globular, platelike, and rodlike particles, which reflects the decreasing relative contribution of long distances to $p(r)$. Hence quantity r_m/D may serve as a criterion of particle shape anisometry.

Quantitative estimates are also possible for very anisometric particles (Glatter, 1979). Consider, for example, the behavior of function $p(r)$ for rodlike particles at $d < r < D$, where d is the cross-section diameter. In this case the number of volume elements spaced r apart is given by integration over the cross sections separated by this distance, i.e.,

$$p(r) = \frac{2}{4\pi} \int_r^D du \iint_{S_c} \rho_c^2 dS_1 dS_2 = \frac{(D - r)\rho_c^2 S_c^2}{2\pi}$$

Therefore function $p(r)$ will be linear for $r > d$, its slope given by

$$dp/dr = -S_c^2 \rho_c^2 / 2\pi \tag{3.46}$$

When $r \approx d$, the behavior of $p(r)$ changes because distances from the same cross section now contribute to the function. The inflection point of $p(r)$ in this region corresponds approximately to $r_i \approx \sqrt{S_c}$, the characteristic size of the cross section (e.g., $r_i \approx R$ for the cylinder of radius R). Typical shapes of $p(r)$ for elongated homogeneous particles are presented in Figure 3.9.

For flat particles, similar considerations prove that function $p(r)$ decreases quadratically for $r > T$, while for $r \approx T$ a transition takes place owing to the additional contribution from the distances along the thickness. The function $f(r) = p(r)/r$ can be used conveniently in these cases (Figure 3.10). Extrapolation of the linear part of the function to $r = 0$ enables the area of the lamella to be estimated as

$$\lim_{r\to 0} f(r) = \rho_t^2 S_l T^2/2 \tag{3.47}$$

In should be noted, however, that in practice it is frequently difficult to specify the linear regions and transition points in function $p(r)$ and $f(r)$, especially for not very anisometric particles, so that the data obtained in such approaches are not always reliable.

The value of R_c or R_t provides another approach when estimating the length L of the rod and thickness T of the lamella. For homogeneous prisms and cylinders one may actually write

$$R_g^2 - R_c^2 = L^2/12 \tag{3.48}$$

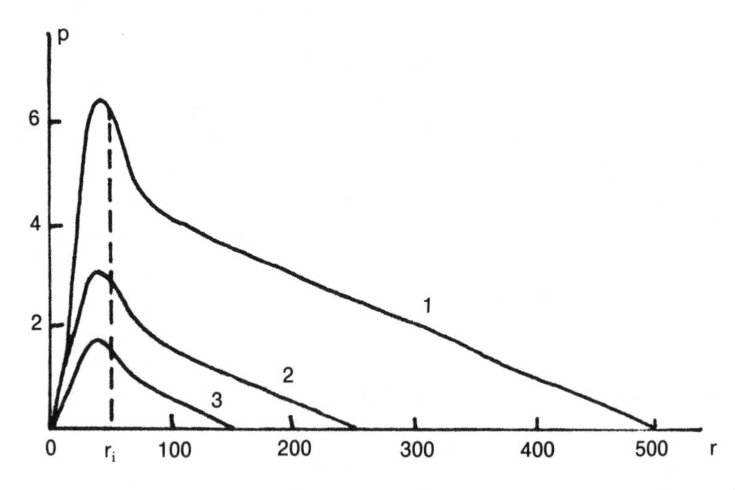

Figure 3.9. Function $p(r)$ for homogeneous prisms with identical cross sections $A \times B = 50 \times 50$ but different heights: (1)–(3) correspond to $L = 500, 250, 150$ (after Glatter, 1979).

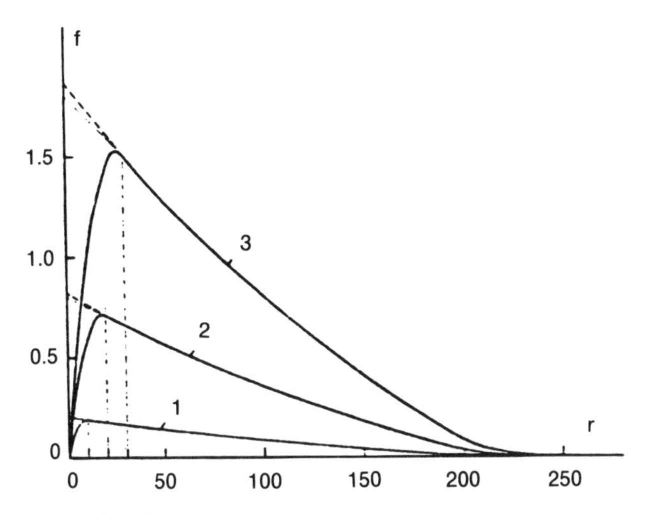

Figure 3.10. Function $f(r)$ for homogeneous prisms with identical basal plane $A \times B = 100 \times 100$ but different thickness: (1)–(3) correspond to $T = 10, 20, 30$ (after Glatter, 1979).

if the particle is elongated, and

$$R_t^2 = T^2/12 \tag{3.49}$$

if it is flattened.

Therefore, a number of geometrical and weight parameters of strongly elongated and flattened particles can be determined from small-angle scattering curves and corresponding characteristic functions. One should, however, bear in mind that all the equations considered are approximate in character, holding only for highly anisometric particles. That is why the parameters of anisometric particles can normally be calculated with poorer accuracy than the invariants considered in Sections 3.3.1–3.3.4.

3.3.6. List of the Basic Equations

The main approaches for calculating integral parameters of particles invariant with respect to the small-angle scattering curve have been examined. For the convenience of the reader, the basic equations for the invariants as well as the conditions governing their applications and (when possible) an approximate estimate of their accuracy are given in Tables 3.2 and 3.3. It is clear that one and the same parameter can frequently be determined in different ways. Such duplicate determi-

Table 3.2. Equations Governing Conditions, and Accuracy of Calculation of Invariants

<table>
<tr><th rowspan="2">Parameters</th><th rowspan="2">D</th><th>R_g</th><th>$I(0)$</th><th rowspan="2">M</th><th colspan="3">Homogeneous approximation</th></tr>
<tr><th>V</th><th>S</th><th>l_m</th></tr>
<tr><td rowspan="2">Reciprocal space</td><td rowspan="2">$\gamma(r) \equiv 0$
or
$F(\omega) \equiv 0$
for $r, \omega > D$</td><td colspan="2">$\ln[I(s)] \approx \ln[I(0)] - s^2 R_g^2/3$</td><td>$[W(0)/W_0] \times \frac{a^2}{(z_p - v\rho_s/N_A)^2 dc N_A}$</td><td>$\frac{2\pi^2 I(0)}{\tilde{Q}}$
$\tilde{Q} = Q(s_{un}) + \frac{c_4}{s_{un}}$</td><td>$\frac{2\pi^3 c_4 I(0)}{\tilde{Q}^2}$</td><td>$\frac{2\pi Q_1(s_{un})}{\tilde{Q}} + \frac{4\pi c_4}{s_{un}^2 \tilde{Q}}$</td></tr>
<tr><td colspan="2">Condition: sufficient number of points in the interval (0, $1/R_g$)
Accuracy up to 1–2%</td><td rowspan="4">Necessary:
absolute measurements, specific partial volume, chemical composition of particle and solvent</td><td colspan="3">$Q(z) = \int_0^z I(s) s^2 ds; \quad Q_1(z) = \int_0^z I(s) s \, ds$
Condition: $I(s_{un})/I(0) \sim 10^{-2}$–$10^{-3}$</td></tr>
<tr><td rowspan="3">Real space</td><td rowspan="3">Condition:
$s_{min} < \pi/D$
Accuracy:
5%</td><td>$\int_0^D r^4 \gamma(r) dr$</td><td rowspan="2">$4\pi \int_0^D r^2 \gamma(r) dr$</td><td>Accuracy 5%</td><td>Accuracy 10%</td><td>Accuracy 10%</td></tr>
<tr><td>$2 \int_0^D r^2 \gamma(r) dr$</td><td>$4\pi \int_0^D r^2 \gamma_0(r) dr$</td><td>$\lim_{r \to 0} 4V \frac{1 - \gamma_0(r)}{r}$</td><td>$2 \int_0^D \gamma_0(r) dr$</td></tr>
<tr><td colspan="2">$s_{min} < \pi/D$, $\Delta s \ll \pi/D$
Accuracy up to 1%</td><td colspan="3">Conditions, accuracy are the same</td></tr>
</table>

Table 3.3. Expressions for Determining the Parameters of Anisometric Particles

	Rodlike particles		Flat particles
R_c	$\ln[sI(s)] \approx \ln[I_c(0)] - s^2R_c^2/2,\ 2\pi/L < s < 1/R_c$	R_t	$\ln[s^2I(s)] \approx \ln[I_s(0)] - s^2R_t^2,\ 2\pi/\sqrt{S_l} \leqslant s \leqslant 1/R_t$
M/L	$\left\{\lim_{s\to 0}[sW(s)]/\pi W_0\right\}\{a^2/(z_p - \bar{v}\rho_s/N_A)^2dcN_A\}$	M/S_l	$\left\{\lim_{s\to 0}[s^2W(s)]/2\pi W_0\right\}\{a^2/(z_p - \bar{v}\rho_s/N_A)^2dcN_A\}$
L	$\pi I(0) \Big/ \lim_{s\to 0}[sI(s)]$	S_l	$2\pi I(0) \Big/ \lim_{s\to 0}[s^2I(s)]$
	Homogeneous approximation		
L_c/S_c	$\pi c_4/Q$	$1/T$	$\pi c_4/Q$
S_c	$\sqrt{-2\pi dp(r)/dr/\rho_c}, \quad r > d_c$	S_l	$\lim_{r\to 0}[p(r)/r]/\rho_t^2T^2$
Section size	Point of transition to linear slope of function $p(r)$	T	Point of maximum of function $f(r)$
L	$[12(R_g^2 - R_c^2)]^{1/2}$	T	$\sqrt{12}R_t$

nation of invariants in usually quite helpful: if the results of different methods coincide, one can at least be sure that the data do not contain any serious errors.

In conclusion we consider the concept of the invariants themselves. It is accepted formally that the calculation of invariants can be performed without any preliminary information about a particle. As a matter of fact, this would be the only case when an ideal, exact scattering curve were known in the whole infinite range of scattering vectors. In reality, smeared sets of experimental data collected at a finite number of discrete points are available. Hence if the measurements were conducted incorrectly (e.g., $\Delta s > \pi/D$ or $s_{min} \ll 1/R_g$), reliable restoration of invariants is impossible by any data-treatment technique. Therefore, usually at least approximate general knowledge about a particle is necessary for data collection and subsequent evaluation of the invariants.

3.4. Scattering by Particles of Simple Shape

Having determined the invariants one may proceed to study the particle's structure. An approximation of a particle by simple geometrical bodes (ellipsoids, prisms, cylinders) is frequently employed as a first step. For instance, in the case of biological macromolecules the approximation provides a rough estimate of their shape, size, and anisometry, allowing certain conclusions to be drawn about their structure in solution.

To begin with we note that small-angle scattering curves are sufficiently sensitive to the shape of a particle, even in simple homogeneous models. Curves $i(s)$ for different homogeneous bodies with the same values of R_g and V are presented in Figure 3.11. One can see that changes in shape lead to distinct differences in the curves, even in the region up to the first subsidiary maximum ($sR_g < 5$).

It is possible, for a number of simple geometrical bodies, to express the small-angle scattering intensity as a function of their parameters (axes, heights, radii) in a convenient form. Corresponding expressions may be derived by calculating the scattering amplitude of a fixed particle (three-dimensional Fourier image of its shape) followed by averaging of the square of the amplitude with respect to the solid angle in reciprocal space. Relations were derived and intensity curves calculated for ellipsoids (Guinier, 1939; Porod, 1948; Mittelbach and Porod, 1962; Soler, 1975), rectangular parallelepipeds (Porod, 1948; Mittelbach and Porod, 1961a), and cylinders (Porod, 1948; Mittelbach and Porod, 1961b). Table 3.4 presents a summary of equations for calculating the intensity of scattering by several simple bodies.

Both visual comparison and automatic searching may be applied to find the best approximation among simple homogeneous (usually three-

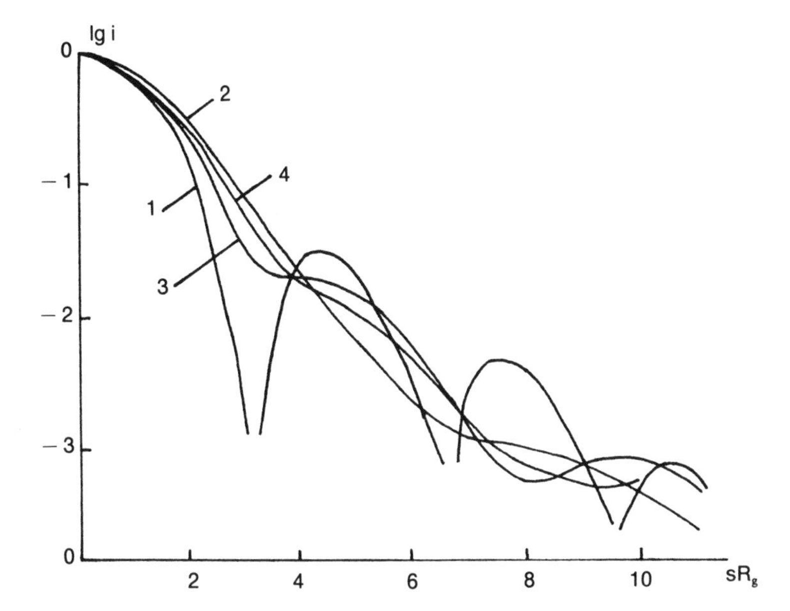

Figure 3.11. Scattering by objects with identical values of R_g and V and various shapes (after Feigin, 1971): (1) spherical shell; (2) triaxial ellipsoid, a: b: c = 0.5: 1: 1.5; (3) four touching ellipsoids, their centers at the vertices of a tetrahedron; (4) the same as (3), but the cavities filled with substance.

parameter) bodies. In Prof. O. Kratky's laboratory (Graz, Austria) many curves have been calculated for scattering from simple bodies. They are plotted in coordinates of lg i versus sa (a being the characteristic size of the particle), thus allowing convenient comparison with experimental data. The use of a computer display as one possible means of such a comparison was described by Hiragi (1979). Qualitatively, one can state that the more anisometric a particle, the flatter, with less distinct maxima, the form of the scattering curve. This is illustrated in Figure 3.12, where the scattering curves for ellipsoids of rotation are shown with different ratios c/a. If the values of the invariants are known, one can estimate the parameters of a model body. Thus, if the values of R_g, V, and D are given and one searches for the best triaxial ellipsoid, the expressions

$$c^0 = D/2 \quad \text{and} \quad a^0, b^0 = ([\alpha \pm (\alpha^2 - 4\beta^2)^{1/2}]/2)^{1/2}$$

($\alpha = 5R_g^2 - D^2/4$, $\beta = 3V/2\pi D$) may be used as a first approximation of its semiaxes.

Table 3.4. Equations for Scattering Intensities of Simple Bodies

Body	Equation
Uniform sphere of radius R	$9\left(\frac{\sin t - t\cos t}{t^3}\right)^2 = \phi^2(t), \qquad t = sR$
Spherical layer with radii $R_1 > R_2$	$(R_1^3 - R_2^3)^{-2}[R_1^3\phi(sR_1) - R_2^3\phi(sR_2)]^2$
Triaxial ellipsoid (semiaxes a, b, c)	$\int_0^1\int_0^1 \phi^2\{s[a^2\cos^2(\frac{1}{2}\pi x) + b^2\sin^2(\frac{1}{2}\pi x)(1-y^2) + c^2y^2]^{1/2}\}dxdy$
Ellipsoid of rotation a: a: va	$\int_0^1 \phi^2[sa(1 + x^2(v^2-1))^{1/2}]dx$
Parallelepiped (edges A, B, C)	$\int_0^1 \Psi_p[s, B(1-x^2)^{1/2}, A]S^2(sBCx/2)dx;\ S(t) = \sin(t)/t$ $\Psi_p(s, B, A) = \frac{2}{\pi}\int_0^{\pi/2} S^2[sA\sin(y/2)]S^2[sB\cos(y/2)]dy$
Right elliptical cylinder with height H, semiaxes of ellipse a, va	$\int_0^1 \Psi_{ec}[s, a(1-x^2)^{1/2}]S^2(sHx/2)dx$ $\Psi_{ec}(s, a) = \frac{1}{\pi}\int_0^{\pi} \Lambda_1^2\left[sa\left(\frac{1+v^2}{2} + \frac{1-v^2}{2}\cos y\right)^{1/2}\right]dy$ $\Lambda_1(t) = 2J_1(t)/t$
Right hollow cylinder with height H, outer radius R_1, inner radius R_2	$\int_0^1 \Psi_{hc}[s, R_1(1-x^2)^{1/2}, R_2(1-x^2)^{1/2}]S^2(sHx/2)dx$ $\Psi_{hc}(s, R_1, R_2) = \frac{1}{1-\gamma^2}[\Lambda_1(sR_1) - \gamma^2\Lambda_1(sR_2)]$ $\gamma = R_2/R_1$
Right circular cylinder of radius R, height H	$4\int_0^1 \frac{J_1^2[sR(1-x^2)^{1/2}]}{[sR(1-x^2)^{1/2}]^2}S^2(sHx/2)dx$
(a) $R = 0$ (infinitely thin rod, height H)	$2\,\mathrm{Si}(sH)/sH - S^2(sH/2), \qquad \mathrm{Si}(t) = \int_0^1 S(x)dx$
(b) $H = 0$ (infinitely thin disk, radius R)	$[2 - \Lambda_1(2sR)]/s^2R^2$

Visual search of the best approximation is an extremely subjective procedure. Automatic search is possible for bodies allowing sufficiently rapid calculation of $I(s)$ at the given parameters (such as the bodies listed in Table 3.4). By expressing the scattering intensity via a hypothetical model as a function of the scattering angle and the chosen parameters

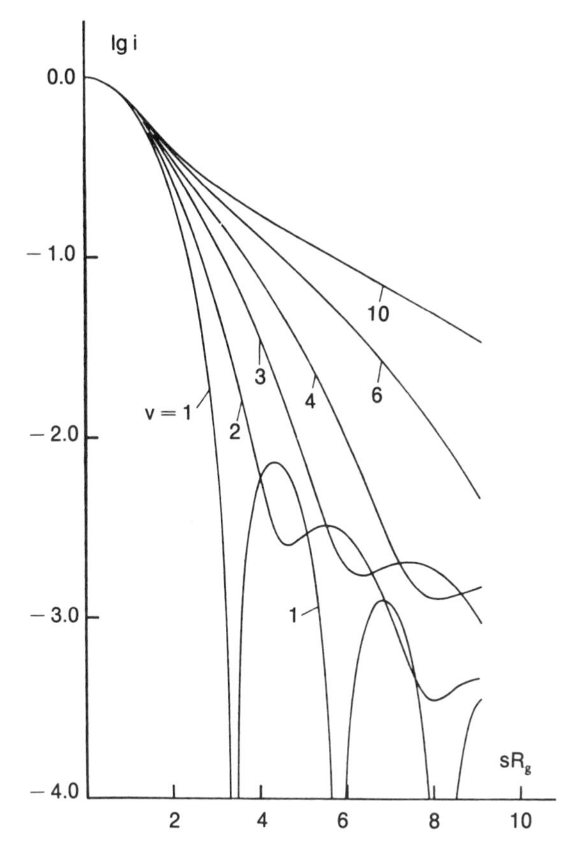

Figure 3.12. Scattering curves for prolate ellipsoids of rotation with ratio of semiaxes $v = c/a$ (after Kratky and Pilz, 1972).

$I(s) = I(s, \mathbf{X})$ [e.g., for an ellipsoid $\mathbf{X} = (a, b, c)$], one can apply a general algorithm described in Section 3.2. From the approximations for different classes of bodies we can choose that providing the best agreement with experiment (namely, with the scattering curve and the values of the invariants). It should be noted, however, that the scattering intensity at sufficiently large s, even in a region of homogeneity, cannot be adequately represented by the scattering curve from a simple body; therefore, the same degree of agreement between the calculated and experimental curves over the whole angular interval of approximations is not possible. Hence a weighting function $W(s)$, which tends to zero as the value of s increases, should be inserted into a function like (3.28). The solution will be strongly influenced by the choice of function, as well as by the choice of the approximation interval itself.

It is also possible to search for simple models in real space. We have already seen (Sections 2.4.2 and 3.3.3) that functions $\gamma(r)$, $p(r)$, and $f(r)$ reflect explicitly the details of a particle shape. Glatter (1979) calculated functions $p(r)$ and $f(r)$ for several simple bodies and their aggregates; sometimes, such modeling appears to be more convenient than searching in s-space. In general, complementary information can be obtained when searching for simple models in real and reciprocal spaces (we saw a similar situation for the calculation of the invariants).

3.5. Modeling Method

Only a rough approximation of particle shape can be achieved in terms of simple geometrical bodies. At the same time, while studying various objects some information on the structure of particles (their symmetry, subunit constitution, and so on) is available. In particular, when solutions of biological macromolecules are investigated a large amount of additional information is frequently known from electron microscopy and different physicochemical measurements. Taking into consideration these data one can search for fine details of particle shape or even structure. Therefore, the following approach (the modeling method) is often used: various plausible models are considered, corresponding scattering intensities are calculated and compared with the experimental curve (as a rule, in the homogeneity region). Although uniqueness of the solution cannot be guaranteed, such an approach in many cases makes it possible to choose the model best corresponding both to the scattering data and to *a priori* information.

3.5.1. Demands on the Technique of Calculation

In the modeling process, a number of scattering curves from different models should be calculated until the best agreement with experimental data is achieved. The models are constructed using the obtained values of the invariants as well as *a priori* information on the particle. The following integral discrepancy factor may serve as an objective criterion for choosing the best model:

$$R_{\mathrm{I}} = \int_{s_{\min}}^{s_{\max}} |I(s) - I_{\mathrm{M}}(s)|\, s^2 ds \bigg/ \int_{s_{\min}}^{s_{\max}} I(s) s^2 ds \tag{3.50}$$

where $I_{\mathrm{M}}(s)$ is the scattering intensity of a model while $(s_{\min}, s_{\max})$ is the interval of fitting.

Such an approach implies considerations of a large number of models (normally, of order 10^1–10^2); therefore the primary problem is how to calculate the scattering intensities for many different bodies as rapidly as possible. Thus, a method of calculating the model intensities should meet two basic demands: (1) rapid calculation and (2) convenient description of a model suitable for a computer.

The exact Debye equation (2.16) is clearly of no use from this point of view. A rougher representation of models is usually used: they are described as consisting of various (identical or nonidentical) subunits.

3.5.2. Subparticle Models

It is natural to extend the approach considered in Section 3.4 and to describe a particle as an aggregate of several three-parameter bodies (subparticles). To this end Müller *et al.* (1979) have used triaxial ellipsoids, elliptical and hollow cylinders, and rectangular prisms. Hence the type, density, coordinates of the mass center, dimensions, and orientation of each subparticle must be specified in order to fully describe the model. Therefore, $11n-6$ parameters are necessary to define a model consisting of n bodies, where the first body is positioned at the origin of coordinates and determines the directions of the coordinate axes. It is very convenient to describe a model in such a way that it can be varied, because varying a model can normally be achieved by appropriate changes in orientation, density, and dimensions of certain subparticles. On the other hand, it is impossible to derive explicit equations for the scattering intensity by such models. It would be necessary to average the scattering intensities by the models numerically with respect to all orientations, involving excessive computer time.

3.5.3. Spheres Method

A faster method of calculation is provided by a modification of the Debye equation. We divide an assumed model into n regions equal in shape. Thus, one can write approximately

$$I_{\mathrm{M}}(s) = g(s) \sum_{i=1}^{N} \sum_{j=1}^{N} W_i W_j \frac{\sin(sr_{ij})}{sr_{ij}} \tag{3.51}$$

where $g(s)$ represents the structure factor of a region averaged over all orientations and W_i is the weight (full scattering length) of the ith region (Rolbin *et al.*, 1971). The accuracy of model presentation depends on the number and dimensions of the regions (spheres are conventionally used for this purpose). For a set of spheres with radius a and equal scattering density one can rewrite equation (3.51) as (Rolbin *et al.*, 1973)

$$I_{\mathrm{M}}(s) = n\phi^2(sa)\left[1 + \frac{2}{n} \sum_{i=1}^{n-1} \sum_{j=i+1}^{n} \frac{\sin(sr_{ij})}{sr_{ij}}\right] \tag{3.52}$$

In order to simplify the definition of the model, Rolbin *et al*. (1971) proposed packing the spheres into a three-dimensional cubic lattice. Essentially less computer time is needed for calculations using this approach than for subparticle models. The main disadvantage of the method is that equation (3.52) holds with sufficient accuracy only when $s < \pi/a$ (for larger angles, the errors due to model representation as a set of spheres become significant). An example comparing the spheres method with the subparticles method is given in Figure 3.13.

The spheres method is now widely applied mainly because of its rapid rate of calculation (allowing a large number of models to be sorted out) and because fine details of its shape can be reproduced with such a model representation. At the same time, the distortions caused by the packing of the spheres are usually of no importance when searching for the shape of a particle (e.g., the quaternary structure of biological macromolecules), if the number of spheres is of order 10^3. The method is simple and may be easily applied even with microcomputers.

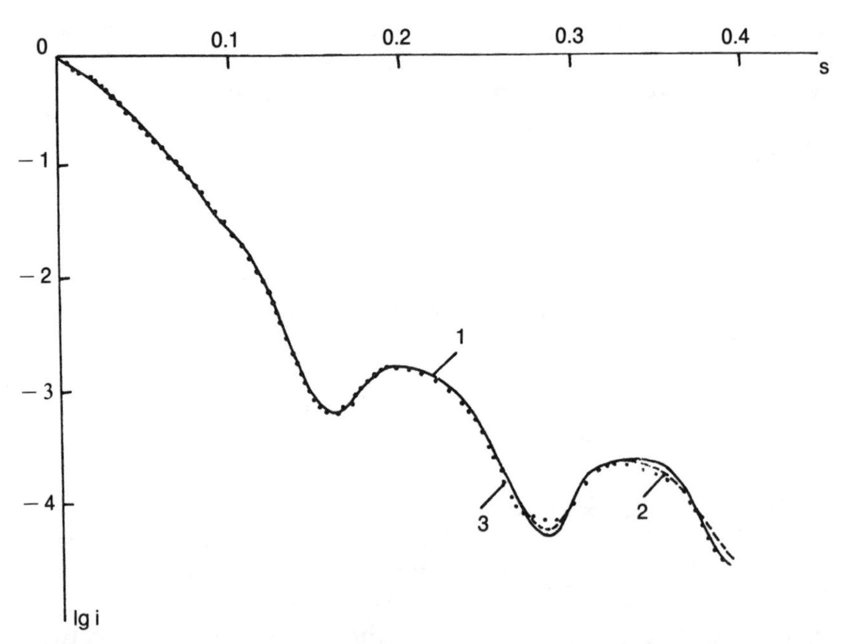

Figure 3.13. Modeling by subparticles method and spheres method: (1) theoretical scattering curve from right circular cylinder with $R = H = 150$; (2) model curve as calculated with subparticles technique, the cylinder divided into three equal ones, $R_I = 0.28\%$, computer time (NORD-100) 92 min; (3) spheres method, number of spheres 2400, $R_I = 3.2\%$, computer time 8 min.

3.5.4. Cube Method

The search for variations between the particle structures in solution and the crystal state is a specific branch of modeling. This task is actual, e.g., for some biological macromolecules, because their structure was determined by single-crystal structure analysis. As a rule, the variations do not involve shape changes, but some intraparticle rearrangements take place, leading to scattering-curve alterations at sufficiently large (on a small-angle scale) angles.

The intensity of small-angle scattering by a particle in solution can be expressed as

$$I(s) = \langle |A_v(\mathbf{s}) - A_s(\mathbf{s})|^2 \rangle$$

where $A_v(\mathbf{s})$ is the amplitude of scattering by the fixed particle in vacuum and $A_s(\mathbf{s})$ is the amplitude of scattering by the particle volume inaccessible to the solvent. Here, one assumes that the volume is filled uniformly with substance of scattering density ρ_s (see Section 2.3.2). If the structure of a particle is known, $A_v(\mathbf{s})$ can be readily calculated according to the general equation (2.1). Two problems arise when it is necessary to evaluate $A_s(\mathbf{s})$: (1) how to define the volume inaccessible to the solvent and (2) how to calculate its structure factor. The cube method (Ninio *et al.*, 1972; Fedorov *et al.*, 1972) solves both problems. The total particle volume is divided into cubes with small (0.5–1.5 Å) edges. Then we solve for each cube, whether or not it belongs to the particle. (The procedure can be carried out by considering the distance between the center of the given cube and the nearest atom of the particle.) Adjacent cubes are joined, if possible, along the z axis to form parallelepipeds. Thus (see Section 1.3.1),

$$A_s(\mathbf{s}) = \rho_s \sum_j \left[8 \frac{\sin(aX)}{aX} \frac{\sin(aY)}{aY} \frac{\sin(l_j Z)}{l_j Z} \right] \exp(i\mathbf{s}\mathbf{r}_j)$$

where the sum is taken over all the parallelepipeds, $2a$ is the edge of the cube, $2l_j$ and $\mathbf{r}_j$ are the length and position of the jth parallelepiped's center of mass, respectively. Such a representation of a particle enables its volume to be filled uniformly, ensuring sufficiently accurate calculation of the scattering intensities at large angles. That is why the method has an advantage over the "effective atomic factor" method, which was proposed by Langridge *et al.* (1960) and resembles the spheres method. Here, it is assumed that the particle consists of a system of spheres with corresponding Van der Waals radii surrounding each atom (for X-ray scattering, hydrogen atoms are omitted). The calculations are readily reduced to an equation similar to the Debye eqution, however, the mistakes at large scattering angles occur due to inhomogeneous filling of the particle volume. Some applications of the cube method are described in Section 4.1.4.

3.5.5. Modeling in Real Space

It is also possible to search for complicated models by comparing characteristic functions. Glatter (1980b) derived the following equation for function $p(r)$ (the particle is represented here by a set of n solid spheres with scattering densities ρ_i and radii R_i):

$$p(r) = \sum_{i=1}^{n} \rho_i^2 p_0(r, R_i) + 2 \sum_{i=1}^{n-1} \sum_{j=i+1}^{n} \rho_i \rho_j \bar{p}(r, d_{ij}, R_i, R_j) \qquad (3.53)$$

Here $p_0(r, R_i)$ is the distance distribution function of the ith sphere [see equation (2.26)], and $\bar{p}(r, d_{ij}, R_i, R_j)$ is the cross term for the ith and jth spheres separated by $d_{ij} = |\mathbf{r}_i - \mathbf{r}_j|$. Expressions for the cross terms with different relations between R_i, R_j, and d_{ij} were also derived. All the equations can be simplified for cases of equal radii and/or densities of the spheres. Further transition to reciprocal space can be undertaken using general equation (2.23). Similar to the spheres method described above, regular packing of the spheres was assumed by Glatter (1980b) to simplify the description of models and calculations.

3.6. Applications of Modeling

Both rough approximations of particles and the fine details of their structure may be sought by the above methods. Several examples illustrating the use of modeling for actual structural studies will be considered here. The investigations of biological macromolecules in solutions is probably the most fruitful field of application of the method, so that studies of biopolymers and their complexes may serve as good examples.

We noted earlier that there is no question of reaching a unique solution using the modeling technique. Successful application of the method depends to a great extent on available *a priori* information. Besides, the results obtained from the small-angle scattering data themselves can also play an important role. Here we speak not only about the use of invariants, but about certain features of the scattering curves as well (such as the occurrence of maxima and minima, and the decreasing nature), which can be interpreted in terms of specific properties of the particle structure. Below, it will be shown how both small-angle scattering and *a priori* information can be taken into account when searching for the models.

3.6.1. Helix pomatia Hemocyanin

Small-angle X-ray investigation of the hemocyanin of *Helix pomatia*, a large protein molecule (Pilz *et al.*, 1970; Pilz *et al.*, 1972), is a good example of refinement of the model shape by taking into consideration more extended portions of the scattering curve. The latter is shown in Figure 3.14. The value $R_g = 164$ Å was determined from the very beginning of the curve. Comparison with theoretical scattering curves from different simple bodies showed the right circular cylinder with ratio $2R/H = 1{:}1.1$ to be the best approximation. By anayzing the height of the first subsidiary maximum (Figure 3.15) one can suggest that the cylinder is hollow, the ratio of the inner to the outer radius being $R_2{:}\ R_1 = 0.45$. Thus, the right hollow cylinder with parameters $H = 360$ Å, $2R_1 = 150$ Å, $2R_2 = 330$ Å (values chosen to yield the experimental value of R_g) seems to be the best approximation at this stage.

Scattering at larger angles should be taken into account for further refinement of the model. There are two pronounced minima s_{01} and s_{02} in the region $s > 0.15$ Å^{-1} that are independent of particle shape; the same minima occur on the scattering curves from separate fragments of the hemocyanin molecule. Therefore they obviously owe their origin to the subunit constitution of the particle. The positions of several first minima of the scattering intensity from globular particles correspond approximately to analogous positions for a solid sphere, namely $s_{0i}R_g =$ 3.48, 6.08, 8.45, . . . [see equation (1.27)]. If the subunits are assumed to be globular in shape, the values of s_{01} and s_{02} yield $R_{gs} = 15.5$ Å, corresponding to a sphere diameter of 40 Å. The models consisting of such subunits were considered in the framework of the overall shape described earlier. The model built up from 360 subunits (Figure 3.16)

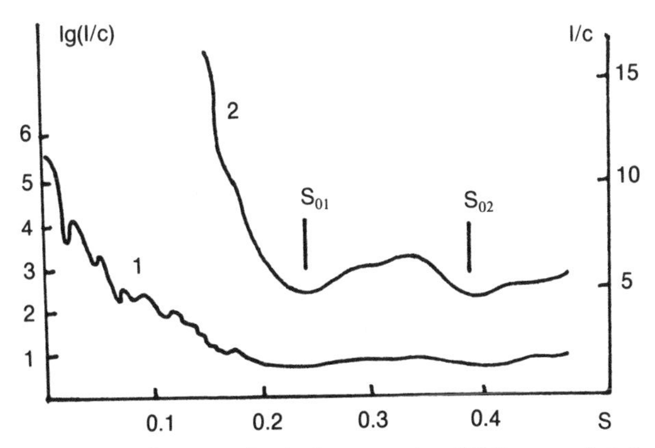

Figure 3.14. X-ray scattering curve for the hemocyanin of *Helix pomatia* (after Pliz *et al.*, 1972): (1) the whole curve on a logarithmic scale; (2) its outer part on an absolute scale.

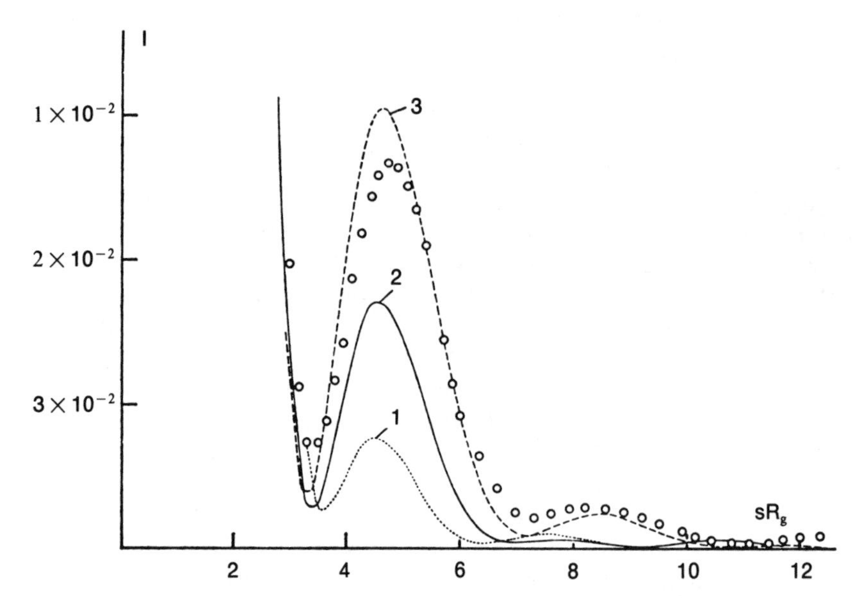

Figure 3.15. First maximum of a scattering curve for the hemocyanin of *Helix pomatia* as compared with theoretical curves of circular cylinders (after Pilz *et al.*, 1970): (1) full cylinder; (2), (3) hollow cylinders with $R_2/R_1 = 0.3$ and 0.5, respectively; circles denote an experimental curve.

best fits the scattering curve and gives fair agreement even for a curve dropping more than three orders of magnitude.

3.6.2. Bacteriophage S_D

Bacterial virus S_D is a large virus infecting *E. coli* bacteria. It contains a double-stranded DNA molecule and, according to electron microscopy data, consists of an isometric polyhedral head and small tail. One of the most accurate shape determinations by modeling was probably carried out when studying this particle (Boyarintseva *et al.*, 1977).

The small-angle X-ray scattering curve of bacteriophage S_D (Figure 3.17) suggests that the overall shape of the phage is almost spherical. The values of the invariants prove to be $R_g = 305$ Å and $V = 2.1 \times 10^8$ Å^3, which determine the general dimensions of a plausible model.

A priori information has provided evidence that the shape of the phage head is nearly isometric (icosahedron and octahedron were the most probable models). Thus, the main problem when searching for the shape was to distinguish among the polyhedrons. Several model calculations have been conducted by Boyarintseva *et al.* (1975). The differences in the scattering curve from regular polyhedrons were proved to be

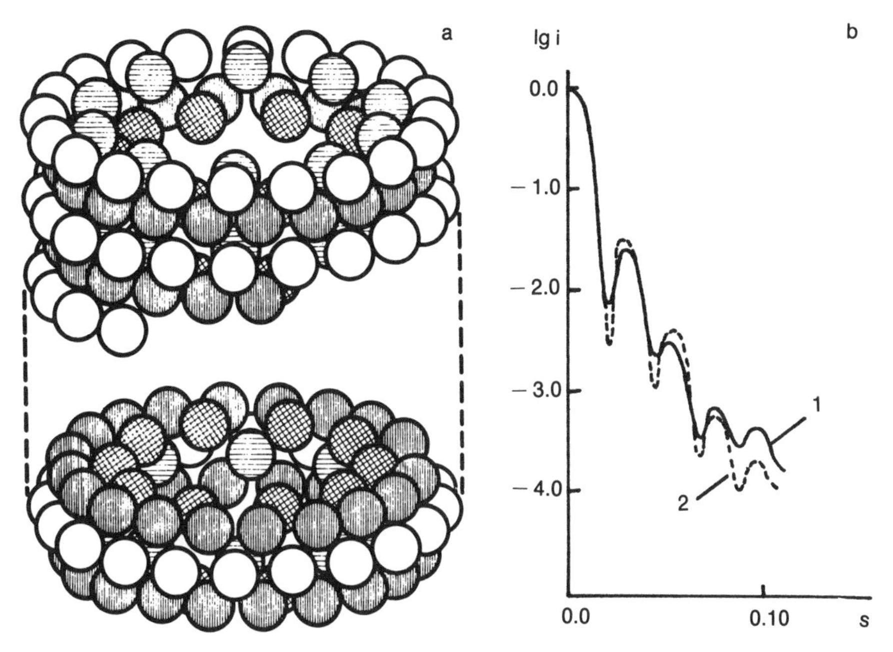

Figure 3.16. Subunit modeling of the hemocyanin of *Helix pomatia* (after Pilz *et al.*, 1972): (a) a model consisting of 360 subunits with diameter 40 Å; (b) scattering curves: (1) experimental, (2) model.

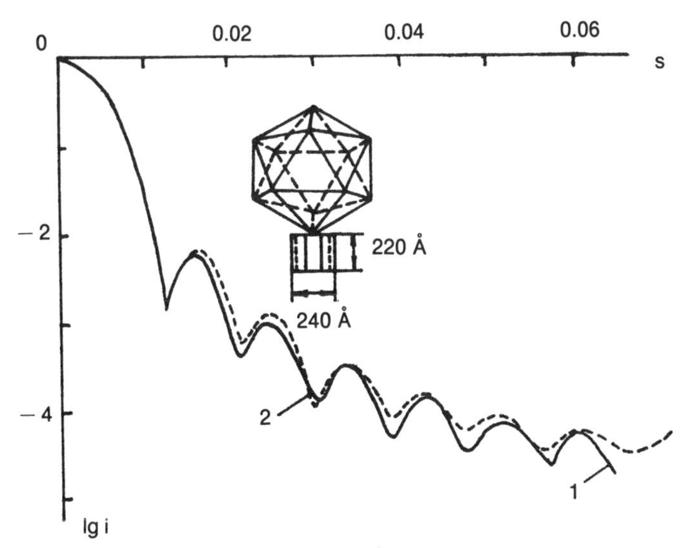

Figure 3.17. Shape modeling of bacteriophage S_D (after Boyarintseva *et al.*, 1977): (1) X-ray experimental scattering curve; (2) scattering by the model.

Table 3.5. Angular Positions[a] of Subsidiary Maxima in the Scattering Curves of Regular Polyhedrons

No. of maximum	Sphere	Icosahedron	Dodecahedron	Octahedron	Cube
1	4.50	4.50	4.50	4.50	4.50
2	7.05	7.05	7.05	7.45	7.45
3	9.60	9.60	9.65	10.35	11.00
4	12.05	12.10	12.30	13.35	14.10
5	16.90	17.35	17.75	—	—
Δ[b]	2.51	2.55	2.64	2.96	3.20

[a] In units of sR_g.
[b] Mean distance between maxima.

conveniently characterized by the angular positions of subsidiary maxima (Table 3.5), these positions being practically the same when adding a small (4–6% in volume) tail.

The calculations allowed some conclusions to be reached about the size and shape of the head. First of all its radius of gyration was evaluated from the position of the first subsidiary maxima (the value proved to be 281 Å). Then, the positions of the first five subsidiary maxima in the experimental curve were compared with the data in Table 3.5. It provided evidence that the shape of the phage head is nearly icosahedral, with the experimental value $\Delta(s_m R_g) = 2.60$.

The intraphage DNA in the S_D phage head is hydrated essentially stronger than the protein component, so that the considered angular region corresponds to the region of homogeneity for the particle. The calculated values of the invariants as well as the above-mentioned conclusions on the head shape were taken into account when modeling in a homogeneous approximation. The spheres method was used to calculate the model scattering intensities. The model shown in Figure 3.17 (icosahedron with edge size 450 Å and a cylindrical tail with volume about 5% related to the head one) gives the best fit. One can see from Figure 3.17 that the model scattering curve is in very good agreement with the experimental curve when the intensity drops almost five orders of magnitude. It enables one to affirm that the model represents the shape of the particle with good precision.

3.6.3. 30 S Ribosomal Subparticle

Modeling of the 30 S subparticle of *E. coli* ribosome (Spirin *et al.*, 1979) is an example of extended application of *a priori* information for constructing a model fitting the data of small-angle scattering; a large amount of such information enabled the construction of inhomogeneous models.

Ribosomes are ribonucleoprotein particles consisting of high-molecular-mass RNA threads and many protein molecules. The 30 S ribosomal subparticle is the subunit of 70 S *E. coli* ribosome and includes the 16 S RNA molecule and 21 proteins. It was proved by various methods (including small-angle scattering contrast variation studies; see Section 4.2.4) that RNA occupies the particle center and the proteins, its periphery. However, such questions as to whether the protein subunits are globular or anisometric, and how they are positioned with respect to the RNA core, were unclear even though many investigations have been undertaken.

Spirin *et al.* (1979) took into account a large quantity of data obtained by various methods. First, they assumed that the shape of the 16 S RNA in ribosome is similar to the V-like one of isolated 16 S RNA in compact form (Vasiliev *et al.*, 1978; this assumption was later confirmed by Serdyuk *et al.*, 1983). Moreover, all the proteins in the 30 S subparticle were supposed to be compact and globular in shape, proceeding from small-angle scattering studies of isolated ribosomal proteins (Serdyuk *et al.*, 1978). When arranging the proteins on the surface of the RNA molecule the data of different physical and physicochemical methods were used. The proposed model of the quaternary structure of the 30 S ribosomal subparticle is shown in Figure 3.18. The model was checked using both X-ray and neutron small-angle scattering data. For the

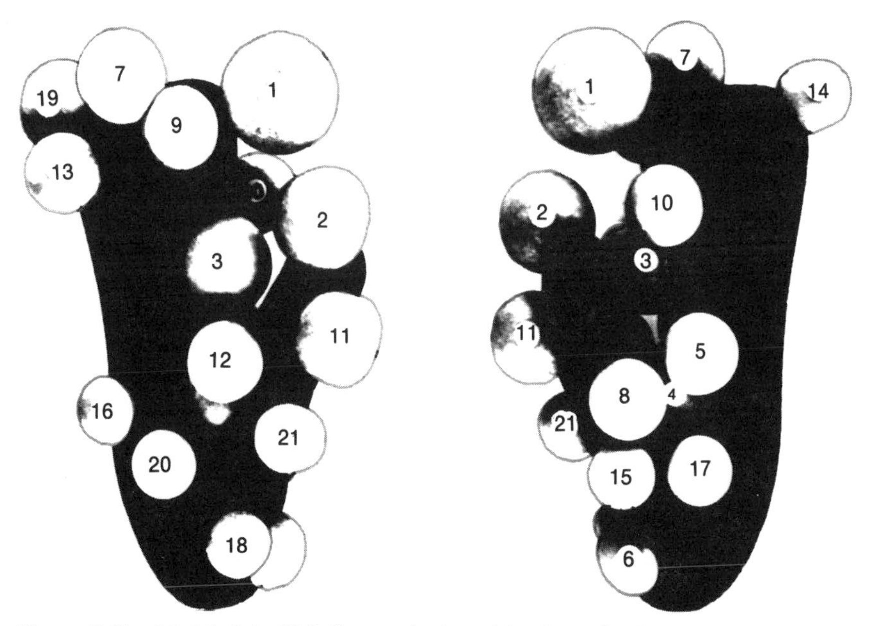

Figure 3.18. Model of the 30 S ribosomal subparticle of *E. coli* (after Spirin *et al.*, 1979). Open circles represent ribosomal proteins (1–21) and the dark area, RNA. Two views from opposite sides are given.

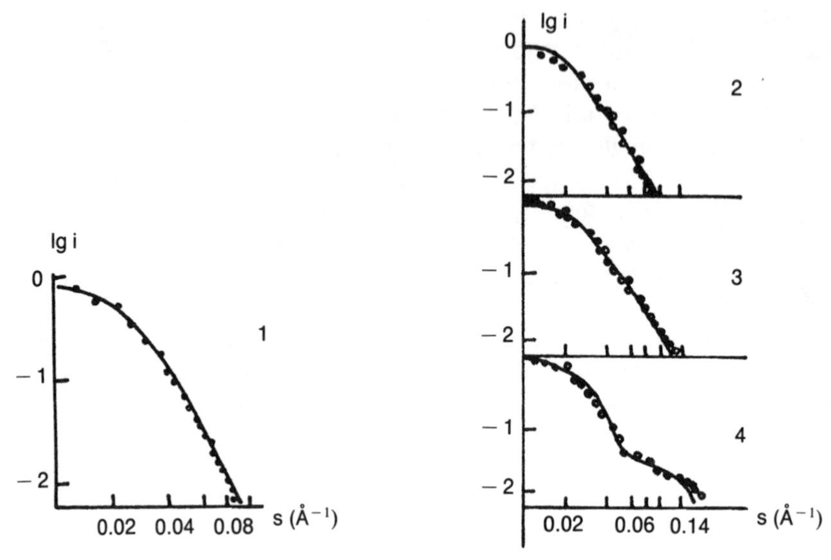

Figure 3.19. Scattering curves from 30 S ribosomal subparticle (after Spirin *et al*., 1979). Dots represent experimental data, solid lines correspond to model curves. (1), (2) refer to X-ray and neutron scattering in H_2O solutions, respectively, and (3), (4) to neutron scattering by RNA and proteins (in solutions with 42% and 70% D_2O), respectively.

calculations, the region occupied by RNA was described by the spheres method (the diameter of a sphere was 10 Å). Also, the proteins were represented as spheres with radii evaluated from their molecular weights. The experimental and theoretical curves of X-ray and neutron scattering by aqueous solutions of 30 S ribosomal subparticles are presented in Figure 3.19 (cases 1 and 2). Cases 3 and 4 correspond to the separate scattering by RNA and proteins. The experimental data were obtained for solutions containing 42% and 70% D_2O, respectively; see Section 4.2.3. One can see that all the experimental and model curves are in good agreement when the intensities drop more than two orders of magnitude. The model cannot, of course, be regarded as very accurate. There are, in particular, contradictions with the results of the triangulation method (see Section 4.3.2). However, the example shows that even inhomogeneous models can be sought in small-angle scattering provided sufficient *a priori* information is available.

3.7. Conclusion

The results of this chapter show that important general information about particles in solution (matrix) can be extracted from small-angle scattering data. Provided an experiment has been carried out correctly, *a priori* information is not necessary to calculate the values of invariants. Geometrical characteristics (R_g, V, l_m, S, D) can be determined from the scattering curves measured in relative units; absolute measurements are required to evaluate weight parameters. There exist special approaches to calculate the parameters of the cross section (for rodlike particles) and thickness (for lamellar particles). The methods of calculating invariants are well developed and allow reliable results to be obtained.

The search for the best approximation among triaxial bodies, making it possible to estimate the general shape and anisometry of a particle, also does not require *a priori* information. However, this kind of information is absolutely necessary to find a complicated model, even in a homogeneous approximation. The modeling method, of course, does not guarantee the uniqueness of a solution. Nevertheless, it is reasonable to believe that a model which fits well both the small-angle scattering curve and the data of other methods describes the particle's structure adequately.

The sampling theorem is very useful when considering problems of information content in small-angle scattering data. On the basis of the theorem proper experimental conditions can be formulated, and the law of error propagation when calculating the invariants can be derived. The errors in the determined values of the invariants prove to be connected to the number of "degrees of freedom," namely, the number of sampling points in the interval (s_{min}, s_{max}).

One arrives at the apparent contradiction by comparing the concepts of the modeling and the sampling theorem. Actually, the number of degrees of freedom for real experiments does not exeed 10^1–10^2. Meanwhile, the methods of modeling described contain essentially a larger number of parameters (relative to, e.g., the spheres method) because when searching for a model some assumptions or others about the structure of a particle are always employed. For example, if we believe that the particle may be well represented by an ellipsoid of rotation, then a two-parameter model must be found. If it is known that the particle has a complicated shape and inhomogeneous density distribution, and some information about the shape and density is available, then complicated models consisting of thousands of spheres or cubes will be considered. It is true that the spatial resolution d achieved by the modeling method cannot be better than $2\pi/s_{max}$. In this sense the possibilities of modeling are limited by the interval of measurements. However, just the opposite situation takes place in research: normally, the model curve starts to deviate strongly from the experimental one at values of s which are essentially smaller than s_{max}. Therefore, the number of independent parameters that can be found from the small-angle scattering curve depends mainly on *a priori* information about the particle rather than on the number of degrees of freedom of the curve.

4

Interpretation of Scattering by Inhomogeneous Particles

The approaches to the interpretation of the small-angle scattering curve considered in the previous chapter (namely, calculation of the invariants and modeling) can be applied, as a rule, only within the framework of a homogeneous approximation. However, they give no information on inhomogeneities which inevitably exist in any real particle and may be associated with its structure as well as solvent penetration. It therefore follows that new methods of interpretation are necessary in order to move forward from analyzing the general parameters to an investigation of the inner structure of particles. The most effective method is to study the change in the particle's (or in its individual segment's) scattering capacity with respect to a solvent, followed by analysis of the corresponding variations in the scattering curves.

After analyzing the general approach to scattering by an inhomogeneous particle with solvent penetration (Section 4.1), we shall deal with methods, widely used in research for studying the inner structure of the particles. The first such method is the contrast-variation technique (Section 4.2), which enables one to decompose scattering caused by particle shape and inhomogeneities. We then describe isomorphous-replacement methods (Section 4.3), where the modification of definite particle segments enables one to seek the distances between them. Some approaches in which the contrast-variation and isomorphous-replacement effects are achieved by means of changes in wavelength or in the type of radiation used are given in Section 4.4.

4.1. Scattering by Inhomogeneous Particles

When considering scattering by particle solutions it was noted earlier that the difference in the scattering curve depends substantially on a solvent density (namely, on the contrast; see Section 2.3.2). This fact is of particular importance in the case of inhomogeneous particles, since here the scattering at low contrast is mainly determined by the distribution of the inner inhomogeneities. Therefore, the solvent influence should be taken into account when analyzing scattering by an inhomogeneous particle.

4.1.1. Solvent Influence

Inhomogeneities within a particle in a solution are defined by two factors. First, there are inhomogeneities in the structure as such due to the difference in density levels of its components ("density" here means "scattering length density"). Second, there are inhomogeneities which owe their origin to solvent penetration into a particle. (We remind the reader that inhomogeneities in the atomic structure of a particle and solvent are of no importance in small-angle scattering.)

The density of a particle in solution is a function that depends on the coordinates and solvent density, so $\rho = \rho(\mathbf{r}, \rho_s)$. (On the assumption that the solvent does not change the particle conformation, it suffices to take into consideration only its density but not its chemical composition.) We now consider in general outline scattering by an inhomogeneous particle in solution (Harrison, 1969; Feigin and Soler, 1975; Soler, 1976).

The full excess scattering length of a particle in solution can be expressed as

$$b_{ef} = \int_V [\rho(\mathbf{r}, \rho_s) - \rho_s] d\mathbf{r} = b_p - V\rho_s$$

where b_p is the full scattering length of a particle and V its partial volume. Assuming b_p and V do not depend on the solvent while particle conformation is preserved, differentiation over ρ_s yields

$$V = -\,\partial b_{ef}/\partial \rho_s = \int_V [1 - \partial \rho(\mathbf{r}, \rho_s)/\partial \rho_s] d\mathbf{r} = \int_V \phi_p(\mathbf{r}, \rho_s) d\mathbf{r} \qquad (4.1)$$

where

$$\phi_p(\mathbf{r}, \rho_s) = 1 - \phi_s(\mathbf{r}, \rho_s) \qquad (4.2)$$

and

$$\phi_s(\mathbf{r}, \rho_s) = \partial \rho(\mathbf{r}, \rho_s)/\partial \rho_s \qquad (4.3)$$

Function $\phi_p(\mathbf{r}, \rho_s)$ and $\phi_s(\mathbf{r}, \rho_s)$, thus introduced, describe the solvent penetration into a particle. The "volume fraction of a particle" $\phi_p(\mathbf{r}, \rho_s)$ shows the contribution of the particle's volume inaccessible to a solvent, in $\rho(\mathbf{r}, \rho_s)$. In the same way $\phi_s(\mathbf{r}, \rho_s)$ represents the "volume fraction of solvent."

We assume that the external parameters (such as pressure, temperature, and pH) being fixed, the volume fraction of a particle and solvent do not depend on ρ_s, i.e., all the solvents penetrate into the particle identically. In this case, one can see from equation (4.3) that the particle density depends linearly on the solvent density and can be expressed as

$$\rho(\mathbf{r}, \rho_s) = \phi_p(\mathbf{r})\rho_p(\mathbf{r}) + \phi_s(\mathbf{r})\rho_s \tag{4.4}$$

where $\rho_p(\mathbf{r})$, the "inner density," is the particle density in vacuum. Figure 4.1 shows schematically a volume element dV consisting of volume dV_p, completely occupied by a particle, and volume dV_s, occupied by the solvent. Functions $\phi_p(\mathbf{r}) = \partial V_p/\partial V$ and $\phi_s(\mathbf{r}) = 1 - \phi_p(\mathbf{r}) = \partial V_s/\partial V$ may be regarded as continuous at small-angle resolution. Thus, for the excess density [equation (2.11)] one has

$$g(\mathbf{r}, \rho_s) = \rho(\mathbf{r}, \rho_s) - \rho_s = \phi_p(\mathbf{r})[\rho_p(\mathbf{r}) - \rho_s] \tag{4.5}$$

which demonstrates explicitly the separation of the two components of

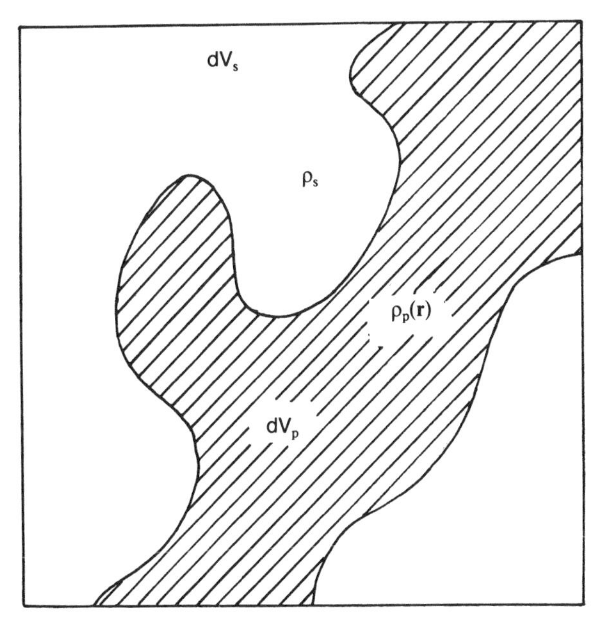

Figure 4.1. Volume element of a particle in solution.

the scattering density: $\rho_p(\mathbf{r})$ determines the structure of a particle in vacuum while $\phi_p(\mathbf{r})$ describes the solvent penetration. The calculation of these functions is possible provided $g(\mathbf{r}, \rho_s)$ is determined for two solvent density values ρ_1 and ρ_2. Then

$$\phi_p(\mathbf{r}) = [g(\mathbf{r}, \rho_2) - g(\mathbf{r}, \rho_1)]/(\rho_1 - \rho_2)$$

and

$$\rho_p(\mathbf{r}) = [\rho_2 g(\mathbf{r}, \rho_1) - \rho_1 g(\mathbf{r}, \rho_2)]/[g(\mathbf{r}, \rho_1) - g(\mathbf{r}, \rho_2)]$$

We note that only relative measurements are required to determine $\rho_p(\mathbf{r})$.

4.1.2. General Equations for Intensity and Invariants

We now consider the general expression for the scattering intensity. Substitution of equation (4.5) in the Debye equation (2.17) gives

$$I(s) = I_0(s) + 2\rho_s I_{01}(s) + \rho_s^2 I_1(s) \tag{4.6}$$

where

$$I_0(s) = \int_V \int_V \phi_p(\mathbf{r}_1)\rho_p(\mathbf{r}_1)\phi_p(\mathbf{r}_2)\rho_p(\mathbf{r}_2)[\sin(sr_{12})/(sr_{12})]d\mathbf{r}_1 d\mathbf{r}_2 \tag{4.7}$$

is the scattering intensity of a body with density $\rho_0(\mathbf{r}) = \phi_p(\mathbf{r})\rho_p(\mathbf{r})$, i.e., by the given particle in vacuum ($r_{12} = |\mathbf{r}_1 - \mathbf{r}_2|$),

$$I_1(s) = \int_V \int_V \phi_p(\mathbf{r}_1)\phi_p(\mathbf{r}_2)[\sin(sr_{12})/(sr_{12})]d\mathbf{r}_1 d\mathbf{r}_2 \tag{4.8}$$

is the term depending on solvent penetration, and

$$I_{01}(s) = \int_V \int_V \phi_p(\mathbf{r}_1)\rho_p(\mathbf{r}_1)\phi_p(\mathbf{r}_2)[\sin(sr_{12})/sr_{12}]d\mathbf{r}_1 d\mathbf{r}_2 \tag{4.9}$$

is the cross term.

Similar dependences on solvent density can also be obtained for the invariants. Thus, by substituting equation (4.5) into the general expression (3.8) for the the radius of gyration, one obtains

$$R_g^2 = R_c^2 + (\bar{\rho}/\Delta\rho)(R_p^2 + L^2 - R_c^2) - (\bar{\rho}/\Delta\rho)^2 L^2 \tag{4.10}$$

where $\bar{\rho} = b_p/V$ is an average particle density, $\Delta\rho$ is a contrast [see equation (2.12)], and R_p and R_c are the radii of gyration of the region inaccessible to the solvent and of the same region with constant density, respectively. For these quantities we can write

$$R_p^2 = b_p^{-1} \int \rho_p(\mathbf{r})\phi_p(\mathbf{r})(\mathbf{r} - \mathbf{r}_{0p})^2 d\mathbf{r} \tag{4.11}$$

and

$$R_c^2 = V^{-1} \int \phi_p(\mathbf{r})(\mathbf{r} - \mathbf{r}_{0c})^2 d\mathbf{r} \tag{4.12}$$

where in both cases $\mathbf{r}_{0p}$ and $\mathbf{r}_{0c}$ are the positions of centers of mass given by

$$\mathbf{r}_{0p} = b_p^{-1} \int \mathbf{r}\phi_p(\mathbf{r})\rho_p(\mathbf{r})d\mathbf{r} \quad \text{and} \quad \mathbf{r}_{0p} = V^{-1} \int \mathbf{r}\phi_p(\mathbf{r})d\mathbf{r}$$

respectively, and $\mathbf{L} = \mathbf{r}_{0c} - \mathbf{r}_{0p}$. It follows from equation (4.10) that, in the general case, the radius of gyration of a particle is a quadratic function of the contrast. For uniform particles $R_p = R_c$, hence $L = 0$, so R_g is independent of the solvent density.

The Porod volume also exhibits dependence on ρ_s. If in equation (3.11) we pass formally to integration over the volume in reciprocal space and make use of the Parseval theorem [equation (1.19)], then we have

$$V(\rho_s) = 2\pi^2 I(0) \left[2\pi^2 \int I(\mathbf{q})d\mathbf{q}\right]^{-1} = I(0)\left[\int g^2(\mathbf{r})d\mathbf{r}\right]^{-1}$$

On substituting equation (4.5) and bearing in mind that $I(0) = \bar{\rho}^2 V^2$, we obtain from this equation that

$$V^{-1}(\rho_s) = V^{-2}\left[\int \phi_p^2(\mathbf{r})d\mathbf{r} + 2(\Delta\rho)^{-1} \int \rho_f(\mathbf{r})\phi_p^2(\mathbf{r})d\mathbf{r} + (\Delta\rho)^{-2} \int \rho_f^2(\mathbf{r})\phi_p^2(\mathbf{r})d\mathbf{r}\right] \tag{4.13}$$

where

$$\rho_f(\mathbf{r}) = \rho_p(\mathbf{r}) - \bar{\rho} \tag{4.14}$$

If the contrast $\Delta\rho$ is much greater than the density fluctuations $\rho_f(\mathbf{r})$, then the Porod volume does not depend on ρ_s but is equal to

$$V_\infty = V^2\left[\int \phi_p^2(\mathbf{r})d\mathbf{r}\right]^{-1}$$

This parameter determines the average value of the volume hydration

$$\bar{\phi}_s = 1 - \bar{\phi}_p = 1 - V/V_\infty$$

4.1.3. Spherically Symmetric Particles

The above approach, although of a general character, is rather inconvenient in practice. Actually, functions $\rho_p(\mathbf{r})$ and $\phi_p(\mathbf{r})$ are related to the corresponding intensities via equations (4.7) and (4.8), which are quite similar to the Debye equation (2.17); their unique restoration is out of the question. Until the present such an approach was applied only to spherically symmetric particles. 'n this case $g(\mathbf{r}, \rho_s) = g(r, \rho_s)$ and the scattering amplitude can be written as

$$A(s, \rho_s) = A_p(s) - \rho_s A_c(s) \tag{4.15}$$

where $A_p(s)$ and $A_c(s)$ are the scattering amplitudes resulting from "particles" with densities $\rho_p(r)\phi_p(r)$ and $\phi_p(r)$, respectively [the connection between the densities and scattering amplitudes is given by equation (1.25)]. Equation (4.15) can be of help when determining the true signs of function $A(s, \rho_s)$ (see Sections 2.4.5 and 5.1), namely, the set of signs should be such that the linear law is fulfilled for different values of ρ_s. The study of Soler (1976), where X-ray measurements were conducted for different solvent densities of the bacteriophage S_D and functions $\rho_p(r)$ and $\phi_s(r)$ where restored in a spherically symmetric approximation, is an example of an application of this approach. One can conclude from the curves shown in Figure 4.2 that the protein component ($\rho_p = 0.41$ eÅ^{-3}, $\phi_s = 10\%$) is concentrated in the regions $0 \leqslant r \leqslant 60$ Å and $320 \leqslant r \leqslant 365$ Å, while DNA ($\rho_p = 0.465$ eÅ^{-3}, $\phi_s = 40\%$) is in the region $60 \leqslant r \leqslant 320$ Å. The obtained level ρ_p for DNA is lower than the theoretical value for "dry" DNA (0.55 eÅ^{-3}). This means that the DNA contains bound water of about 0.6 g H_2O per 1 g of "dry" DNA.

4.1.4. Large-Angle Scattering

Another aproach for taking into account the solvent influence, "large-angle scattering" (Fedorov *et al*., 1974), is noteworthy. This method can be applied if the structure of a particle is known in the crystalline state, and is now used for protein molecules. After calculating the small-angle scattering curve using the atomic coordinates of a protein, one can compare it with experiment. The "large-angle" regions of the curves ($0.2 \leqslant s \leqslant 1$ Å^{-1}) turn out to be sensitive to small changes in the structural organization of a molecule (e.g., to the packing features of polypeptide chains in proteins; see Fedorov *et al*., 1976). On comparing these curves, various conclusions can be drawn about the identity or differences in particle structure in a crystal and in solution.

By means of modeling (such as the "cube method," see Section 3.5.4) one can search for structural changes in a particle caused by solvent

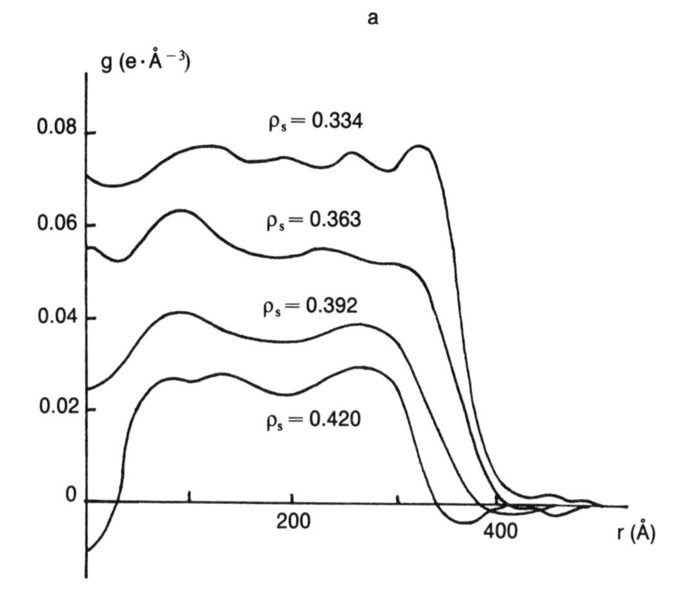

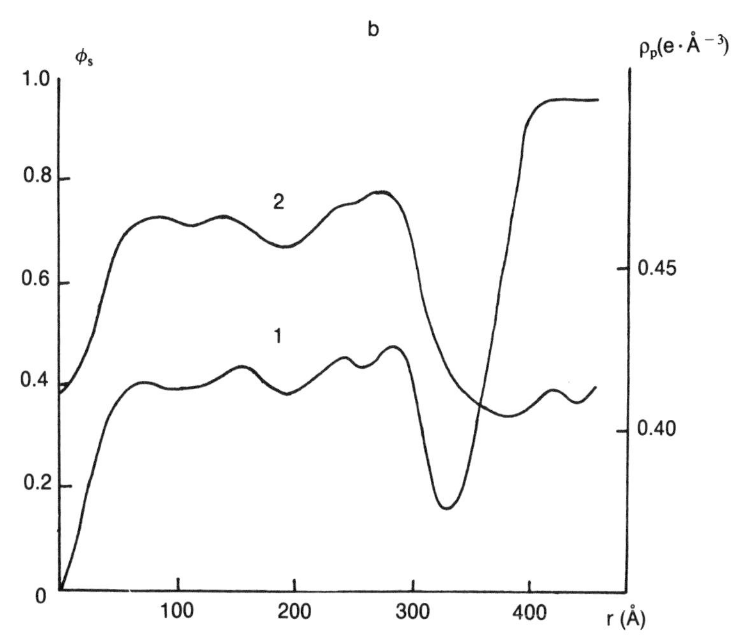

Figure 4.2. Determination of the inner density and hydration of bacteriophage S_D: (a) excess scattering densities of phage S_D in various solvent densities; (b) calculated functions: (1) $\phi_s(r)$, (2) $\rho_p(r)$.

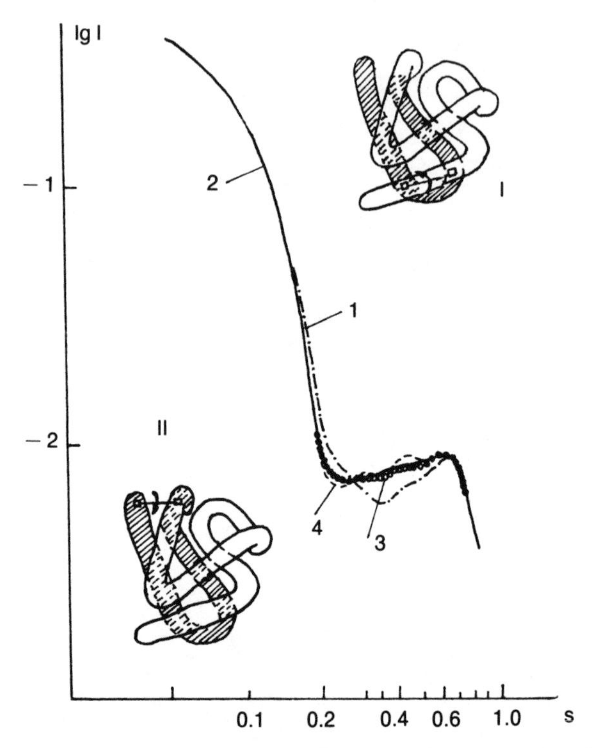

Figure 4.3. Theoretical scattering curves of Mb-SW and its structural modifications (after Fedorov and Denesyuk, 1978): (1) Mb-SW in a crystal; (2) experimental small-angle scattering curve; (3), (4) two modifications obtained by shifting or rotating parts of the peptide chain by 5° to 10° about axes I and II, respectively.

penetration. Fedorov and Denesyuk (1978) used this method to compare the small-angle scattering curves from sperm-whale myoglobin (Mb-SW) in the crystalline state (calculated from atomic coordinates) and in solution (experiment). The two curves differ substantially (Figure 4.3, curves 1 and 2). The analysis of possible structural chain rearrangements of the Mb-SW molecule shows that best agreement with experiment is given by a small (about 2 Å) increase in distance between the "hairpin" GH and the outer segment of the molecule (Figure 4.3, curves 3 and 4).

The "improved cube method" (Müller, 1983) extends the possibilities of this technique and combines the advantages of both the cube method and the "effective atomic factors" method (see Section 3.5.4). The cube edge is equal to 0.5 Å, while the possibility of water-molecule penetration into a particle is evaluated in terms of the movement of the molecule as a sphere with van der Waals radius 1.4 Å along the surfaces of van der Waals spheres surrounding the atoms in the particle. Müller

(1983) used such an approach to perform model calculations for different DNA and RNA conformations. It was found possible to study some fine effects, such as variations in the base pair angles in the DNA double helix.

4.2. Contrast Variation

We have seen above that, in the general case, when one cannot predict the degree and nature of solvent penetration, it is very difficult to obtain new structural information using solvent density variations. Therefore practical use of this approach requires simplifying assumptions about function $\phi_p(\mathbf{r})$.

4.2.1. Basic Functions

The model of a particle whose density does not depend on the solvent was found to be the most fruitful in practice. In this case

$$\phi_p(\mathbf{r}) = \begin{cases} 1, & \text{if } \mathbf{r} \text{ lies within the particle} \\ 0, & \text{if } \mathbf{r} \text{ lies outside the particle} \end{cases} \tag{4.16}$$

is the shape function of the particle. It is thus convenient to write the excess density of a particle as follows (Stuhrmann and Kirste, 1965):

$$g(\mathbf{r}, \rho_s) = \Delta\rho\,\phi_p(\mathbf{r}) + \rho_f(\mathbf{r}) \tag{4.17}$$

The method based on representation (4.16) and (4.17) is mainly related to Prof. H. Stuhrmann's (FRG) research and was called "contrast variation." With the above assumptions equation (4.6) for the scattering intensity takes the form

$$I(s) = I_s(s) + \Delta\rho I_{cs}(s) + (\Delta\rho)^2 I_c(s) \tag{4.18}$$

where $I_c(s)$ is the scattering by a particle with density given by equation (4.16), i.e., by the particle "shape," $I_s(s)$ is the scattering by particle inhomogeneities [function $\rho_f(\mathbf{r})$], and $I_{cs}(s)$ is the cross term which, according to the Cauchy–Buniakovsky inequality, satisfies the condition

$$|I_{cs}(s)| \leqslant 2[I_c(s)I_s(s)]^{1/2}$$

for any s.

The scattering intensity components introduced here are called the basic functions. If one measures $I(s)$ for n values of the contrast $\Delta\rho$, then each experimental point s_i will yield a system of n linear equations (4.18) for the three independent variables $I_c(s_i)$, $I_s(s_i)$, and $I_{cs}(s_i)$ that can be solved for $n \geqslant 3$. Therefore, no less than three scattering curves for different contrast are available; hence the basic functions can be uniquely determined. Such an expansion makes it possible to separate the contribution from a particle shape and its inhomogeneities to the scattering intensity, illustrated in Figure 4.4. All the above methods of invariant calculation and modeling by homogeneous bodies may be applied to function $I_c(s)$; the typical inhomogeneities in the inner structure of a particle are more explicitly apparent in function $I_s(s)$ than in $I(s)$. Similar separation holds in real space too. Actually, substitution of equation (4.17) into expression (2.21) for the correlation function yields

$$\gamma(r) = \gamma_s(r) + \Delta\rho\gamma_{cs}(r) + (\Delta\rho)^2\gamma_c(r) \tag{4.19}$$

where $\gamma_c(r) = V\gamma_0(r)$ is the characteristic function of a particle shape,

$$\gamma_s(r) = \frac{1}{4\pi}\int_\Omega d\omega \int_V \rho_f(\mathbf{u})\rho_f(\mathbf{u}+\mathbf{r})d\mathbf{u}$$

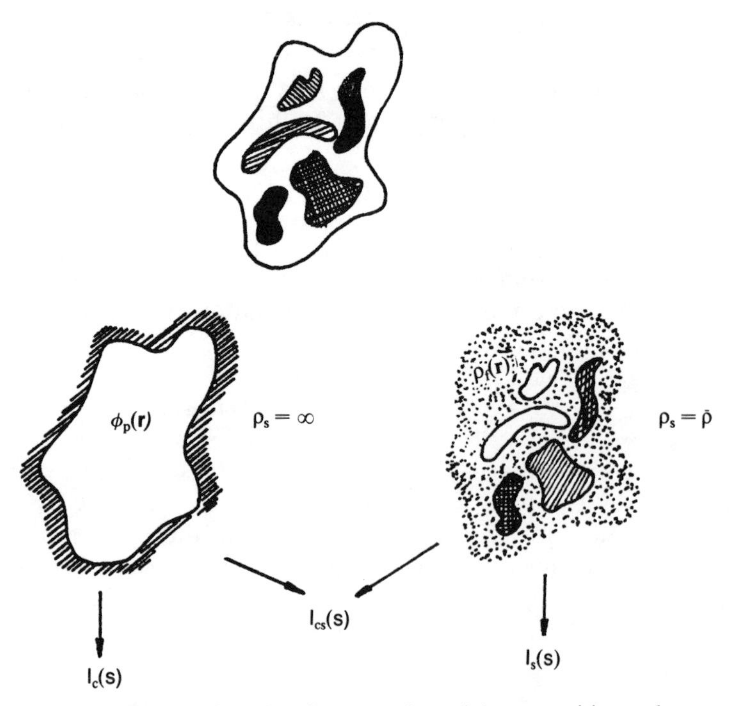

Figure 4.4. Contribution of particle shape and inner inhomogeneities to the scattering intensity.

is the correlation function of inhomogeneities, and $\gamma_{cs}(r)$ is the cross term. It is evident that functions $\gamma_c(r)$, $\gamma_s(r)$, and $\gamma_{cs}(r)$ are related to the corresponding basic intensities by the same transform (2.23).

4.2.2. Dependence of Invariants on Contrast

The dependence of the radius of gyration on the contrast is most important in research. By introducing equation (4.16) into equations (4.11) and (4.12) we obtain equation (4.10) in the form (Stuhrmann and Kirste, 1967; Luzzati *et al*., 1976)

$$R_g^2 = R_c^2 + \alpha \Delta\rho - \beta/(\Delta\rho)^2 \tag{4.20}$$

where

$$\alpha = V^{-1} \int \rho_f(\mathbf{r}) r^2 d\mathbf{r} \tag{4.21}$$

and

$$\beta = V^{-2} \int\int \rho_f(\mathbf{r}_2)\rho_f(\mathbf{r}_1)\mathbf{r}_1\mathbf{r}_2 d\mathbf{r}_1 d\mathbf{r}_2 \tag{4.22}$$

Measurements of R_g with three or more different contrasts enable the coefficients of this square law to be determined, where R_c is understood as the radius of gyration of a particle with density $\phi_p(\mathbf{r})$ (the "shape" radius of gyration). It is noteworthy that for homogeneous particles $R_g = R_c$ does not depend on the contrast; α describes the relative arrangement of higher and lower density regions with respect to the center of mass of a shape, so α is positive when denser regions are close to the periphery, and vice versa. The value of β represents an estimate of the distance between the center of mass of a particle and that of the distribution of inhomogeneities; $\beta = 0$ when they coincide; otherwise $\beta \neq 0$ (the particle center of mass is displaced with varying ρ_s).

We consider the practically important case of a particle consisting of two components with different scattering densities ρ_1 and ρ_2. In this case one can initially select the solvent density equal to the density of one component and hence eliminate its contribution to the scattering intensity and observe the scattering by the other component. Such a study was first carried out in the X-ray investigations of apoferritin by Fischbach and Anderegg (1965). The radius of gyration can be expressed as (Serdyuk and Fedorov, 1973; Moore *et al*., 1974)

$$R_g^2 = R_1^2 x_1 + R_2^2(1 - x_1) + L_{12}^2 x_1(1 - x_1) \tag{4.23}$$

where R_1 and R_2 are the radii of gyration of the components with respect to their centers of mass, L_{12} is the distance between these centers, and

$$x_1 = \Delta\rho_1 V_1/(\Delta\rho_1 V_1 + \Delta\rho_2 V_2) \tag{4.24}$$

is the relative fraction of scattering from the first component, V_1, V_2, and $\Delta\rho_1 = \rho_1 - \rho_s$, $\Delta\rho_2 = \rho_2 - \rho_s$ being the volumes and excess densities of the components, respectively. Having determined R_g for three different values of ρ_s, one can calculate the values of R_1, R_2, and L_{12}, which describe the relative arrangement of the components within the particle. It is obviously more appropriate when studying two-component particles to use equation (4.23) rather than the general relation (4.20).

The dependence of zero-angle scattering on the contrast is also of great importance. It is clear by the definition of $\rho_f(\mathbf{r})$ that $I_s(0) = I_{cs}(0) = 0$; hence

$$I(0) = (\Delta\rho)^2 I_c(0) = (\Delta\rho)^2 V^2 \tag{4.25}$$

i.e., $[I(0)]^{1/2}$ is the linear function of the contrast, the angle of inclination of the line being determined by the volume of a particle. A matching point, where $\Delta\rho = 0$, and therefore $I(0) = 0$, is an important characteristic of a particle. At this point $\bar{\rho} = \rho_s$, so the average density level within the particle is readily given by its determination.

Equation (4.13) enables the Porod volume to be expressed in the form

$$V^{-1}(\rho_s) = V^{-1} + (\Delta\rho V)^{-2} \int_V \rho_f^2(\mathbf{r}) d\mathbf{r}$$

As soon as $V(\rho_s)$ has been determined for two values of ρ_s, the particle volume and mean-square density fluctuation can be evaluated.

4.2.3. Contrasting Techniques

The contrast-variation method, separating information on the particle as a whole and on its inner inhomogeneities, makes it possible both to determine the particle shape more accurately and to search for some features of its inner structure. Therefore, the method is widely used in small-angle scattering, particularly for investigations of biological objects. In X-ray scattering the solvent density changes are achieved by using solutions of biologically inactive substances (such as glycerine and saccharose), while in neutron scattering, D_2O is substituted for a certain amount of water in the solvent. Figure 4.5 and Table 4.1 show possibilities of contrast variation for both types of radiation, where data of Stuhrmann and Miller (1978) and Ostanevitch and Serdyuk (1982) are used. It is seen that H_2O/D_2O mixtures give better possibilities

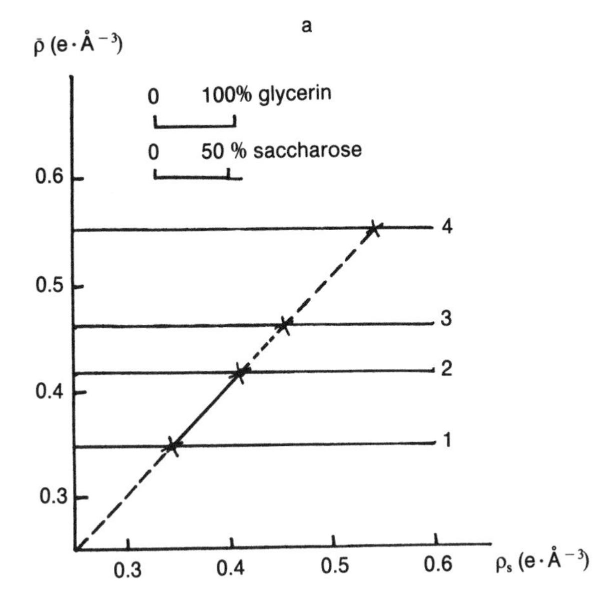

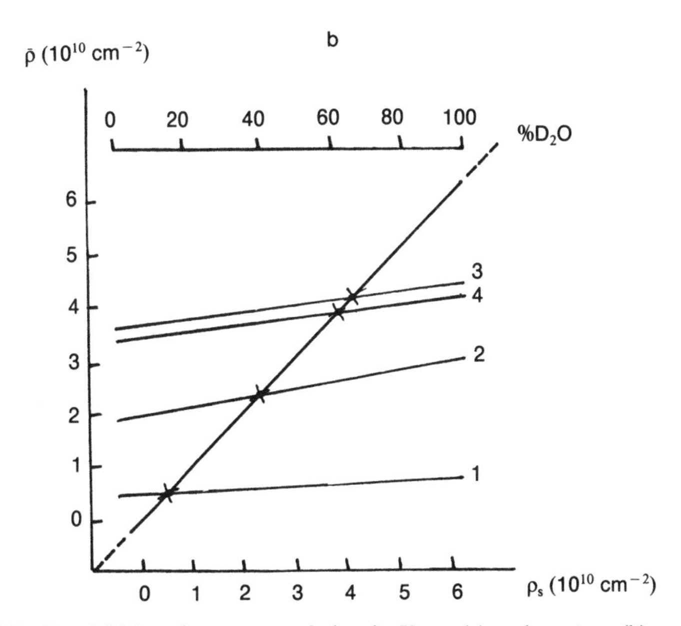

Figure 4.5. Possibilities of contrast variation in X-ray (a) and neutron (b) scattering. Lines (1)–(4) correspond to low-density lipoproteins, proteins, RNA, and DNA. The scattering density of solutes in H_2O/D_2O mixtures increases slightly due to H/D exchange. Matching points are denoted by crosses.

Table 4.1. Scattering Densities of Some Solvents and Components of Biological Macromolecules

Substance	X rays ($Å^{-3}$)	X rays: Solvent for matching	Neutrons ρ (10^{10} cm^{-2}): In H_2O	In D_2O	At matching point	D_2O for matching
Solvents						
H_2O	0.334	—	−0.56	—	—	—
D_2O	0.334	—	6.34	—	—	—
Glycerin	0.41	—	1.25	—	—	—
50% Saccharose in H_2O	0.40	—	1.25	—	—	—
Components of biopolymers						
Low-density lipoprotein	0.347	10% saccharose; 17% glycerin	0.47	0.67	0.48	15%
Protein	0.42	65% saccharose	1.9	2.95	2.33	42%
RNA	0.46	No	3.6	4.4	4.15	68%
DNA	0.55	No	3.4	4.2	3.9	64%

for contrast variation, matching any component of biological macromolecules.

The solvent composition frequently affects the structure of a particle, and relationship (4.17) is not fulfilled. This situation complicates the practical application of the contrast-variation method. In X-ray investigations, the method can be applied mainly to sufficiently large ($M > 10^5$ daltons) particles whose shape will not be altered substantially by the joining of salt, saccharose, or glycerin molecules. The problem of H–D exchange arises in the case of neutron scattering, because some hydrogen atoms in a particle can be replaced by deuterium atoms, and that leads to a change in $\rho_f(\mathbf{r})$. The highest degree of H–D exchange can be estimated assuming complete replacement of the hydrogen by deuterium atoms in polar groups, which are easily accessible for solvent penetration, and hence for isotopic exchange. Since the ratio of the number of hydrogen atoms in polar groups to their total number is determined only by the chemical composition of the given particle component, then, assuming the exchanged hydrogens are proportional to the volume fraction of D_2O in the solvent, one can calculate the scattering density of the component as a function of the D_2O content. These functions are given by the straight lines in Figure 4.5b, while their intersections with the lines corresponding to the solvent density denote the matching points of the components.

An attempt was made by Ibel and Stuhrmann (1975) to account for H–D exchange by introducing into equation (4.17) the "degree of exchange" function, $0 \leqslant \phi_e(\mathbf{r}) \leqslant 1$, which describes the positions and scattering lengths of the exchanged atoms:

$$g(\mathbf{r}, \rho_s) = \Delta\rho[\phi_p(\mathbf{r}) - \phi_e(\mathbf{r})] + \rho_f(\mathbf{r}) \tag{4.26}$$

Contrast variation may in principle provide some information on $\phi_e(\mathbf{r})$. Thus, one can write for $I(0)$

$$[I(0)/\Delta\rho]^{1/2} = \int_V [\phi_p(\mathbf{r}) - \phi_e(\mathbf{r})] d\mathbf{r} = (1-u)V \tag{4.27}$$

where the parameter $0 \leqslant u \leqslant 1$ provides a general estimate of the exchange effects. It is clear, however, that without simplifying assumptions pertaining to $\phi_e(\mathbf{r})$ we shall return to a general expression like (4.5), which is difficult to use in practice. If the centers of mass of $\phi_e(\mathbf{r})$ and $\phi_p(\mathbf{r})$ are assumed to coincide, one obtains a relation similar to (4.20). Equations (4.21) and (4.22) will involve $(1-u)V$ instead of V, and the expression

$$R_{ce}^2 = (R_c^2 - uR_e^2)/(1-u) \tag{4.28}$$

instead of R_c^2, where R_e is the "radius of gyration" of the distribution $\phi_e(\mathbf{r})$ and allows one to describe the arrangement of the exchanged hydrogens in a particle, i.e., the nature of solvent penetration into it.

The solvent influence in contrast variation can be estimated quantitatively by considering zero-angle scattering. Actually, since the scattering length per molecular-mass unit is defined by the chemical composition of a particle

$$z_p = \sum_i x_i b_i \Big/ \sum_i x_i A_i \tag{4.29}$$

where x_i, b_i, and A_i are, respectively, the number, scattering length, and atomic mass of each kind of atom in a particle, then, having determined the value of $\bar{\rho}_p$ at the matching point, one can estimate the partial specific volume of a particle by

$$\bar{v}_z \approx z_p N_A / \bar{\rho}_p$$

The value of $\bar{v}$, on the other hand, is normally established by other methods. The coincidence of $\bar{v}_z$ and $\bar{v}$ should be indicative of the fact that the solvent does not penetrate into the particle. If these values are different, then the expression

$$u = (\bar{v}_z - \bar{v})/\bar{v}$$

provides an approximate estimate of the exchange effects. In neutron scattering, the number of exchanged hydrogens can be evaluated by including the summation in equation (4.29) over deuteriums and seeking to attain the coincidence of $\bar{v}_z$ and $\bar{v}$.

It should be noted that X-ray and neutron contrast-variation techniques give different values for the particle volume V: in the former case the volume inaccessible to the solvent [i.e., to rather large (compared to the H_2O molecule) salts, saccharose, and similar molecules], in the latter case the "dry" volume, inaccessible to D_2O. (The same situation prevails for H_2O.) Hence the X-ray and neutron techniques may supplement each other.

4.2.4. Applications of the Contrast-Variation Technique

We shall now consider some actual examples of structural small-angle studies using contrast variation. It may well be that the most interesting application of the method in X-ray scattering is the study of human-serum low-density lipoprotein (LDL). These particles, close to spherically symmetric, consist of the lipoprotein core and outer protein shell. They have been studied for a long time by various methods, including contrast variation in both X-ray (Tardieu *et al*., 1976) and neutron (Stuhrmann *et al*., 1975b) scattering, but some of the characteristic features of their structure remain to be clarified. Luzzati *et al*. (1979) carried out the X-ray measurements on LDL, the solvent density being varied from 0.336 up to 0.446e Å^3 by adding NaBr in the solution. Nearly 14 solutions were measured with different contrasts, at the two temperatures 21 °C and 41 °C corresponding to different phase states of the LDL molecule (reversed temperature transition takes place at 39 °C).

The resulting dependences of R_g and $I(0)$ on the contrast are shown in Figure 4.6. The values of $\bar{\rho}_p$ and $\bar{v}$ were determined from the plots of $[I(0, \Delta\rho)]^{1/2}$. The calculated value of $\bar{v}$ (at 21 °C) agreed well with densitometry data, thus confirming the suggestion about particle density being independent of solvent density. Parameters R_c and α were evaluated from function $R_g^2(\Delta\rho)$. The α value proved, as expected, to be positive (components with higher density, namely proteins, are on the periphery of the particle), while the ratio $R_c^3/V = 0.132$ differed from the corresponding value for a solid sphere (0.111). This suggested that the shape of the LDL particle, although sufficiently isometric, differs essentially from spherical.

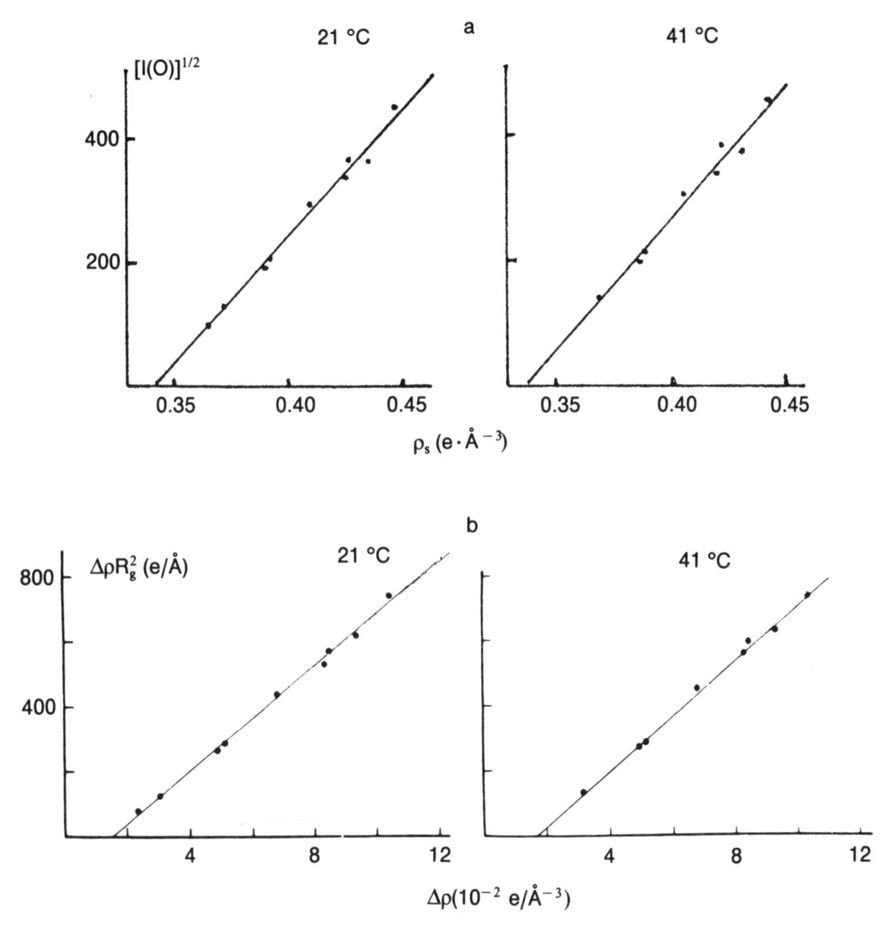

Figure 4.6. Forward scattering intensity (a) and radius of gyration (b) as functions of contrast for two temperature states of LDL (after Luzzati *et al.*, 1979).

The basic functions (shown in Figure 4.7) as well as corresponding correlation functions were calculated from experimental data. The largest particle dimension obtained, $D = 295 \pm 5$ Å, also confirmed that the particle is not spherical. On the other hand, the distance distribution function $p(r)$, calculated from functions $\gamma_c(r)$, $\gamma_s(r)$, and $\gamma_{cs}(r)$, were found at the matching points of protein to be very close to such a function for a solid sphere. Hence, one can conclude that the central lipid core possessed nearly spherical shape, with diameter equal to 208 ± 5 Å. The LDL particle was therefore found to consist of an almost spherical lipid core and protein subunits located isometrically on its surface.

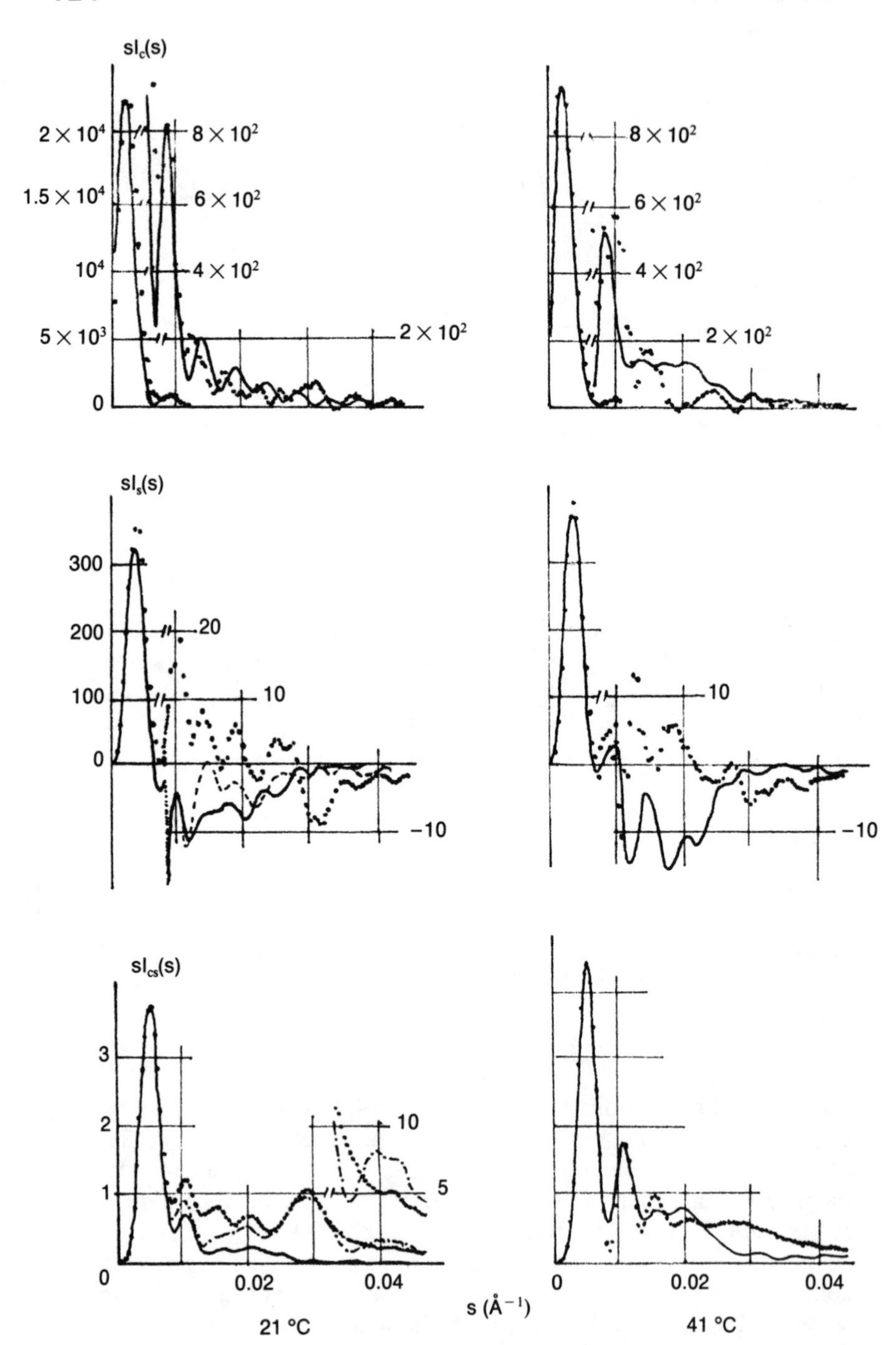

Figure 4.7. Basic functions of LDL for $T = 21$ °C and 41 °C (after Luzzati *et al.*, 1979). Dots denote experiment, solid lines, the model.

According to electron microscopy data (Gulik-Krzywicki *et al.*, 1979), the LDL particles possess tetrahedral symmetry (point group 23). This fact and the above results were used to propose models of LDL constitution at 21 °C and 41 °C (Figure 4.8) and are in good accord with the basic function (see Figure 4.7). These models also agree with physical and chemical data on LDL, and demonstrate characteristic changes taking place at 39 °C thermal phase transition.

The contrast variation in neutron scattering is most frequently used owing to the possibility of wide-range variations in the solvent density by adding heavy water to the solution. In particular many studies are associated with the name of H. Stuhrmann, the developer of the contrast-variation method itself, for instance, the above-mentioned studies of lipoproteins (Stuhrmann *et al.*, 1975b), ferritin (Stuhrmann *et al.*, 1976b), lysozyme (Stuhrmann and Fuess, 1976), fibrinogen (Marguerie and Stuhrmann, 1976), and 70 S ribosome and its 30 S and 50 S subparticles (Stuhrmann *et al.*, 1976a; Stuhrmann *et al.*, 1978).

Ibel and Stuhrmann (1975) studied myoglobin samples and compared neutron contrast-variation data with corresponding X-ray data and with the small-angle curves, calculated by atomic coordinates. Quantity $I(0)$ as a function of the solvent densities for both X-ray and neutron experiments is shown in Figure 4.9. Calculations of the partial specific volume give 0.765 cm³/g for X-ray and 0.74 cm³/g for neutron studies. (H–D exchange was taken into account in the latter case on the assumption that at the matching point of 40% D_2O 105 out of 207 hydrogen atoms in the molecule were replaced by deuterium atoms.) Densitometry experiments give the value $\bar{v} = 0.741$ cm³/g. These results confirm the account of H–D exchange being correct (for X-ray experiments the value of $\bar{v}$ changes due to an addition of glycerin to the

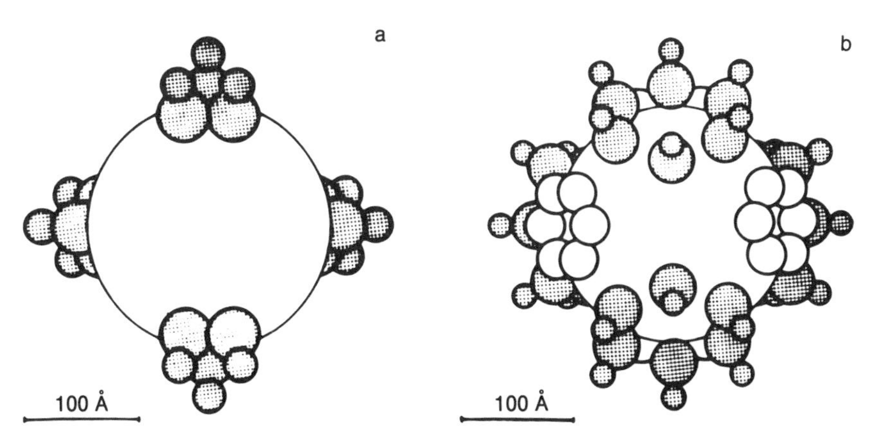

Figure 4.8. Models of LDL for $T = 21$ °C (a) and 41 °C (b) (after Luzzati *et al.*, 1979). Protein particles are shown as shaded areas.

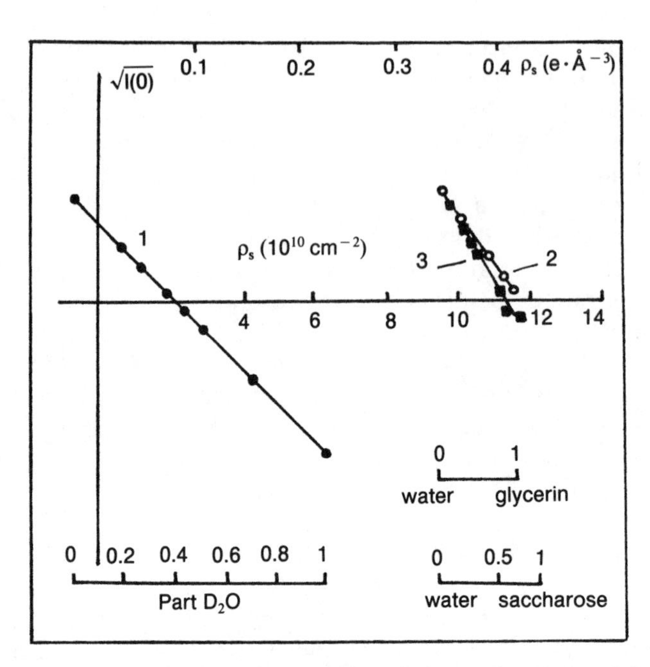

Figure 4.9. Zero-angle scattering of myoglobin solutions (after Ibel and Stuhrmann, 1975): (1) neutron scattering; (2) X-ray scattering (water/glycerol mixtures); (3) X-ray scattering (sugar solutions).

solution). On the other hand, if H–D exchange were not taken into account, this would lead to a considerable underestimate of the particle volume ($u \approx 0.2$), while in X-ray measurements $u < 0.05$, i.e., the solvent almost does not penetrate into the molecule.

The radius of gyration as a function of the contrast is given in Figure 4.10. The values of R_g^2 at infinite contrast are (234 ± 6) Å^2 and (240 ± 10) Å^2 for neutron and X-ray techniques, respectively. Calculations from the known structure give 237 Å^2, which means that, using both X rays and neutrons, the "shape" radius of gyration can be evaluated with sufficient accuracy. The use of equations (4.26)–(4.28) for neutron data gives $R_c^2 = 236$ Å and $R_e^2 = 300 \pm 60$ Å; therefore, H–D exchange takes place in the main on the particle surface. The values of α for neutron and X-ray studies are $(3.5 \pm 1.0) \times 10^{-5}$ and $(3.0 \pm 1.5) \times 10^{-5}$, respectively (the theoretical values are 2.8×10^{-5} and 2.5×10^{-5}). Meanwhile, one fails to determine β from X-ray data because the contrast range is too small [for neutrons, $\beta = (3 \pm 1) \times 10^{-12}$ with the theoretical value 1.5×10^{-12}].

Consideration of the basic functions, where curves $I_c(s)$ and $I_s(s)$ are shown in Figure 4.11, leads to the following conclusions. X-ray and

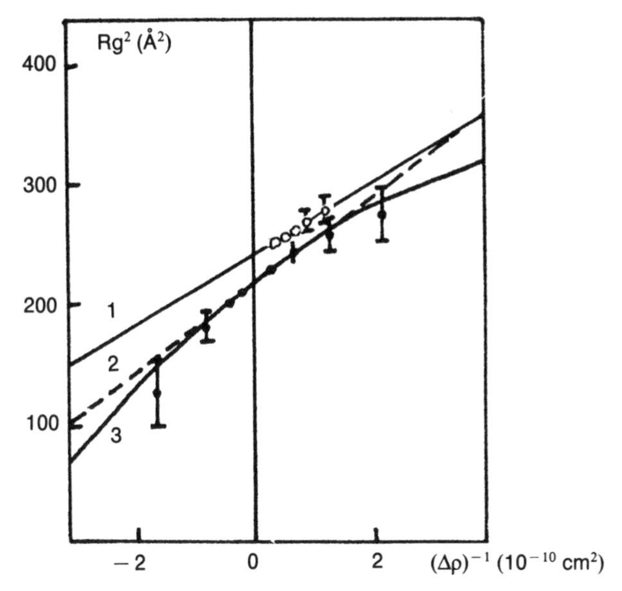

Figure 4.10. Radius of gyration of myoglobin as a function of contrast (after Ibel and Stuhrmann, 1975). Open circles denote X-ray data and shaded circles, neutron data. (1), (2) straight lines with slope α calculated near $(\Delta\rho)^{-1} = 0$; (3) calculated from the myoglobin model (neutrons).

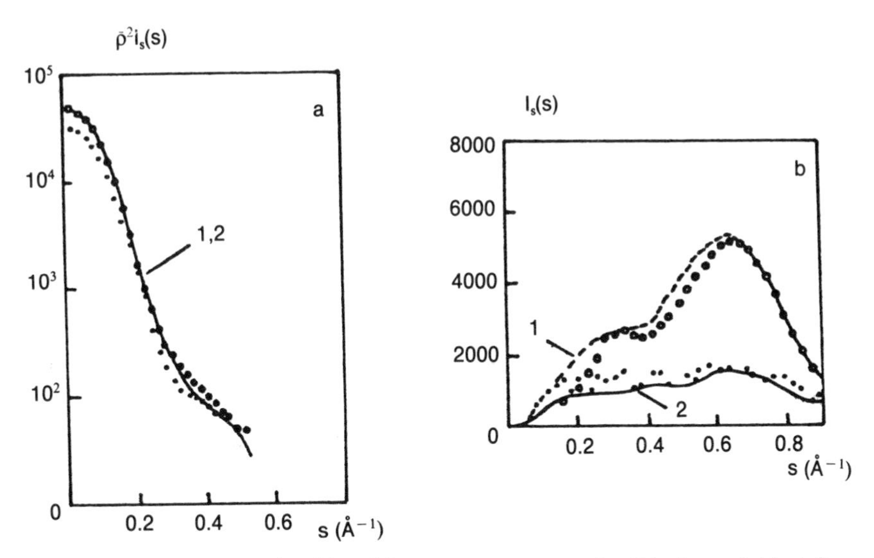

Figure 4.11. Shape scattering (a) and inner structure scattering (b) of myoglobin (after Ibel and Stuhrmann, 1975). Open circles refer to X-ray scattering, shaded circles to neutron scattering. Dotted and solid curves refer to X-ray and neutron scattering, respectively, as calculated from the myoglobin model. The curves of $I_c(s)$ are given for the contrast $\Delta\rho = 10^{10}\ \mathrm{cm}^{-2}$.

neutron curves of $I_c(s)$ are practically identical (the neutron curve is located somewhat lower due to H–D exchange) and agree well with the theoretical data. At the same time, the theoretical X-ray and neutron curves of $I_s(s)$ differ substantially from one another as well as the accuracy of the experimental data. At very small angles the neutron data are more precise, while at larger angles the X-ray data are more accurate. The lower accuracy of neutron data is caused by the lower source intensity and by strong incoherent scattering from hydrogen atoms (see Section 1.5). On the whole, the conclusion reached in the paper is that, owing to the wide contrast range, the neutron contrast variation permits one to obtain more reliable results at the initial angles ($s < 0.2$ Å^{-1}) while at $s > 0.3$ Å^{-1} the X-ray contrast variation is preferable.

Another interesting example is the study of the 50 S subparticle of *E. coli* ribosome (Stuhrmann *et al.*, 1976a). Measurements were carried out for 11 solvent densities (from pure H_2O to pure D_2O) at angles satisfying 4×10^{-3} Å^{-1} $\leqslant s \leqslant 0.5$ Å^{-1}. The initial parts of the curves were measured at different concentrations and extrapolated to infinite dilution. The samples were investigated in two different buffers at various temperatures. A number of structure parameters, in particular the dependence of R_g on the contrast (Figure 4.12), were calculated precisely using

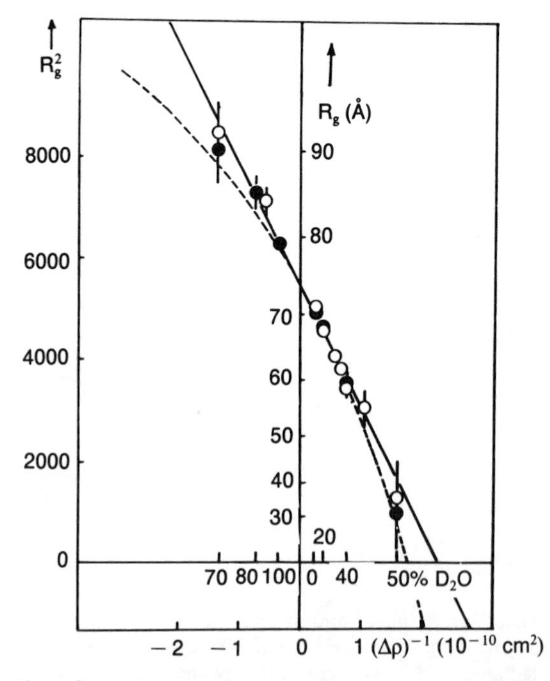

Figure 4.12. Radius of gyration as a function of contrast for the 50 S ribosomal subparticle (after Stuhrmann *et al.*, 1976a). Open and shaded circles correspond to measurements in different solvents.

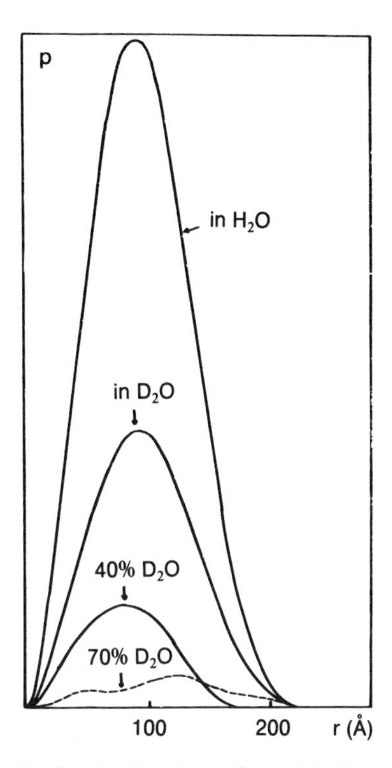

Figure 4.13. Distance distribution functions of the 50 S ribosomal subparticle at various contrasts (after Stuhrmann *et al*., 1976a).

these data. The authors investigated the relative arrangement of the two components of ribosome: protein and RNA. The experiments involving 40% D_2O (with the protein matched) were analyzed with this aim. The calculated curves of $p(r)$ for different D_2O concentrations are presented in Figure 4.13. The radius of gyration and diameter of the protein component were found equal to 90 ± 2 Å and 220 Å, which corresponds to the diameter of the whole particle; the respective figures for RNA are 59 ± 1 Å and 170 Å. The distance between the centers of mass of the two components was estimated using the curvature of the function $R_g(1/\Delta\rho)$ with the aid of equations (4.23) and (4.24), and was found equal to 20 Å. It thus follows that the 50 S subparticle consists of the compact core (RNA) surrounded by the protein shell. Further analysis of the resulting curve $I_c(s)$ enabled the shape of this particle to be determined by a direct method (see Section 5.4.2).

Experiments on the relative arrangement of particle components are also possible in more complicated cases. Koch *et al*. (1978) investigated the 70 S ribosome, consisting of the 50 S and 30 S subparticles. Values of R_g as functions of reciprocal contrast are shown in Figure 4.14. The

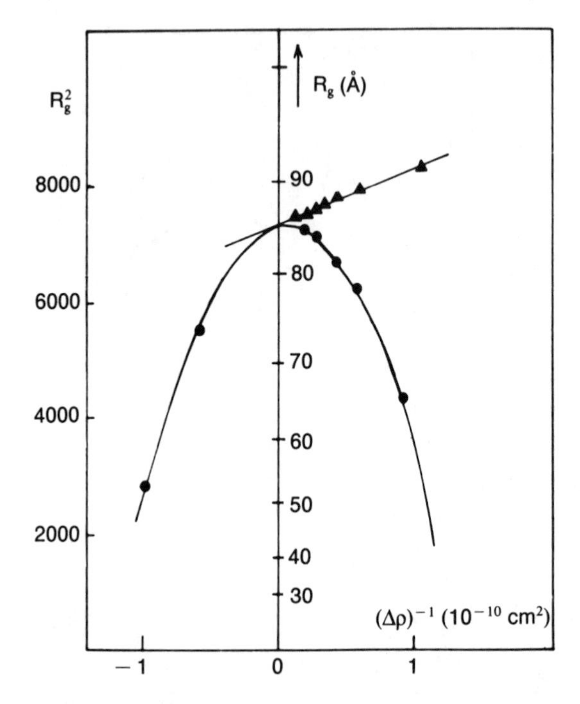

Figure 4.14. Dependence of R_g on contrast for 70 S ribosome (after Koch *et al.*, 1978). The straight line refers to both subparticles deuterated, and the parabola to only the 30 S subparticle deuterated.

straight line corresponds to the sample with both 50 S and 30 S subparticles deuterated (they were extracted from the ribosome after being grown in heavy water), while the parabola matches the native 50 S subparticle and deuterated 30 S subparticle. In the first case, the center of mass is not displaced by a change in contrast ($\beta = 0$; the same holds for native particles; see Stuhrmann *et al.*, 1978). At the same time, a well-defined parabolic dependence appears so that the distance can be calculated between the centers of mass of 50 S and 30 S subparticles ($L = 88 \pm 15$ Å). We note that, as expected, the values of R_g at infinite contrast are the same for both samples ($R_c = 85 \pm 2$ Å).

Kinetic (time-resolved) experiments should also be mentioned, where H–D exchange becomes an object of investigation rather than a complicating factor. Actually we have already seen that functions $I(0)$ and R_g depend substantially on the degree of H–D exchange. Therefore, if one carries out a series of sufficiently rapid experiments immediately after addition of D_2O to H_2O solution, it is possible, while analyzing the temporary variations in these functions, to evaluate the rate of exchange as well as the sequence of deuteration of various segments of the particle (Schelten *et al.*, 1972; Stuhrmann and Miller, 1978). Such experiments can be performed only if a high-flux reactor and an instrument equipped with a position-sensitive detector are available (see Section 8.4).

4.3. Isomorphous-Replacement Methods

The contrast-variation method involves a change in the scattering length of a particle as a whole relative to the solvent, thus enabling the total contribution to be separated from various intrinsic inhomogeneities of the particle. As seen earlier, the interpretation of this contribution and the determination of the relative arrangement of inhomogeneities is a rather complex task. Therefore, the following method is often used: one changes the excess scattering density not of the whole particle, but of its separate segments. Then, the above formalism can be employed to separate the contribution of these segments to the scattering intensity, and thereby obtain information on their relative arrangement.

4.3.1. Heavy-Atom Labels

We first consider the case where n point scatterers (labels) with scattering lengths b_i are added to a particle:

$$\rho_{\mathrm{m}}(\mathbf{r}) = \rho(\mathbf{r}) + \sum_{i=1}^{n} b_i \delta(\mathbf{r} - \mathbf{r}_i)$$

where $\mathbf{r}_i$ are the label coordinates. Thus, the scattering intensity is given by

$$I_{\mathrm{m}}(s) = I(s) + \sum_{i=1}^{n} \sum_{j=1}^{n} b_i b_j \frac{\sin(sr_{ij})}{sr_{ij}} + W(s) \qquad (4.30)$$

where $I(s)$ is the scattering from a particle without labels and $W(s)$ is the cross term (the solvent scattering is supposed to be eliminated). The second term on the right-hand side of equation (4.30) conveys information on the distances between labels. For a given number of labels, these distances can be estimated by analyzing the positions of its maxima and minima.

Hence for the simplest case of two equal labels b_{m} this function can be written as $2b_{\mathrm{m}}^2[1 + \sin(sr_{12})/sr_{12}]$, the maxima corresponding to the roots of the equation $\mathrm{tg}(sr_{12}) = sr_{12}$, i.e., $sr_{12} = 0$, 4.493, 7.854, 10.904, and so on.

Such an approach was proposed to determine characteristic distances in particles using small-angle X-ray scattering (Kratky and Worthman, 1947; Vainshtein *et al*., 1970). In order that the scattering of labels should be noticeable against the background of the scattering by a particle, the values of b_m should be sufficiently high. Therefore compounds with heavy atoms were used as labels, whence the method draws its name. Isomorphous attachment of labels was performed to "mark" the definite points of a particle. Since the cross term $W(s)$ decreases very rapidly with s [see, for example, the model calculation by Feigin *et al*. (1978)], the difference between scattering from labeled and native particles seems to be a satisfactory estimate for the scattering from labels.

The investigation of histidinedecarboxylase (HDC) exemplifies application of the method (Feigin *et al*., 1978; Vainshtein *et al*., 1980). The heavy-atom labels with four mercury atoms were attached to three fast-titrating SH groups and the scattering curves from labeled HDC_m and native HDC measured. Two different approaches were used to calculate scattering by labels in various series of experiments: (1) subtraction of the HDC from the HDC_m curve (Figure 4.15a), and (2) calculation of the difference curve for scattering by a solution with the same concentration of HDC_m and HDC (subtraction of solvent scattering is not required here). The obtained difference curves are presented in Figure 4.15b, which shows that the sets of their maxima coincide well enough. If the labels are placed at the vertices of an equilateral triangle with side 69 ± 3 Å, one obtains the best agreement with these data.

The heavy-atom label method therefore permits one to obtain extra information about the organization of the particles under investigation. One should, however, bear in mind that the method is inapplicable to particles of large molecular weight for which scattering from labels cannot be evaluated from the experimental data with sufficient accuracy. According to Feigin *et al*. (1978) the condition $N_m > 4 \times 10^{-3} N_p$ should be fulfilled, where N_m is the number of electrons in a label and N_p is the number of electrons in a particle.

4.3.2. The Triangulation Method

We now consider the general case when a change takes place in the scattering length of particle segments. Suppose two segments 1 and 2 are marked in a particle, the distance between their centers of mass being d_{12} (Figure 4.16a). Then, by somehow changing the scattering length of these segments, the scattering intensity can be expressed in the form

$$I_b(s) = I_a(s) + I_1(s) + I_2(s) + C_{a1}(s) + C_{a2}(s) + C_{12}(s)$$

where $I_a(s)$ is the scattering intensity from a native particle, $I_1(s)$ and $I_2(s)$

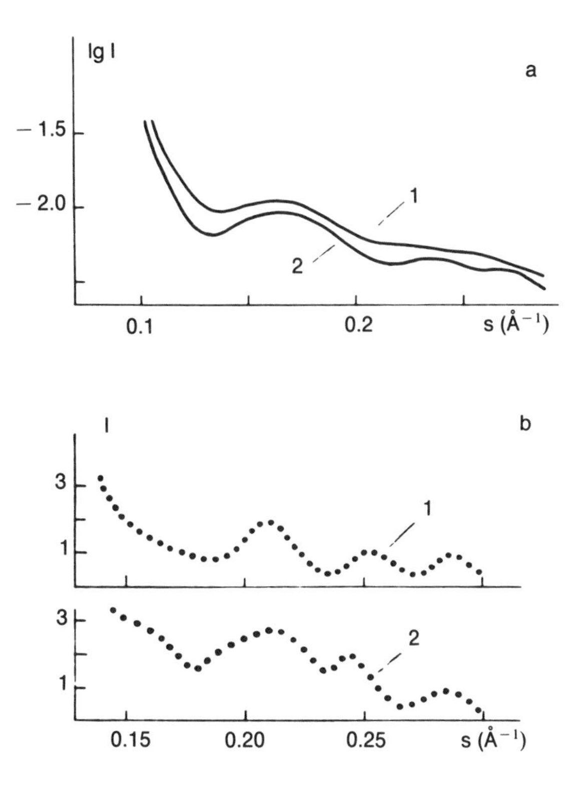

Figure 4.15. Determination of the distances between heavy-atom labels in HDC (after Vainshtein *et al.*, 1980): (a) scattering curves from labeled (1) and native (2) particles; (b) scattering from labels. Curves (1) and (2) correspond to different techniques of calculation; see the text.

are the scattering from segments 1 and 2 with excess densities $g_1(\mathbf{r})$ and $g_2(\mathbf{r})$ equal to the differences between the scattering length densities of these segments in the modified and native states, $C_{a1}(s)$ and $C_{a2}(s)$ represent the interference between scattering from the modified segments and a particle, and $C_{12}(s)$ corresponds to interference scattering from the particles shown in Figure 4.16d.

By modifying the two segments separately, we obtain (see Figure 4.16c,d)

$$I_c(s) = I_a(s) + I_1(s) + C_{a1}(s) \quad \text{and} \quad I_d(s) = I_a(s) + I_2(s) + C_{a2}(s)$$

Hence

$$C_{12}(s) = I_a(s) + I_b(s) - I_c(s) - I_d(s) \tag{4.31}$$

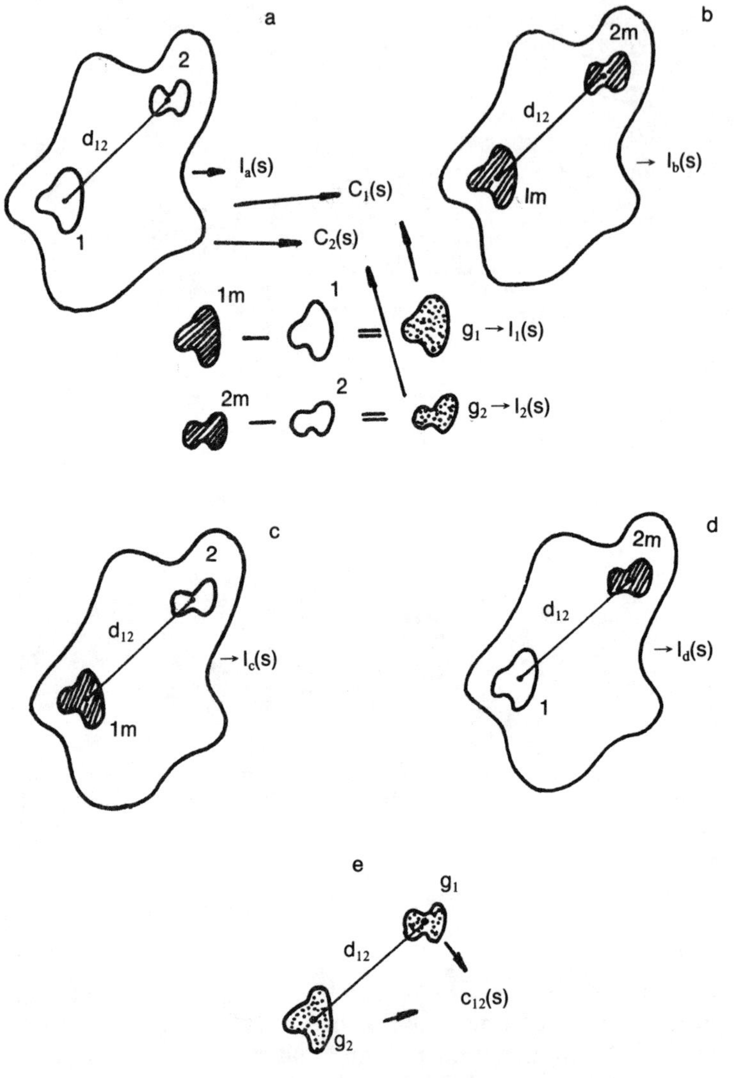

Figure 4.16. Determination of the distance between two modified segments.

It thus follows that, having measured particle scattering for four different states, one can separate the term which depends only on the difference between the scattering densities of definite particle segments in their native and modified states, and on the distance between their centers of mass (Engelman and Moore, 1972). In practice, it is preferable to measure the scattering from equimolar solutions of particles a and b, c and d, rather than functions $I_a(s)$, $I_b(s)$, $I_c(s)$, and $I_d(s)$. It was shown by

Hoppe (1972, 1973) that the difference $I_{a+b}(s) - I_{c+b}(s)$ also provides the function $C_{12}(s)$. One important advantage of such an approach is that the interference effects between the modified segments in different particles become negligible, so one can perform experiments with highly concentrated solutions.

The positions of the maxima of $C_{12}(s)$, as in the case of heavy-atom labels, are connected to the distance d_{12}. We note that equation (4.31) also holds for this method, so making it possible to eliminate the cross term $W(s)$ in equation (4.30). If the distances between several fixed segments of a particle are known, one can determine their three-dimensional arrangement using the triangulation method (Langer *et al.*, 1978).

Actually, if n points ($n \geqslant 3$) are given in space, $3n - 6$ coordinates describe their relative arrangement. These coordinates can be determined if $4n - 10$ interpoint distances are known. On the other hand, there are $n(n-1)/2$ distances between n points. Therefore if the relationship

$$n(n-1)/2 \geqslant 4n - 10 \geqslant 3n - 6$$

holds (this is so for $n \geqslant 4$), one can in principle restore the three-dimensional relative arrangement of segments using intersegment distances. If the latter are known exactly, the coordinates can be obtained (up to enantiomorphic structure) by simple geometrical construction (see Engelman *et al.*, 1975).

Engelman, Moore, and co-workers carried out an extensive study of the 30 S subparticle of *E. coli* ribosome with the aid of the triangulation method. Deuteration of selected protein pairs was used to achieve an appropriate change in their scattering lengths (the total number of proteins in the particle equals 21). The distances between the proteins were determined, after which their three-dimensional relative arrangement was constructed. A great deal of biochemical work as well as high-precision intensity measurements were required in the course of this study. Besides, difficulties arose when determining the values of d_{12}, because their simple calculation using the positions of the maxima of $C_{12}(s)$ is possible only if d_{12} is much greater than the dimensions of the modifying segments. The distance distribution function $p_{12}(r)$ corresponding to $C_{12}(s)$ was used to estimate d_{12} more precisely. By considering the explicit expression for $C_{12}(s)$, namely

$$C_{12}(s) = \int_{V_1}\int_{V_2} g_1(\mathbf{r}_1)g_2(\mathbf{r}_2)\frac{\sin(s\,|\mathbf{d}_{12} + \mathbf{r}_2 - \mathbf{r}_1|)}{s\,|\mathbf{d}_{12} + \mathbf{r}_2 - \mathbf{r}_1|}\,d\mathbf{r}_1 d\mathbf{r}_2$$

where $\mathbf{r}_1$ and $\mathbf{r}_2$ are vectors drawn from the centers of mass of segments 1 and 2, and $\mathbf{d}_{12}$ is the vector connecting these centers, one sees that the function

$$p_{12}(r) = r \int_0^\infty C_{12}(s) \sin(sr) s \, ds$$

yields the length distribution of vectors joining the points in segments 1 and 2. Langer *et al*. (1978) chose the coordinate of the center of mass of this distribution as an estimate of d_{12}; however, in this case the result depends on the shape and relative arrangement of the segments (Moore *et al*., 1977). A more general approach was suggested by Moore and Weinstein (1979). Using the theorem of parallel axes, one can easily show that the "radius of gyration" corresponding to the distribution $p_{12}(r)$, i.e., its second moment, is given by

$$R_{12}^2 = M_{12} = \int_0^\infty p_{12}(r) r^2 dr \left[\int_0^\infty p_{12}(r) dr \right]^{-1} = R_1^2 + R_2^2 + d_{12}^2$$

where R_1 and R_2 are the radii of gyration of the particle with densities $g_1(\mathbf{r})$ and $g_2(\mathbf{r})$ about their centers of mass. Values of M_{ij} can be used to restore the three-dimensional arrangement of the segments. This increases the number of independent variables to $4n - 6$ (the quantities R_i are added), but the number of values of M_{ij} that can be determined still remains sufficiently large. Moore and Weinstein (1979) suggested a method for seeking the coordinates of the centers of mass and radii of gyration of the segments under investigation. They proposed minimizing the functional

$$\phi = \sum_{i,j} [M_{ij} - (d_{ij}^2 + R_i^2 + R_j^2)]^2 / \sigma_{ij}^2$$

where σ_{ij} is the error in M_{ij} while

$$d_{ij} = (x_i - x_j)^2 + (y_i - y_j)^2 + (z_i - z_j)^2$$

By solving the corresponding system of normal equations, one finds the required values of the coordinates and radii of gyration. It should be noted, however, that this system is nonlinear, therefore an iteration procedure should be used to find the solution. The method of Moore and Weinstein (1979) works for $n > 8$.

The triangulation method was further developed by Ramakrishnan and Moore (1981), who describe the general technique for neutron scattering data analysis, which enables one to calculate functions $p_{12}(r)$ and their moments with an estimate of the accuracy of the results. The accompanying paper (Ramakrishnan *et al*., 1981) presents results on triangulation of the 30 S subparticle of *E. coli* ribosome. As many as 43 difference scattering curves were measured and corresponding moments calculated. A typical set of data is shown in Figure 4.17. The coordinates as well as the radii of gyration of 12 proteins were determined. Ramakrishnan *et al.* (1984) presented a map of 15 proteins, and the protein mapping was almost completed by Moore *et al.* (1986), who gave the positions of 19 proteins on the basis of 83 pairwise distances (only S2 and

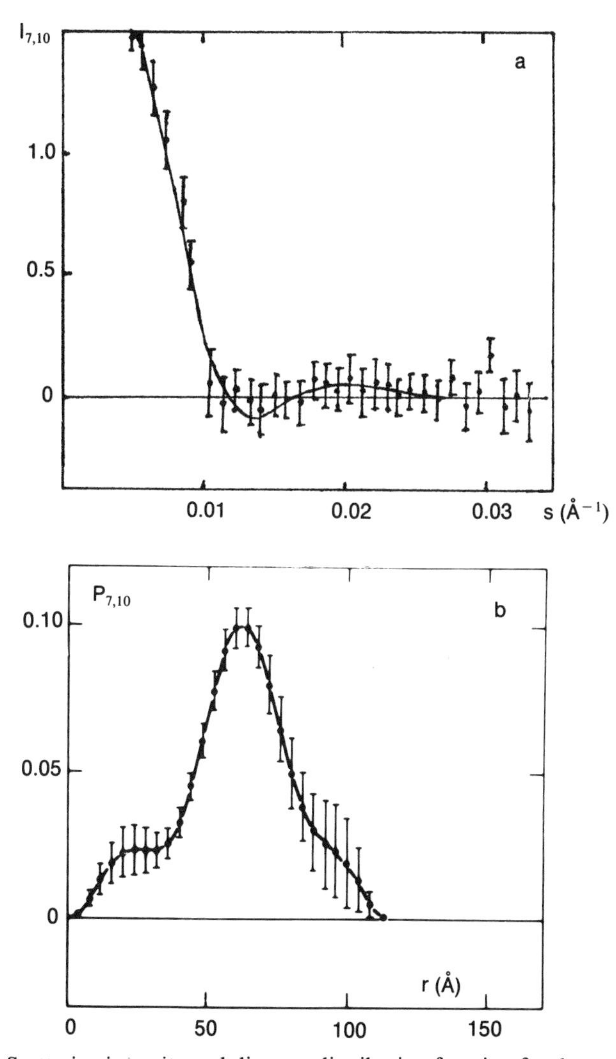

Figure 4.17. Scattering intensity and distance distribution function for the protein pair S7–S10 (after Ramakrishnan *et al.*, 1981): (a) intensities, where dots represent experimental data with errors and the solid line corresponds to inverse calculations using function $P_{7,10}$; (b) function $P_{7,10}$, where associated errors are estimated following Ramakrishnan and Moore (1981).

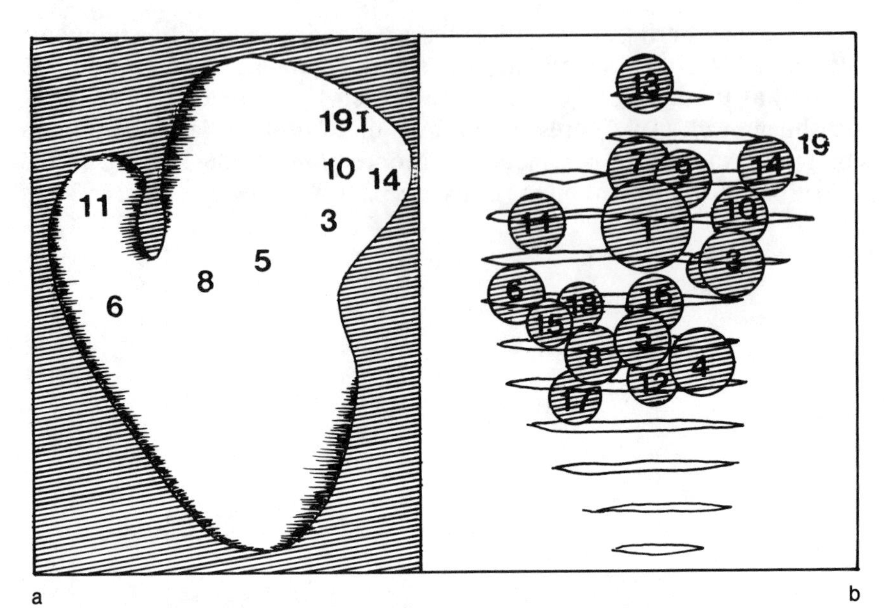

Figure 4.18. Superposition of the protein map in 30 S *E. coli* ribosome according to the triangulation method on the electron microscopic model of Kahan *et al*. (1981): (a) space-filling model of the electron microscopic structure with the locations of antigenic determinants marked; (b) positions of proteins merged with the electron microscopic contours (after Moore *et al*., 1986).

S21) remained to be found). A complete map was presented by D. Engelman *et al*. at the Seventh Small-Angle Scattering Conference (Prague, 1987). Figure 4.18 shows the arrangement of the obtained protein molecules merged with the immunoelectron microscopy data of Kahan *et al*. (1981).

Probably the most serious problem in the practical application of the triangulation method is that the radii of gyration of the proteins are small compared to the distances in the 30 S subparticle. The result is that the values of R_i are determined with large errors. Speaking of the triangulation technique as such it should be noted that the method can be applied only to objects in which one can effectively change the scattering lengths of selected segments without affecting the others. Moreover, application of the method to its full extent is rather complex, both experimentally and computationally.

4.4. Variation in the Applied Radiation

The methods examined above employed an appropriate modification of an investigated sample to change the scattering density of particle or its segments relative to the solvent. Sometimes hard preparative work is required, especially in the methods of heavy-atom labels and triangulation. Moreover, serious difficulties arise due to possible structural rearrangements of a particle under such a modification, such as H–D exchange and distortions caused by heavy atoms. There is, however, the possibility of obtaining the necessary information on the inner structure of a particle without changing an object, that is, to somehow or other change the radiation employed.

4.4.1. Combined Use of Various Types of Radiation

We first consider the question of whether it is useful to use different radiations when investigating the same sample. As we know, X rays are scattered by "free" electrons (Section 1.4) and neutrons by nuclei (Section 1.5). Let g_x and g_n be the excess scattering densities for the two types of radiation, respectively. The method of light scattering (diffraction of electromagnetic radiation in the region of visible light) is also widely used in studies of highly disperse systems. Light is scattered by electrons from outer atomic shells, the excess scattering density being proportional to an increment in the refractive index:

$$g_l = (\partial n/\partial c)/\bar{v} = n - n_0$$

where $\bar{v}$ is the partial specific volume of a substance, n is the refractive index of a solution, and n_0 is the refractive index of the solvent. It is clear

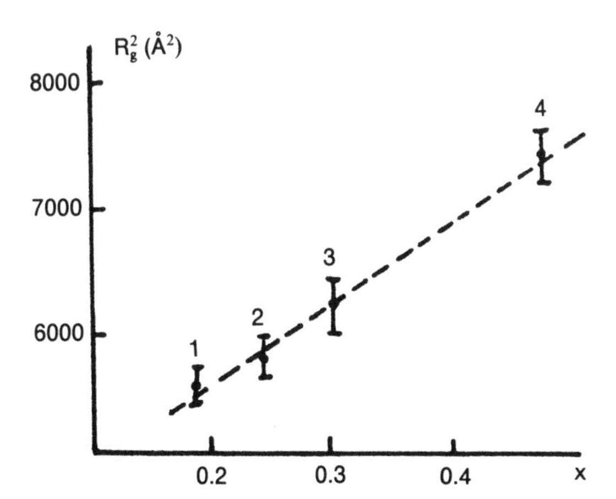

Figure 4.19. Radius of gyration of the 50 S ribosomal subparticle as a function of the volume scattering fraction of protein, obtained by using various radiations (after Serdyuk and Grenader, 1975): (1) X-ray scattering; (2), (3) neutron scattering in H_2O and D_2O, respectively; (4) light scattering.

Table 4.2. Excess Scattering Densities of Biological Objects in Aqueous Solutions

	Radiation, density		
Component	X rays (eÅ^{-3})	Neutrons (10^{10} cm^{-2})	Light
Protein	0.09	1.9	0.26
RNA	0.23	3.6	0.34
Phospholipids	0.01	0.7	0.16
Polysaccharides	0.22	2.5	0.25

that since the particle consists of several components with different values of g_x, g_n, and g_l, the X-ray, neutron, and visible-light scattering measurements yield information analogous to measurements with three different solvent densities during contrast variation. Such an approach was proposed by Serdyuk and Fedorov (1973) and Serdyuk (1974) for studying the structure of synthetic block copolymers and two-component biological macromolecules in solution. The possibilities of this technique are determined by the relative changes in the excess scattering densities of different particle components while replacing one radiation with another. The values of these densities for some components of biological macromolecues in aqueous solutions are presented in Table 4.2.

Serdyuk and Grenader (1975) used this technique combined with equations (4.23) and (4.24) to investigate the relative arrangement of RNA and proteins in the 50 S subparticle of *E. coli* ribosome. (Application of the contrast-variation method to this object was described in Section 4.2.4.) Figure 4.19 shows the measured radius of gyration as a function of the relative scattering fraction of the protein component. The calculated radii of gyration of RNA and the protein component are 65 and 104 Å, while the distance between their centers of mass does not exceed 25 Å. These data, being somewhat different from those obtained by Stuhrmann *et al.* (1976a), also provide evidence that the central region of the 50 S subparticle is filled by RNA.

Hence the combined use of different radiations allows contrast variation without changing the object. We note, however, that the range of contrasts, determined for two-component particles by the lower and upper limits of the variable x_1 in equation (4.24), proves to be essentially smaller than for the contrast variation in neutron scattering.

4.4.2. Anomalous (Resonant) Scattering

The use of anomalous scattering (mainly in X-ray studies) provides another possibility of changing the excess scattering density of a particle or its components. It was shown in Section 1.4 that the atomic scattering amplitudes depend on the X-ray wavelength in such a way that the anomalous corrections become significant only near the absorption edges of an atom. Provided there are atoms in the sample that scatter X rays anomalously at some wavelength λ_k, the variation in wavelength near λ_k will lead to a change in the scattering lengths of the segments containing these atoms. Hence the resonant atoms may be found both in a particle and in a solvent. These possibilities will be dealt with below (see Stuhrmann, 1981a,b; Stuhrmann and Notbohm, 1981).

1. Anomalous scatterers are distributed uniformly over the entire solvent. In this case the solvent density will depend on the wavelength in line with equation (1.30), and variation in λ near the absorption edge will lead to a somewhat "complex contrast variation." Equation (4.18) is replaced by

$$I(s) = [(\Delta\rho_0 - \Delta\rho')^2 + \Delta\rho''^2]I_c(s) + (\Delta\rho_0 - \Delta\rho')I_{cs} + I_s(s)$$

where the meaning of the basic functions remains unchanged; $\Delta\rho_0$ is the contrast far away from the absorption edge, while $\Delta\rho'$ and $\Delta\rho''$ are the real and imaginary parts of the anomalous corrections to the solvent density. The radius of gyration is then given by

$$R_g^2 = R_c^2 + \frac{(\Delta\rho_0 - \Delta\rho')\alpha}{(\Delta\rho_0 - \Delta\rho')^2 + \Delta\rho''^2} - \frac{\beta}{(\Delta\rho_0 - \Delta\rho')^2 + \Delta\rho''^2}$$

At high contrasts $|\Delta\rho_0| \gg |\Delta\rho''|$; these expressions reduce to equations (4.18) and (4.20) with $\Delta\rho$ replaced by $\Delta\rho_0 - \Delta\rho'$. Such an approach allows contrast variation to be carried out with one and the same solvent.

Stuhrmann and Gabriel (1983) studied ferritin molecules in 30% CsCl solution near the L_3-absorption edge of cesium ($\lambda_k = 2.47$ Å). The absorption spectrum of the solvent was measured in order to calculate function $\Delta\rho'(\lambda)$ by equation (1.31) (see Figure 4.20a). In Figure 4.20b, the obtained dependence of R_g on contrast is compared with the results of neutron experiments. It is evident that, though the contrast range for anomalous scattering data is rather narrow, the value of R_c can be determined with sufficient accuracy.

2. Resonant atoms are distributed nonuniformly over the solvent. This case confirms, e.g., the investigations of polyelectrolyte solutions, where ion clouds are formed around the charged segments of molecules. Anomalous scattering allows one to study the configuration of ions in the solution which surrounds the particles, and thus to obtain information on the particle structure (Stuhrmann, 1981b; Templeton *et al*., 1980).

3. Resonant atoms are bound to definite segments in particles (the former are lacking in the solvent). This is, perhaps, the most significant

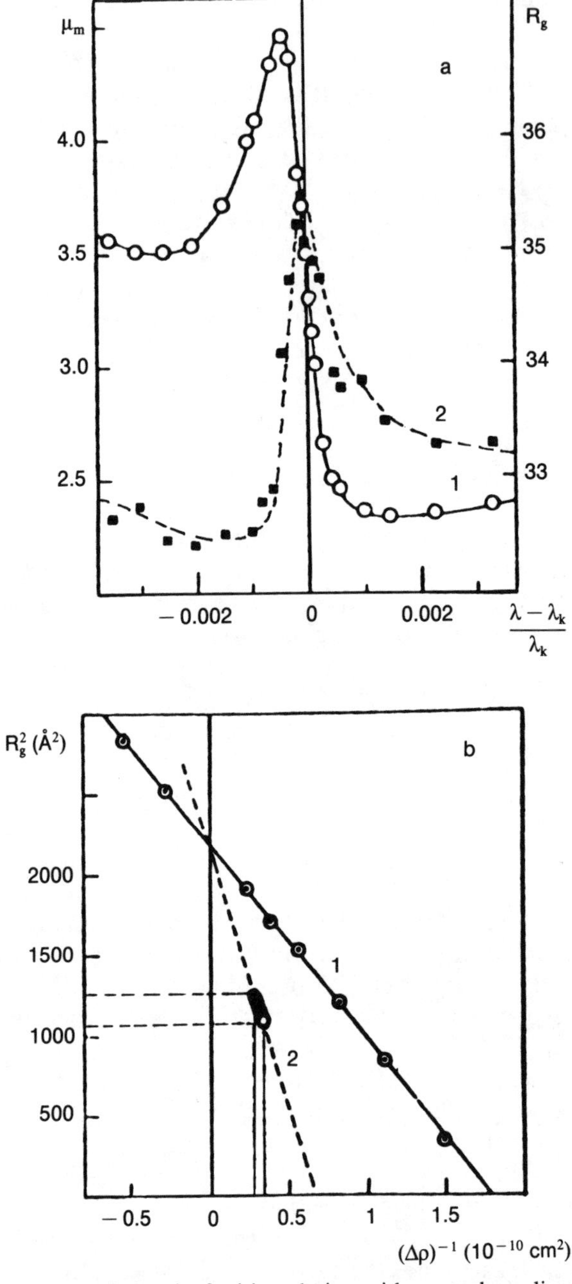

Figure 4.20. Contrast variation in ferritin solution with anomalous dispersion (after Stuhrmann and Gabriel, 1983): (a) extinction of a solution (1) and the radius of gyration of ferritin (2) near the L_3-absorption edge of Cs ($\lambda = 2.47$ Å); (b) the radius of gyration of ferritin versus the reciprocal contrast for neutron (1) and anomalous dispersion (2).

case. The scattering density of a particle can be expressed as

$$\rho(\mathbf{r}) = \rho_0(\mathbf{r}) + (\rho' + i\rho'')\rho_a(\mathbf{r})$$

where $\rho_0(\mathbf{r})$ is the density far away from the resonance, $\rho_a(\mathbf{r})$ describes the positions of the anomalous scatters, and ρ' and ρ'' are the anomalous corrections. Hence the small-angle scattering intensity is

$$I(s) = I_0(s) + \rho' I_{a0}(s) + [\rho'^2 + \rho''^2] I_a(s) \qquad (4.32)$$

where $I_0(s)$ is the scattering intensity far away from the absorption edge,

$$I_{a0}(s) = 2 \int\int \rho_0(\mathbf{r}_1)\rho_a(\mathbf{r}_2) \frac{\sin(sr_{12})}{sr_{12}} d\mathbf{r}_1 d\mathbf{r}_2$$

is the cross term, and $I_a(s)$ is the anomalous scattering intensity. The similarity between relationship (4.32) and equations (4.18) and (4.30) is evident. The distribution of resonant atoms in a particle (e.g., of metal atoms in metal-containing proteins) is described by a sum of δ-functions, similar to that described in Section 4.3.1, a change in wavelength corresponding to isomorphous replacement with the introduction of heavy-atom labels, no sample treatment being involved.

It should be noted, however, that in this case the anomalous scattering intensity is very weak, thus experimentally still unobservable. Figure 4.21 presents scattering intensity curves of hemoglobin near the K absorption edge for iron ($\lambda = 1.743$ Å). It is clear that the cross term amounts to about 1% of the total scattering, the intensity of anomalous scattering being so weak that it cannot be determined from the experimental data. The curve of $I_a(s)$ shown was calculated from the proposed model. Therefore the distances between resonant atoms should be sought on the basis of an examination of the cross term, which is undoubtedly a more complex procedure than the analysis of scattering from labels.

Not only metals, but also phosphorus or sulfur atoms can be used as resonant labels in any investigation of biological objects. The content of these atoms in the latter may be rather high: sulfur enters the composition of dísulfide bridges in proteins; phosphorus is an element of the RNA and DNA helixes. So while investigating proteins and nuclear acids, anomalous scattering from phosphorus and sulfur atoms may be very helpful.

Thus anomalous scattering provides new possibilities for investigating various systems. Experiments involving these methods may be performed by using synchrotron radiation, which enables wide wavelength variation at high source power.

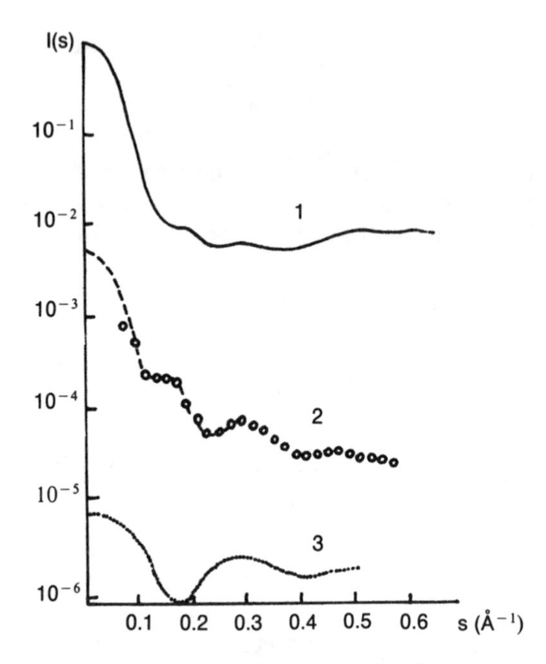

Figure 4.21. Anomalous dispersion study of hemoglobin (after Stuhrmann, 1981): (1) $I_0(s)$; (2) $I_{a0}(s)$; (3) $I_a(s)$ (the latter is calculated from the model).

4.5. Conclusion

The methods for investigating scattering by inhomogeneous particles may be split conditionally into two groups. The first (the "integral" one) includes approaches dealing with changes in the contrast, i.e., the scattering density of the whole particle is changed with respect to the solvent. In the classical contrast-variation method this is achieved at the cost of a change in solvent density; a similar effect may be obtained by changing the type or (if the solvent contains anomalous scatters) wavelength of the radiation employed. The results obtained by these methods permit one, first, to establish whether the particle in question is homogeneous or not, at least within the framework of small-angle resolution. Moreover, it appears possible to separate the "homogenous" and "inhomogeneous" components of the scattering intensity and the scattering invariants in order to determine scattering by the "shape" of a particle, scattering by its individual components, and to obtain some general features of the intrinsic inhomogeneity distribution. The contrast-variation method allows standard approaches to be used widely in the study of inhomogeneous particles (cf. Chapter 3). In this way the general parameters of these particles can be determined.

Table 4.3. Methods for Investigating Inhomogeneous Biological Particles

Technique	Advantages	Shortcomings	Suitable objects
Contrast variation			
X rays (adding salts, glycerin, saccharose)	Simple preparation of samples	Narrow contrast range. Applicable to large particles	Large viruses, macromolecular complexes
Neutrons (adding D_2O)	Wide range of contrast. Simple preparation of samples	Structural changes caused by H–D exchange	Two- and multicomponent particles
Joint use of different types of radiation	No sample modification	Narrow contrast interval. Limited range of application	Two-component particles (lipoproteins, ribosomes)
Isomorphous replacement			
X-ray (heavy-atom labels)	Simple preparation	Possible structural changes caused by the attached labels	Proteins (in particular, enzymes)
Neutrons (triangulation)	Possibility of restoring three-dimensional structure	Complicated preparation, experiment, and data treatment. Applicable only to certain classes of particles	Particles with distinct quaternary structure and subunits
Anomalous dispersion (X-ray)	No sample modification	Weak scattering effects. Cross term to be treated	Proteins (especially metal-containing), nucleotides

The other ("differential") group of methods is related to isomorphous replacement. Hence, a change in the excess scattering density takes place not for the whole particle, but for its segments. Heavy-atom labels, subunit deuteration, as well as anomalous dispersion are employed to this end. Thus, one can determine the distances between the modified segments and, when applying triangulation, their shape and three-dimensional arrangement within a particle. Hence this group of methods, unlike the previous one, serves for analyzing specific features of the particle structure.

It is noteworthy that the methods used to investigate inhomogeneous particles have their analogs in single-crystal structure analysis. For instance, the contrast-variation technique was used back in 1952 in X-ray work on single crystals of hemoglobin by Bragg and Perutz (1952), and is now applied in neutron analysis of single crystals at low resolution (see, for example, Saibil *et al*., 1976). Methods employing heavy-atom derivatives are used in phase determination when examining biological crystals (Blandell and Johnson, 1976). Other approaches used in structure analysis may also be applied in small-angle scattering (such as Mössbauer radiation; see Hermes *et al*., 1980).

Each of the reviewed methods has its own merits and its own disadvantages, reflected schematically in Table 4.3. It is up to the researcher to decide which path to follow.

5

Direct Methods

Only several general parameters of particles can be obtained directly from the methods of interpretation considered in the previous chapters. It is not always possible to study their inner structure by varying the experimental conditions; moreover, these studies involve very difficult preparative work and data evaluation. Another class of techniques will be examined in this chapter, namely, direct methods which enable one to restore the particle structure, i.e., the scattering density distribution $\rho(\mathbf{r})$, provided some assumptions have been made about this distribution.

It is clear that the problem of obtaining a three-dimensional function $\rho(\mathbf{r})$ from a one-dimensional function $I(s)$ has an infinite variety of solutions. Hence, by saying "direct method" we mean an approach that results in a maximum possible narrowing of the class of sought solutions, subject to some justified restrictions. These restrictions may be determined by geometrical considerations, the chemical composition of a particle, its physical properties, and so on. In Section 5.1, cases will first be examined in which the scattering density distribution can be described as a function of one variable. Thus for several types of particle the concept of scattering amplitude, the moduli of which are measured experimentally, can be introduced (as for single-crystal structure analysis). The sign problem should be solved to restore the density distribution in these cases, i.e., true signs of the scattering amplitudes should be determined. Section 5.2 deals with techniques for solving the problem.

The mathematical apparatus of spherical harmonics as applied to small-angle diffraction is presented in Section 5.3. This apparatus proves very helpful when formulating and solving the inverse problem, namely, how to restore function $\rho(\mathbf{r})$ from function $I(s)$, because it enables explicit expressions to be written for manifold functions $\rho(\mathbf{r})$ corresponding to the given function $I(s)$. Methods for evaluating the multipole components of scattering density, which have been developed for different types of particle, are discussed in Section 5.4. They sometimes allow one to reconstruct the scattering density $\rho(\mathbf{r})$ with resolution up to 10–20 Å.

5.1. One-Dimensional Density Distributions

In the general case, a three-dimensional function $\rho(\mathbf{r})$ is required to describe the scattering density distribution within a particle. If the three-dimensional function $I(\mathbf{s})$ (scattering intensity by a fixed particle) were known, the phase problem (Section 2.1.1) would be solved for restoring function $\rho(\mathbf{r})$. In small-angle scattering, however, only the averaged, one-dimensional function

$$I(s) = \frac{1}{4\pi} \int_{\Omega} I(\mathbf{s}) d\Omega$$

is available, so the Fourier image of the scattering density — function $A(\mathbf{s})$ — cannot be determined. If, however, the scattering density is a function of one variable, $\rho(\mathbf{r}) = \rho(\xi)$, then the scattering amplitude and intensity also depend on one variable: $A(\mathbf{s}) = A(\eta)$ and $I(\mathbf{s}) = I(\eta)$. After solving the phase problem for function $I(\eta)$, one can find $A(\eta)$ and then restore function $\rho(\xi)$. In contrast to the three-dimensional case functions $\rho(\xi)$ and $A(\eta)$ are not necessarily a pair of mutual Fourier images, since variables ξ and η are used instead of $\mathbf{r}$ and $\mathbf{s}$.

Function $I(s)$, available from small-angle scattering data, does not, generally speaking, coincide with $I(\eta)$. There are only a few cases when one can write explicitly a transform relating the distribution $\rho(\xi)$ to the scattering amplitude $A(s)$ whose moduli are determined experimentally.

First, the case of the spherically symmetric particle should be mentioned ($\xi_s = r$, $\eta_s = s$). Here, the scattering amplitude $A(s) = \pm [I(s)]^{1/2}$ is related to the scattering density by equation (1.25) while the inverse transformation can be written as

$$\rho(r) = \rho_s(r) = \frac{1}{2\pi^2} \int_0^{\infty} A_s(s) \frac{\sin(sr)}{sr} s^2 ds \tag{5.1}$$

Therefore, the radial density distribution for spherically symmetric particles can be restored, if the signs of the scattering amplitude have been determined.

Another relation can be derived for the case of rodlike particles. If the particle is cylindrically symmetric with respect to the long axis, the density along the axis being constant, then function $\rho(\mathbf{r})$ can be represented as $\rho(\xi_c)$, where ξ_c is the distance from the axis. Hence one can write

(Fedorov and Aleshin, 1966; Fedorov, 1971)

$$\rho_c(\xi_c) = \frac{1}{2\pi} \int_0^\infty A_c(s) J_0(s\xi_c) s \, ds \tag{5.2}$$

where $A_c(s) = \pm [I_c(s)]^{1/2}$ [see equation (2.36)]. Therefore, if the signs of the cross-sectional scattering amplitude are determined, the density distribution in the cross section can be calculated.

Flat (lamellar) particles represent another important case. If the density ρ_t is assumed to vary only in the direction perpendicular to the lamella plane ($\xi_t = z$) and, moreover, to be an even function, namely $\rho_t(\xi_t) = \rho_t(-\xi_t)$, then the equation

$$\rho_t(\xi_t) = \frac{1}{\pi} \int_0^\infty A_t(s) \cos(s\xi_t) ds \tag{5.3}$$

can be derived, where $A_t(s) = \pm [I_t(s)]^{1/2}$ (Lesselauer *et al.*, 1971; Laggner *et al.*, 1979). Here the "thickness" scattering amplitude defines the density distribution along the lamella thickness.

The above cases often have application in research as approximations. The problem to be solved is a special sort of phase problem: the true signs of functions $A(s)$, $A_c(s)$, or $A_t(s)$ must be found. The corresponding density distribution functions can then be calculated using transform (5.1), (5.2), or (5.3). A graphical approach is often used in order to determine the true signs of the scattering amplitudes. By assuming that $A(s)$ can change its sign only at the minima of $I(s)$, different signs are assigned to the maxima of $I(s)$. An alternative set of signs for the successive maxima of $I(s)$ is chosen conventionally as most probable (Fischbach and Anderegg, 1965; Fedorov *et al.*, 1977). Two possibilities of constructing function $A(s)$ are presented in Figures 5.1 and 5.2. In Figure 5.1 functions $|A(s)|$ and $-|A(s)|$ are plotted together and a smooth curve with alternative signs of maxima is drawn, while Figure 5.2 requires an empirical "background" to be subtracted, hence taking into account deviations from the assumed symmetry. Different undesirable factors, such as experimental errors near the minima of $I(s)$ or deviations from the assumed symmetry, can be removed "by hand" using the above techniques. To check the chosen signs, models are usually constructed with the help of the obtained distributions and then the scattering intensities of the models are compared with the experimental curve (see Figure 5.2b,c).

This approach, however, suffers from some shortcomings. First, only the beginning of the scattering curve, where the maxima are distinct and

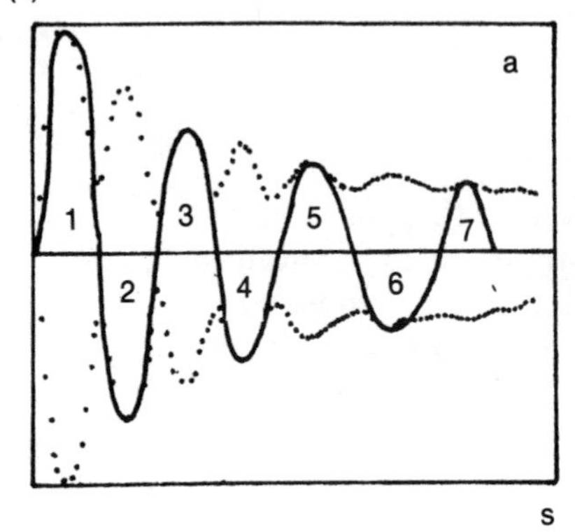

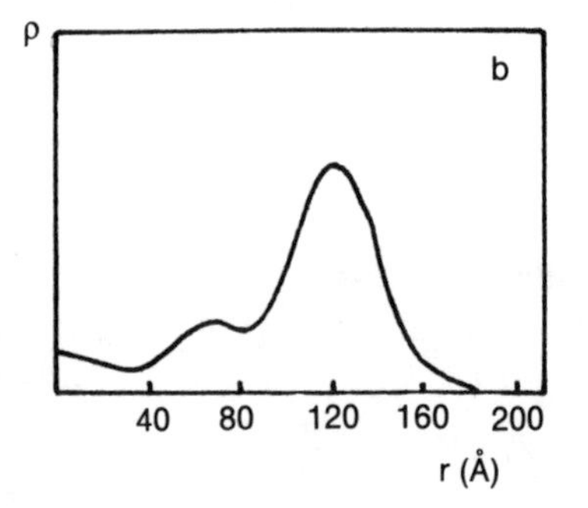

Figure 5.1. Construction of scattering amplitudes by extrapolating the minima of $I(s)$ to zero [scattering data for wild cucumber mosaic virus; after Anderegg (1965)]: (a) plotting function $A(s)$; (b) restored density distribution $\rho(r)$.

measured with good statistics, can be treated in such a manner. Second, the reliability of the results is not very high, because one cannot be sure that the chosen set of signs is correct. It is seen from Section 4.1.3 that the contrast-variation method can be of advantage when searching for a true set of signs (Mateu *et al*., 1972; Müller *et al*., 1974). In a number of cases direct solution of this problem is possible.

5.2. Solving the One-Dimensional Sign Problem

The possibility of solving the one-dimensional sign problem for Fourier transforms has been examined by Hosemann and Bagchi (1962). It was shown that if function $\rho(\xi)$ is real and symmetric (antisymmetric) and has finite support, i.e., the function differs from zero in some finite interval: $\rho(\xi)\equiv 0$ for $|\xi| > R < \infty$, then it can be restored (up to the factor ± 1) from the function $I(\eta) = \{\mathscr{F}[\rho(\xi)]\}^2$; in other words, there

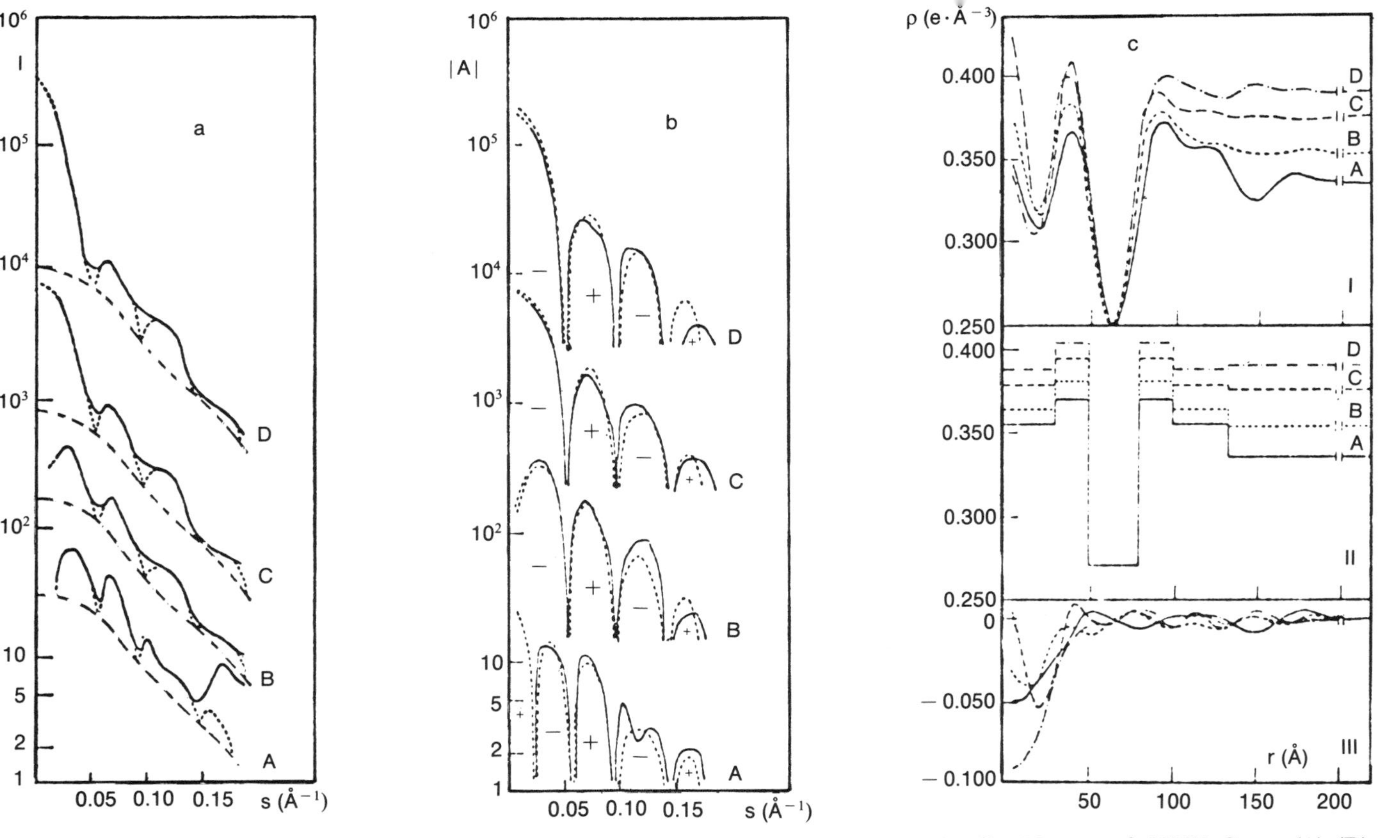

Figure 5.2. Construction of amplitude $A(s)$ by "background" subtraction [low-density lipoprotein; after Mateu *et al.* 1972)]. Curves (A)–(D) correspond to solvent densities $\rho_s = 0.3335$, 0.354, 0.376, 0.391 $eÅ^{-3}$. (a) Intensity curves (solid lines), background (dashed lines), and extrapolated intensities (dotted lines); (b) functions $|A(s)|$ (solid lines) and scattering from models $|A_M(s)|$ (dashed lines); (c) radial density distributions: I, experimental; II, constructed models; III, Fourier images of functions $[A(s) - A_M(s)]$.

exists a unique set of signs which result in a function $\rho(\xi)$ with such a support. These considerations are fulfilled for the cases considered in Section 5.1, since the dimensions of particles are finite. The diameters of a spherically symmetric particle and of the cross section of a rodlike particle as well as the thickness of a lamella can be readily estimated using the corresponding correlation functions $\gamma(r)$, $\gamma_c(\xi_c)$, and $\gamma_t(\xi_t)$ (see Section 2.4).

5.2.1. Use of Correlation Functions

The correlation functions can be used for restoring $\rho(\xi)$ without an explicit solution of the sign problem. To that end, the corresponding integral equation relating $\rho(\xi)$ to $\gamma(\xi)$ must be solved. For the one-dimensional Fourier transform the equation can be written as the self-convolution

$$\gamma(\xi) = \mathscr{F}^{-1}[I(s)] = \int_{-\infty}^{\infty} \rho(u+\xi)\rho(u)du \tag{5.4}$$

For function $\rho(\xi)$ with restricted support this equation, similar to the sign problem, has a unique solution (up to the factor ± 1). Several methods have been proposed for solving equation (5.4) (see Pape, 1974; Bradaczek and Luger, 1978; Pape and Kreutz, 1978). In particular, the convolution square-root technique of Glatter (1981) should be mentioned, where $\rho(\xi)$ is assumed to be a superposition of equal step functions and equation (5.4) is reduced to a system of nonlinear equations. The system is solved by an iteration procedure with an additional stabilization routine. The approach, although somewhat sophisticated, is very stable with respect to experimental error and requires minimum *a priori* information.

Self-convolution equations are widely used in structure analysis. One should bear in mind, however, that formal equation (5.4) is valid only for a one-dimensional Fourier transform. At the same time, we have already seen that the type of transform changes when passing from variables **r** and **s** to variables ξ and η. The "thickness" scattering amplitude for lamellar particles is a Fourier image of the symmetric function $\rho_t(\xi_t)$ which differs from zero in the interval $(-T/2; T/2)$, so equation (5.4) holds for $\gamma_t(\xi_t)$. In the cases of cylindrical and spherically symmetric particles, however, functions $\gamma_c(\xi_c)$ and $\gamma(r)$ result from two- and three-dimensional averaging over the corresponding self-convolutions of functions $\rho_c(\xi_c)$ and $\rho(r)$ (see Sections 2.4.5 and 2.4.2); therefore equation (5.4) does not occur. Thus, there is no simple convolution theorem for Hankel transforms corresponding to the case of rodlike particles (Sneddon, 1951). Writing $s^2I(s) = [sA(s)\ sA(s)]$ and substituting equation (1.25) into equation (2.22), one gets for spherically symmetric particles

$$\gamma(r)=\frac{8}{r}\int_0^\infty \frac{\sin(sr)}{sr}\,ds\int_0^R \mu\rho(\mu)\sin(s\mu)d\mu\int_0^\infty \eta\rho(\eta)\sin(s\eta)d\eta$$

$$=\frac{8}{r}\int_0^R \mu\rho(\mu)d\mu\int_0^\infty \eta\rho(\eta)d\eta\int_0^\infty \frac{\sin(sr)\sin(s\mu)\sin(s\eta)}{s}\,ds$$

where the condition $\rho(r)\equiv 0$ for $r>R$ is used and the order of integration is changed. The inner integral over s can be evaluated:

$$I(a,b,c)=\int_0^\infty \frac{\sin(ax)\sin(bx)\sin(cx)}{x}\,dx=\frac{\pi}{8}(1-\varepsilon)$$

where $a\geqslant b\geqslant c$, $\varepsilon=\text{sign}(a-b-c)$; see Prudnikov *et al.* (1981). Separate consideration of the cases $r\geqslant\mu\geqslant\eta$, $\mu\geqslant r\geqslant\eta$, and $\eta\geqslant\mu\geqslant r$ yields

$$I(r,\mu,\eta)=\begin{cases}\pi/4, & |r-\mu|\leqslant\eta\leqslant r+\mu\\ 0, & \eta<|r-\mu|,\quad \eta>r-\mu\end{cases}$$

and finally

$$\gamma(r)=\frac{2\pi}{r}\int_{\max(0,R)}^R \mu\rho(\mu)d\mu\int_{|r-\mu|}^{\min(R,r+\mu)} \eta\rho(\eta)d\eta \tag{5.5}$$

Feigin *et al.* (1981) obtained the same equation by direct averaging over the three-dimensional self-convolution integral (2.23) and proposed a numerical stepwise method of solving this equation for function $\rho(r)$. The method gives reliable results for exact input data; however, statistical errors were disregarded by the authors.

Therefore, the self-convolution equation must be modified when applied to small-angle scattering problems. Glatter (1981) covered the cases of cylindrical and spherical particles by deriving special equations with allowance for the corresponding average of the Patterson function. An improvement in the convolution square-root technique was presented by Glatter and Hainisch (1984), where unequal step representation of $\rho(\xi)$ is used. Various systematic deviations in the scattering curve (e.g., constant background) can be taken into account with this technique.

5.2.2. Box-Function Refinement

This iteration technique is probably the simplest and most reliable method for finding the signs of a scattering amplitude. Such methods are widely used to solve the sign (or phase) problem when the intensity is a continuous funtion (Makowski, 1981), for instance, when noncrystallographic symmetry is taken into account in crystal structure analysis (Crowther, 1967, 1969).

The basic features of the method are as follows. We assume functions $\rho(\xi)$ and $A(\eta)$ are connected by an integral transformation, such as the sine-Fourier transform $A(\eta)=\mathscr{F}_s[\rho(\xi)]$, and that condition $\rho(\xi)\equiv 0$ for $\xi > R$ is fulfilled. Function $|A(\eta)| = [I(\eta)]^{1/2}$ is given, while function $\rho(\xi)$ must be restored by the inverse transform $\rho(\xi)=\mathscr{F}_s^{-1}[\pm |A(\eta)|]$. The following iteration procedure is constructed. A function $\rho_k(\xi)$ is chosen as the kth approximation, which obeys the requirement of finiteness, that is, $\rho_k(\xi)\equiv 0$ for $\xi > R$. Thus a function $A_k(\eta)=\mathscr{F}_s[\rho_k(\xi)]$ is specified that does not, generally speaking, coincide with $A(\eta)$. A modified function $\tilde{A}_k(\eta)$ can be obtained by assigning the signs of $A_k(\eta)$ to the moduli of $A(\eta)$:

$$\tilde{A}_k(\eta)=\operatorname{sign}[A_k(\eta)]\,|A(\eta)| \tag{5.6}$$

The function $\tilde{\rho}_{k+1}(\xi)=\mathscr{F}_s^{-1}[\tilde{A}_k(\eta)]$ corresponds to the given moduli of amplitude, but it is not finite. The function

$$\rho_{k+1}(\xi)=\Pi(\xi-R)\tilde{\rho}_k(\xi) \tag{5.7}$$

where $\Pi(\xi-R)$ is step function (1.26), is thus chosen as the next approximation. Then the process is repeated until the signs stop changing. The convergence of the procedure can be tested by R factors of discrepancy in reciprocal and real space:

$$R_I=\int_0^R [I(\eta)-A_k^2(\eta)]^2 \Big/ \int_0^R I^2(\eta)d\eta \tag{5.8}$$

and

$$R_\rho=\int_0^R [\rho_{k+1}(\xi)-\rho_k(\xi)]^2 d\xi \Big/ \int_0^R \rho_k^2(\xi)d\xi \tag{5.9}$$

For the sine- (cosine-) Fourier transform it was proved by Svergun *et al.* (1984) that such a method converges to the true solution (up to the factor ± 1) whatever the initial approximation.

In the general case of a one-dimensional Fourier transform $A(\eta)=\mathscr{F}[\rho(\xi)]$ for a finite function $\rho(\xi)$ which differs from zero in an interval $R_1 \leqslant \xi \leqslant R_2$, a similar iteration procedure can be carried out. Equation (5.6) ensures the phase assignment while the function

$$\Pi(\xi, R_1, R_2)=\begin{cases} 1, & R_1 \leqslant \xi \leqslant R_2 \\ 0, & \xi < R_1,\quad \xi > R_2 \end{cases}$$

is substituted in equation (5.7) instead of $\Pi(\xi - R)$. The number of possible solutions, with the same support and same moduli of a Fourier image, depends on the number of complex zeros of the analytical continuation of function $A(\eta)$ (Walther, 1962; Fiddy and Ross, 1979). The iteration procedure will converge to one of these solutions depending on the initial approximation.

It is therefore clear that application of the "box-function refinement" method for lamellar particles (the cosine-Fourier transform), and also for spherically symmetric particles, where functions $r\rho(r)$ and $sA(s)$ are related by the sine-Fourier transform, is substantiated mathematically and allows a true solution to be found. Svergun *et al.* (1983, 1984) showed that the method is stable with respect to experimental errors, namely, the uncertainty of the value of R, the statistical noise in $I(s)$, and the termination of $I(s)$ for $s > s_{max}$. Moreover, it is possible to refine the value of R using the method: a series of calculations with different values of R yields a sharp increase in the factor R_I of the restored solution when the value of R employed becomes smaller than its true value (Svergun *et al.*, 1983). The iteration process converges rapidly (normally, no more than 5 or 6 iterations are required to achieve factor R_ρ of order 10^{-3}–10^{-4}). Svergun *et al.* (1984) indicated that the signs of function $\cos(sR)$ are suitable as the initial approximation for minimizing the computing times. A model application of the method as well as its stability is illustrated in Figure 5.3 for the sine-Fourier transform.

A similar approach can also be applied for rodlike particles (the Hankel transform). Model examples show that the method also allows reliable restoration of function $\rho_c(\xi_t)$.

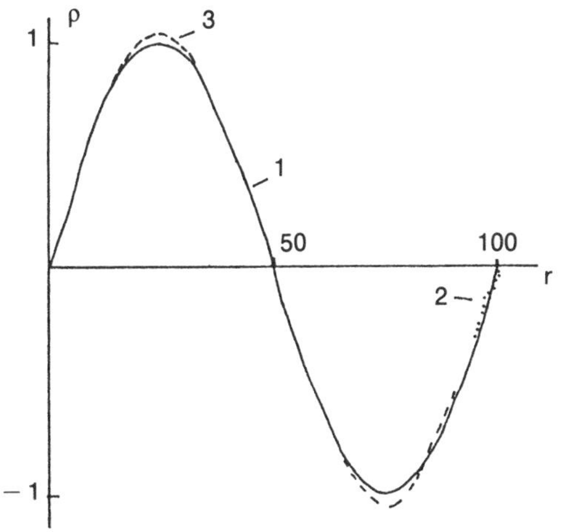

Figure 5.3. Restoration of the model density distribution (after Svergun *et al.*, 1984): (1) function $\rho(\xi)$, $R = 100$; (2) restored function for exact data, $s_{max} = 0.5$; (3) restoration when inserting errors into all input parameters [$I(s)$ includes statistical noise with $\sigma = 20\%$, $R = 90$, $s_{max} = 0.2$].

5.3. Multipole Theory of Small-Angle Diffraction

Conventional Fourier transforms are not very suitable for developing direct methods in small-angle scattering. There are only few cases for which the concept of scattering amplitude as a Fourier image of scattering density can be introduced. We shall now consider another approach for describing small-angle diffraction. This approach in which spherical harmonics are used was first applied by Harrison (1969) and later developed comprehensively by Stuhrmann (1970a). It proves to be of greater advantage for the whole theory of small-angle scattering and, in particular, for direct methods.

Any bounded three-dimensional (particle density) function $\rho(\mathbf{r})$ can be represented as a series

$$\rho(\mathbf{r}) = \sum_{l=0}^{\infty} \sum_{m=-l}^{l} \rho_{lm}(r) Y_{lm}(\omega) \tag{5.10}$$

where

$$\rho_{lm}(r) = \int_{\omega} Y^*(\omega)\rho(\mathbf{r})d\omega \tag{5.11}$$

are radial coefficients and $Y_{lm}(\omega) = Y_{lm}(\theta, \omega)$ are spherical harmonics (Abramowitz and Stegun, 1964). In reciprocal space the scattering amplitude $A(\mathbf{s})$ of a fixed particle can be expressed in the same way:

$$A(\mathbf{s}) = \sum_{l=0}^{\infty} \sum_{m=-l}^{l} A_{lm}(s) Y_{lm}(\Omega) \tag{5.12}$$

where

$$A_{lm}(s) = \int_{\Omega} Y^*(\Omega)A(\mathbf{s})d\Omega$$

On the other hand, the scattering amplitude can be written (symmetrically with respect to $\mathbf{s}$ and $\mathbf{r}$; see Section 1.2) as

$$A(\mathbf{s}) = (2\pi)^{-3/2} \int_{V} \rho(\mathbf{r})\exp\{i\mathbf{sr}\}d\mathbf{r} \tag{5.13}$$

while the following expansion is valid for function $\exp(i\mathbf{sr})$:

$$\exp\{i\mathbf{sr}\} = 4\pi \sum_{l=0}^{\infty} \sum_{m=-l}^{l} i^l j_l(sr) Y_{lm}(\Omega) Y^*_{lm}(\omega) \tag{5.14}$$

where $j_l(sr)$ are the spherical Bessel functions

$$j_l(z)=\left(\frac{\pi}{2z}\right)^{1/2}J_{l+1/2}(z)=z^l\left(-\frac{1}{z}\frac{d}{dz}\right)^l\frac{\sin(z)}{z}$$

On substituting equation (5.14) into equation (5.13) one obtains

$$A(\mathbf{s})=\sum_{l=0}^{\infty}\sum_{m=-l}^{l}\left(\frac{2}{\pi}\right)^{1/2}i_l\int_V\rho(\mathbf{r})j_l(sr)Y^*_{lm}(\omega)d\mathbf{r}\,Y_{lm}(\Omega)$$

or, taking equation (5.11) into account,

$$A(\mathbf{s})=\sum_{l=0}^{\infty}\sum_{m=-l}^{l}\left(\frac{2}{\pi}\right)^{1/2}i_l\left[\int_0^{\infty}\rho_{lm}(r)j_l(sr)r^2dr\right]Y_{lm}(\Omega)$$

Comparison with equation (5.12) yields

$$A_m(s)=\left(\frac{2}{\pi}\right)^{1/2}i^l\int_0^{\infty}\rho_{lm}(r)j_l(sr)r^2dr \tag{5.15}$$

The inverse transformation

$$\rho_{lm}(r)=\left(\frac{2}{\pi}\right)^{1/2}(-i)^l\int_0^{\infty}A_{lm}(s)j_l(sr)s^2ds \tag{5.16}$$

can be derived in the same way.

Furthermore, substitution of the expansion

$$\frac{\sin(s\,|\mathbf{r}-\mathbf{r}'|)}{s\,|\mathbf{r}-\mathbf{r}'|}=4\pi\sum_{l=0}^{\infty}\sum_{m=-l}^{l}j_l(sr)j_l(sr')Y^*_{lm}(\omega)Y_{lm}(\omega')$$

(see Nikiforov and Uvarov, 1974) into the Debye equation (2.17) enables one to obtain

$$\begin{aligned}I(s)&=4\pi\left[\sum_{l=0}^{\infty}\sum_{m=-1}^{l}\int_0^{\infty}\int_0^{4\pi}\rho(\mathbf{r})Y^*_{lm}(\omega)d\omega r^2j_l(sr)dr\right]\\&\quad\times\left[(-1)^m\int_0^{\infty}\int_0^{4\pi}\rho(r')Y^*_{lm}(\omega')d\omega' r'^2j_l(sr')dr'\right]\\&=4\pi\sum_{l=0}^{\infty}\sum_{m=-l}^{l}(-1)^m\int_0^{\infty}r^2\rho_{lm}(r)j_l(sr)dr\int_0^{\infty}r'^2\rho_{lm}(r')j_l(sr')dr'\end{aligned}$$

where equation (5.11) and the equation $Y^*_{lm}(\omega)=(-1)^m Y_{lm}(\omega)$ have been used. If use is made of equation (5.15), one finally arrives at the equation

$$I(s)=2\pi^2 \sum_{l=0}^{\infty} \sum_{m=-l}^{l} |A_{lm}(s)|^2 \quad (5.17)$$

Therefore, the small-angle scattering intensity can be represented mathematically as a sum of squares of Hankel transforms of the corresponding orders from the multipole components of the scattering density. Physically it means that, using this approach, the latter is separated into the components

$$\rho_{lm}(\mathbf{r})=\rho_{lm}(r)Y_{lm}(\omega) \quad (5.18)$$

producing additive contributions to the total scattering intensity. This description is very convenient. Actually, all the previously used representations of $\rho(\mathbf{r})$ as a sum of components (see, for example, the contrast-variation technique; Section 4.2) inevitably led to the cross terms appearing in the scattering curve, which complicated practical applications. In the present case the specific scattering amplitude $A_{lm}(s)$ corresponds to each component (5.18) and no cross terms appear. Therefore the multipole expansion seems to be a reasonable way of describing the isotropic small-angle scattering intensity.

The spherical harmonics $Y_{lm}(\theta, \varphi)$ are functions defined on the surface of a sphere. General equations as well as explicit expressions for a number of harmonics are given in Table 5.1. The zero harmonic Y_{00} (monopole) displays a homogeneous density distribution over the whole surface of the sphere (Figure 5.4a). For $l=1$ (dipole terms) the densities $\rho_{lm}(\mathbf{r})$ exhibit two poles with opposite signs; terms with $l=2$ (quadrupole) have four poles, and so on (Figure 5.4b,c). The harmonics with $m=0$ (axial) have axial symmetry (they do not depend on the azimuthal angle φ) and terms $m=l$ (sectorial) divide the whole surface into $2l$ sectors of alternate signs, being increasingly confined to the equatorial plane at higher index (Figure 5.5).

The angular resolution provided by harmonics depends on their orders. With increasing l, structures (5.18) on the surface become more detailed (i.e., the typical distance between angular inhomogeneities decreases). According to the basic diffraction principles it would shift the maxima of the intensity functions $|A_{lm}(s)|$ toward larger scattering angles. This situation is ensured mathematically by increasing the order of the corresponding Hankel transform (5.15). An analysis of this equation permits one more interesting conclusion to be reached. If we compare transforms (5.15) of two radial functions $\rho_{lm}(r)$ and $\rho_{nk}(r)$, then it is evident that if $l=n$ and the functions themselves are equal, then

Table 5.1. Spherical Harmonics $Y_{lm}(\theta, \varphi)$

General Equation

$$Y_{lm}(\theta, \varphi) = \left[\frac{2l+1}{4\pi} \cdot \frac{(l-|m|)!}{(l+|m|)!}\right]^{1/2} \cdot P_l^{|m|}(\cos\theta)\exp(im\varphi), \qquad 0 \leqslant \theta \leqslant \pi, \quad 0 \leqslant \pi \leqslant 2\pi$$

$P_l^m(\cos\theta)$ are associated Legendre functions of the first kind

Zonal Harmonics

$$Y_{l0}(\theta, \varphi) = Y_{l0}(\theta) = \left(\frac{2l+1}{4\pi}\right)^{1/2} P_l(\cos\theta); \quad P_l(\cos\theta) \text{ are Legendre polynomials}$$

Sectorial Harmonics

$$Y_{ll}(\theta, \varphi) = \left[\frac{2l+1}{4\pi(2l)!}\right]^{1/2} \cdot 1 \cdot 3 \cdot 5 \cdots (2l-1) \cdot \sin^l\theta \cdot \exp(il\varphi)$$

l, m	$Y_{lm}(\theta, \varphi)$	l, m	$Y_{lm}(\theta, \varphi)$
0,0	$\left(\frac{1}{4\pi}\right)^{1/2}$	2, ± 2	$\frac{3}{2}\left(\frac{5}{96\pi}\right)^{1/2}(1-\cos 2\theta)\exp(\pm 2i\varphi)$
1,0	$\left(\frac{3}{4\pi}\right)^{1/2}\cos\theta$	3,0	$\frac{1}{16}\left(\frac{7}{\pi}\right)^{1/2}(5\cos 3\theta + 3\cos\theta)$
1, ± 1	$\left(\frac{3}{8\pi}\right)^{1/2} \cdot \sin(\theta)\exp(\pm i\varphi)$	3, ± 1	$\frac{3}{8}\left(\frac{7}{48\pi}\right)^{1/2}(\sin\theta + 5\sin 3\theta)\exp(\pm i\varphi)$
2,0	$\frac{1}{8}\left(\frac{5}{\pi}\right)^{1/2} \cdot [3\cos(2\theta) + 1]$	3, ± 2	$\frac{15}{4}\left(\frac{7}{480\pi}\right)^{1/2}\cos\theta(\cos\theta - \cos 3\theta)\exp(\pm 2i\varphi)$
2, ± 1	$\frac{3}{2}\left(\frac{5}{24\pi}\right)^{1/2} \cdot \sin(2\theta)\exp(\pm i\varphi)$	3, ± 3	$\frac{15}{4}\left(\frac{7}{2880\pi}\right)^{1/2}(3\sin\theta - \sin 3\theta)\exp(\pm 3i\varphi)$

$A_{lm}(s) = A_{nk}(s)$ for any m and k. Consequently, one and the same partial scattering amplitude can be produced by several different densities (5.18). A similar situation is observed in quantum mechanics. It is well known that the spherical harmonics are the angular-momentum eigenfunctions (Edmonds, 1957). Thus, the stationary states of a system corresponding to the same value of l have the same energy; only the magnitude of the momentum projection along the z direction depends on the value of m.

These special features of functions $A_{lm}(s)$ are illustrated qualitatively in Figure 5.4, where radial functions $\rho_{lm}(r) = \delta(r - R)$, $R = \text{const}$, have

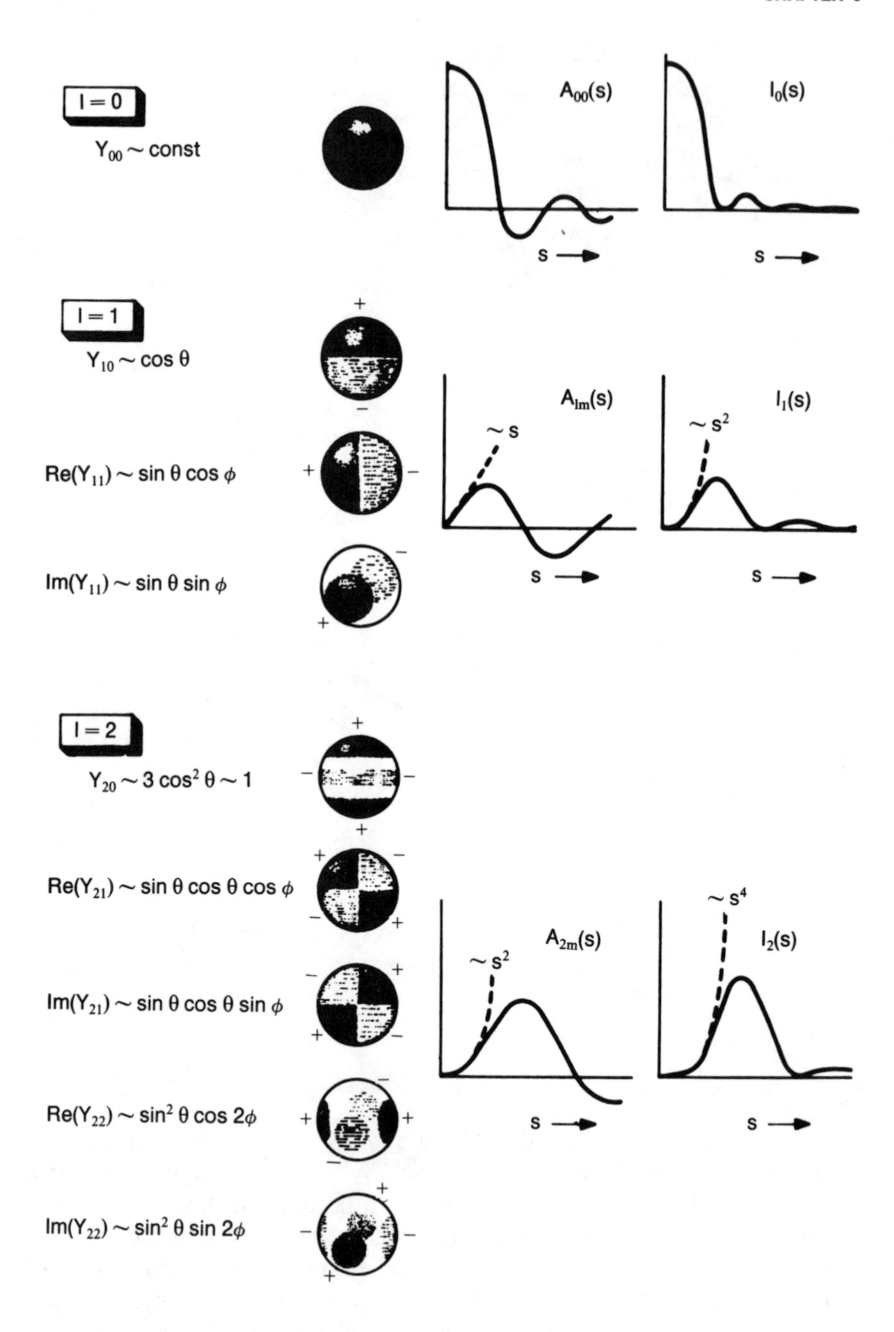

Figure 5.4. Multipole components of a structure and their amplitudes $A_{lm}(s)$ and intensities $I_l(s)$ (after Stuhrmann and Miller, 1978).

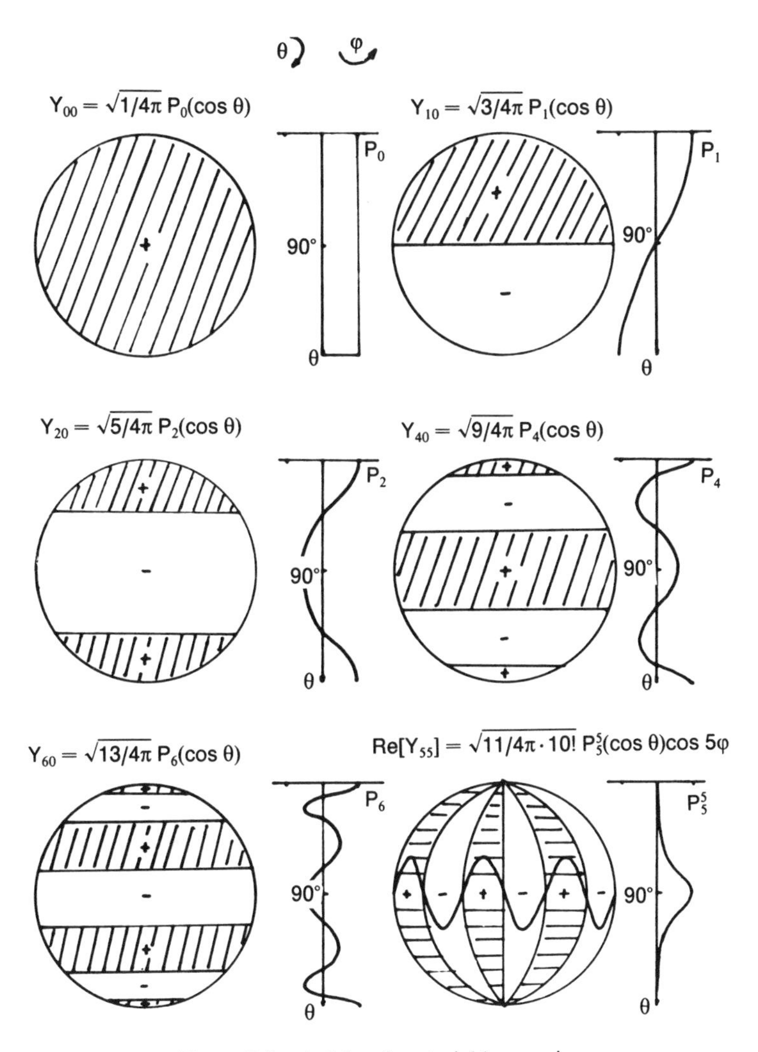

Figure 5.5. Axial and sectorial harmonics.

been selected. Bessel functions can be represented in the series form

$$j_l(sr) = \sum_{p=0}^{\infty} d_{lp}(sr)^{l+2p} = \sum_{p=0}^{\infty} \frac{(-1)^p (sr)^{l+2p}}{2^p p! [2(l+p)+1]!!} \tag{5.19}$$

(see Abramowitz and Stegun, 1964); hence one can see that functions $A_{lm}(s)$ will increase with s^l for small s (in particular, all the functions, besides $A_{00}(s)$, are equal to zero at $a = 0$). Equations (5.10) and (5.17) allow one to demonstrate explicitly the structural ambiguities inherent in small-angle scattering. Actually, equation (5.10) can be rewritten as

$$\rho(\mathbf{r}) = \sum_{l=0}^{\infty} \rho_l(\mathbf{r}) \tag{5.20}$$

where

$$\rho_l(\mathbf{r}) = \sum_{m=-l}^{l} \rho_{lm}(r) Y_{lm}(\omega) \tag{5.21}$$

Rotation of density $\rho(\mathbf{r})$ specified by Euler angles α, β, γ is obtained by transforming $Y_{lm}(\omega)$ by the equation

$$Y_{lm}(\omega') = \sum_{m=-l}^{l} D^l_{mm'}(\alpha, \beta, \gamma) Y_{lm'}(\omega) \tag{5.22}$$

where $D^l_{mm'}(\alpha, \beta, \gamma)$ (D-functions) are matrix elements of the rotational operator (Brink and Satchler, 1968). The symmetrical properties of the D-functions (see Edmonds, 1957) enable one to show that the function

$$I_l(s) = \sum_{m=-l}^{l} |A_{lm}(s)|^2 \tag{5.23}$$

is conserved under arbitrary rotation of density (5.21) according to equation (5.22). Therefore, a manifold of functions $\rho(\mathbf{r})$ generated by independent rotations of partial structures $\rho_l(\mathbf{r})$ will give the same function $I(s)$. It is obvious that, although functions $\rho_{lm}(r)$ do not change, functions $\rho(\mathbf{r})$ may differ substantially (see Figure 5.6). Another class of

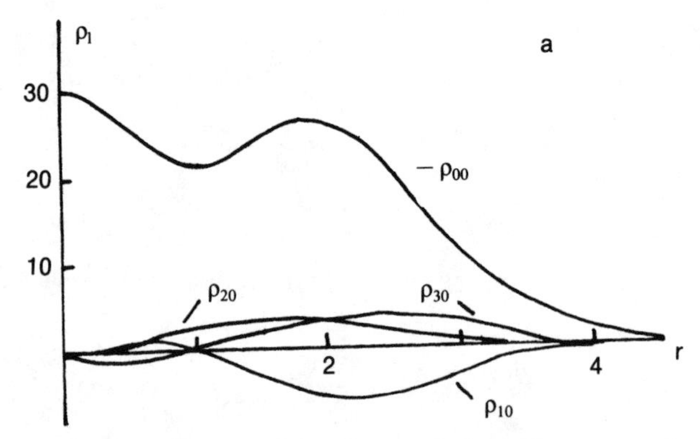

Figure 5.6. Different structures giving the same intensity curve (after Stuhrmann, 1970a): (a) radial functions ($l = 0, 1, 2, 3$); (b) structures constructed from corresponding partial densities with their rotations [the cross sections in the plane $z = 0$ are given; structure (1) is axially symmetric]; (c) curve $I(s)$ and partial intensities $I_{l0}(s)$.

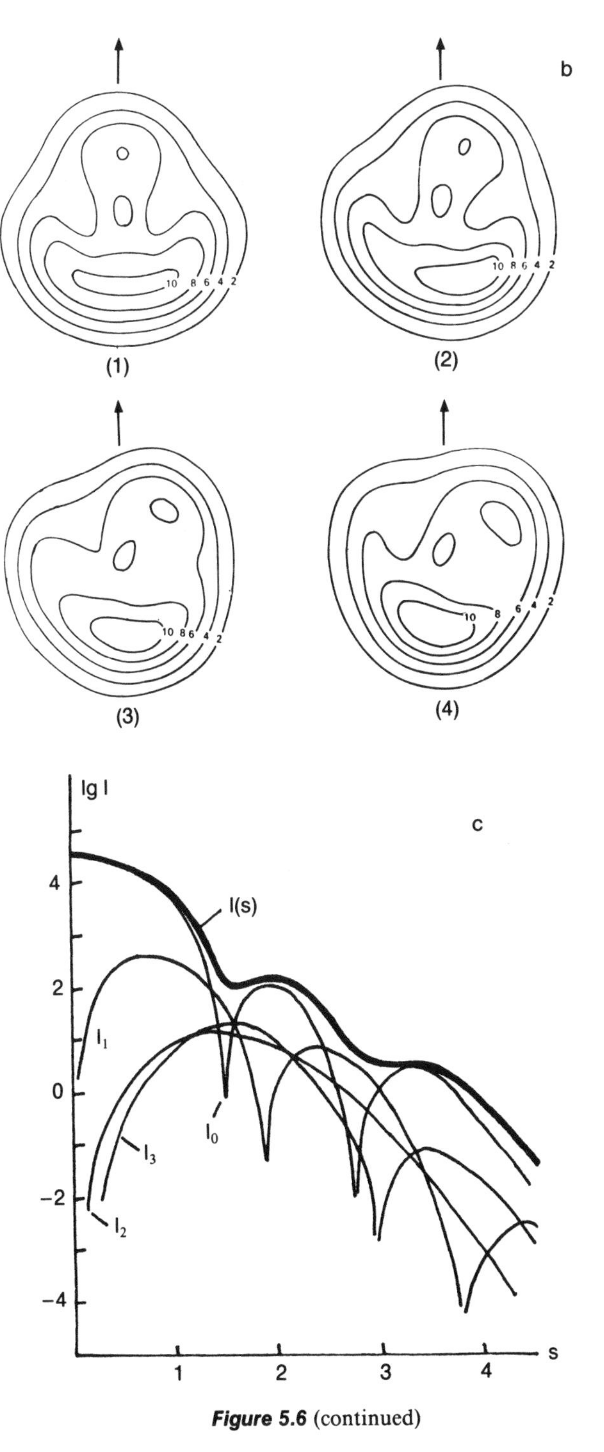

Figure 5.6 (continued)

functions may be obtained by displacing the origin of coordinates in real space. The multipole components vary under this operation, but function $\rho(\mathbf{r})$ remains congruent and neither $I(s)$ nor the intensity scattered by a fixed particle $I(\mathbf{s})$ changes.

Therefore, multipole theory helps one to describe the whole variety of structures corresponding to the given curve $I(s)$. Two problems should be solved for the theory to be used as a method of structure determination. First, it is necessary to somehow or other extract the partial amplitudes $A_{lm}(s)$ from the sum (5.17); second, after calculating radial functions $\rho_{lm}(r)$, the function $\rho(\mathbf{r})$ should be constructed with an account of all the ambiguities listed above. The latter problem can, in principle, be solved provided *a priori* information about the particle is available. However, arbitrary functions $\rho_{lm}(r)$ may be obtained when separating the summands from sum (5.17). Therefore, the problem of how to evaluate the partial ampitudes $A_{lm}(s)$ is very important and, to solve it, there must be some additional information. The possibilities of practical application of multipole theory depend mainly on the solution to this problem. With this in mind several different approaches are examined below.

5.4. Determination of Multipole Components

Methods for separating partial amplitudes require some kind of *a priori* information about the particle, and these methods can be applied only with certain types of particle. There are different methods for different types of particle; however, one characteristic feature is inherent in all of them, namely, it should be assumed that the structure can be described satisfactorily by a finite (small enough) number of multipole components, so that the truncated series (5.17) can be used. Such an assumption is justified for compact particles with globular shape: in this case series (5.10) converges rapidly and higher multipoles may be omitted within the framework of the small-angle scattering resolution. At the same time, for highly anisometric particles considerable distortions may be introduced by the truncation itself. Therefore, it is clear initially that the spherical-harmonics technique is applicable mainly for globular particles.

5.4.1. Isometric Particles

It is very convenient to use the multipole expansion for interpretation of scattering by particles of isometric shape (such as viruses, chromatin monomers, or lipoproteins). Since the particles are nearly spherical, the monopole term predominates. Higher multipoles produce the isometric shape of a particle, differing substantially from zero for $r \approx R$, where R is the radius of a particle in the spherical approximation. We can therefore write approximately

$$\rho(r) \approx [1/(4\pi)^{1/2}]\rho_{00}(r) + \sum_{l=1}^{\infty} \rho_l \delta(r - R) H_l(\omega) \tag{5.24}$$

where

$$H_l(\omega) = \sum_{m=-l}^{l} a_{lm} Y_{lm}(\omega)$$

Hence the scattering intensity is given by

$$I(s) \approx 2\pi^2 A_{00}^2(s) + 4\pi \sum_{l=0}^{\infty} c_l^2 R^4 j_l^2(sR)$$

where

$$c_l = \rho_l \sum_{m=-l}^{l} a_{lm}$$

Particle symmetry leads to "selection rules" for the terms in the spherical harmonic expansion (5.24). Thus, for the icosahedral case, the permitted orders are given by $l = 6p + 10q + 15r$, where p, q, and r are integers (see Finch and Holmes, 1967) and the first nonzero term is $l = 6$. It is possible, in principle, for the order to be higher, depending on symmetry properties. The orders of nonzero terms as well as their contributions (the values of c_l^2) can be estimated by analysis of deviations from the spherical structure, the maximum contribution of the given term corresponding to the first maximum of the corresponding Bessel function $j_l(sR)$ (see Figure 5.4 and Figure 5.8 below). Such an approach was used to investigate the isometric structure of protein shells of spherical viruses (Harrison, 1969; Jack and Harrison, 1975).

5.4.2. *Shape of Uniform Particles*

The condition of particle homogeneity suffices for the multipole components to be separated in the scattering curve. This case is of practical importance, because the homogeneous approximation is widely used in research and, moreover, the "shape" scattering curve can be determined using the contrast-variation technique (see Section 4.2.1).

For a uniform particle, one can assume after suitable normalization of the scattering curve that

$$\rho_c(\mathbf{r}) = \phi_p(\mathbf{r}) = \begin{cases} 1, & 0 \leqslant r \leqslant F(\omega) \\ 0, & r > F(\omega) \end{cases}$$

where $F(\omega)$ is a unique function describing the particle shape, $0 \leqslant \theta \leqslant \pi$, and $0 \leqslant \varphi \leqslant 2\pi$. It is evident that in this case expansion (5.17) cannot be arbitrary — certain relationships between the amplitudes $A_{lm}^{c}(s)$ should exist. These relationships could be used as restrictions on separation of the multipole components (Stuhrmann, 1970b,c; Stuhrmann *et al.*, 1975a).

If $F(\omega)$ is expressed in series form

$$F(\omega) = \sum_{l=0}^{\infty} \sum_{m=-l}^{l} f_{lm} Y_{lm}(\omega) \tag{5.25}$$

where

$$f_{lm} = \int_{\omega} \rho_c(\mathbf{r}) Y_{lm}^{*}(\omega) d\omega$$

and equation (5.15) is taken into account, one obtains

$$A_{lm}^{c}(s) = i^{l} \left(\frac{2}{\pi}\right)^{1/2} \int_{r=0}^{F(\omega)} \int_{\omega} Y_{lm}^{*}(\omega) d\omega j_l(sr) r^2 dr$$

Substitution of $j_l(sr)$ by expansion (5.19) results in the expansion

$$A_{lm}^{c}(s) = i^{l} \left(\frac{2}{\pi}\right)^{1/2} \sum_{p=0}^{\infty} \frac{d_{lp}}{l+2p+3} f_{lm}^{(l+2p+3)} s^{l+2p} \tag{5.26}$$

where

$$f_{lm}^{(q)} = \int_{\omega} F^{q}(\omega) Y_{lm}^{*}(\omega) d\omega$$

Introduction of equation (5.26) into equation (5.17) yields

$$I_c(s) = 2\pi^2 \sum_{n=0}^{\infty} a_n s^{2n}$$

where

$$a_n = \sum_{l=0}^{n} \sum_{p=0}^{n-l} \sum_{m=-l}^{l} \frac{d_{lp} d_{l,n-p} f_{lm}^{(l+2p+3)} f_{lm}^{*(2n+l-2p+3)}}{(l+2p+3)(2n+l-2p+3)} \tag{5.27}$$

Therefore the coefficients of the power-series expansion of $I_c(s)$ prove to be related to the values of f_{lm}, which determine the particle shape. It is convenient to evaluate a_n by means of the correlation function $\gamma_c(r)$. On replacing function $(\sin sr)/sr$ in equation (2.23) by the Mclaurin series, one has

$$I_c(s) = 4\pi \sum_{n=0}^{\infty} \frac{(-1)^n}{(2n+1)!} \int_0^D \gamma_c(r) r^{2n+2} dr \, s^{2n}$$

and therefore

$$a_n = \frac{(-1)^n}{4\pi(2n+1)!} \int_0^D \gamma_c(r) r^{2n+2} dr$$

Having calculated coefficients a_n from the experimental curve $I_c(s)$, the values of f_{lm}, and hence the particle shape, can be evaluated. System (5.27) is, however, nonlinear, so it can be solved only approximately. In the course of studying the 50 S ribosomal subparticle Stuhrmann *et al.* (1975a) applied an iteration procedure: starting from an initial shape, successive approximations were performed to find a solution. Agreement between the obtained and experimentally determined coefficients a_m, as well as the stability of the solution with respect to the initial approximation, confirmed the validity of the procedure. About 100 computer runs were performed with very different initial structures and different harmonics in equation (5.25). The determined shape of the 50 S subparticle is shown in Figure 5.7; harmonics with $l = 0, 1, 2, 3$ were used and a spatial resolution of about 100 Å. About the method as such, one can say that system (5.26) permits a unique solution if the number of unknowns does not exceed six (Stuhrmann, 1970b,c)

5.4.3. Separation of Bessel Functions

The methods considered for extracting multipole components dealt in fact with restoration of the particle shape (this can often also be achieved with the help of modeling). However, one can search for decomposition (5.17) with more general assumptions about the particle structure, i.e., about function $\rho(\mathbf{r})$.

Behavioral features of functions $A_{lm}(s)$ examined in Section 5.3 may serve as a basis for separating the summands from expansion (5.17). Stuhrmann and Fuess (1976) and Marguerie and Stuhrmann (1976) attempted to perform this decomposition by means of stepwise refinement of the structure. They fitted successively multipole components with $l = 0, 1, 2, \ldots$ on those sections where the contribution from a given multipole was subtracted and the next one fitted to the rest of the curve, and so on. This technique, however, is rather empirical and permits no

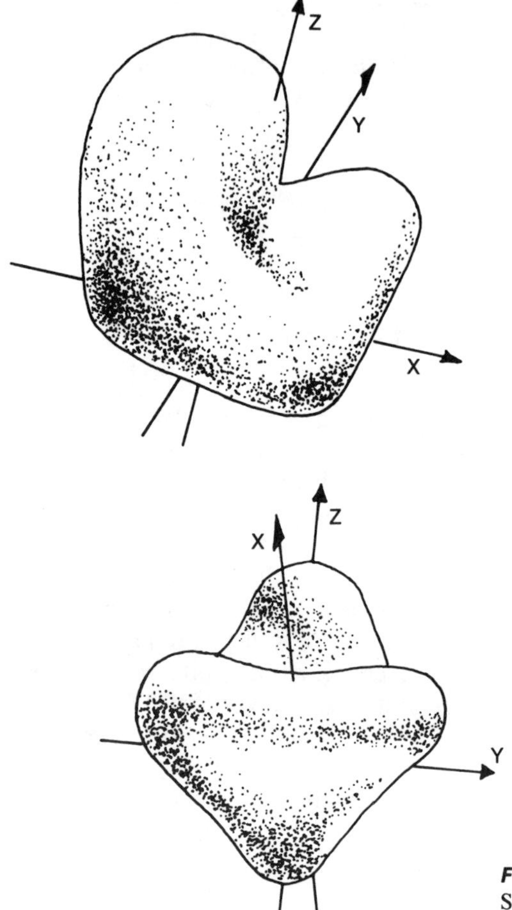

Figure 5.7. Shape restoration of 50 S ribosomal subparticle (after Stuhrmann *et al*., 1977).

reliable results. A more general approach taking into account both the properties of function $A_m(s)$ and additional information about the particle was proposed by Svergun *et al*. (1981, 1982). The main features of the approach will now be examined.

The condition that each function $\rho_{lm}(r)$ possesses a finite support is one of the most significant physical requirements governing a possible solution. Actually, since any particle is bounded in space, there exists such a value of R with the property that $\rho(\mathbf{r}) = 0$ [and therefore $\rho_{lm}(r) = 0$] for $r > R$. Essential restrictions are imposed on function $A_{lm}(s)$ by this condition (we saw something similar in Section 5.2 for the Fourier transform). It is noteworthy that, assuming R to be the distance between the center of mass of the particle and the most removed point of the particle, the ambiguity of the solution caused by possible displacements of the origin of coordinates is eliminated (because it means that the origin coincides with the center of mass). However, other ambiguities still exist. In particular, it is impossible to distinguish in the scattering curve among harmonics with different m and the same l. Therefore, only special types of particles permit the condition of finiteness of the radial functions to be used. In other words, it is necessary that a function $\rho_l(r) = \rho_{lk}(r)$ should exist for each l, which makes a considerably greater contribution to partial density (5.21), and hence to partial intensity (5.23), than other radial functions with the same index l. In this case we can write

$$\rho_l(\mathbf{r}) = \rho_l(r) Y_{lk}(\omega)$$

and

$$I_l(s) = |A_l(s)|^2$$

where $\rho_l(r)$ are finite functions related to $A_l(s)$ by transform (5.15). The case of an axially symmetric particle is most illustrative. Here $\rho_{lm}(r) = 0$ for $m \neq 0$, thus $k = 0$ and $\rho_l(r) = \rho_{l0}(r)$ for each l. Construction of density $\rho(\mathbf{r})$ using functions $\rho_l(\mathbf{r})$ is also simplified in this case, since any rotation of the partial density through an angle not equal to π disturbs the axial symmetry.

If the particle density is represented satisfactorily by a finite number $L + 1$ of harmonics, the problem of the separation of the partial amplitudes can be formulated as

$$I_l(s) = 2\pi^2 \sum_{l=0}^{L} |A_l(s)|^2 \tag{5.28}$$

spectra (5.16) of functions $A_l(s)$ being restricted by the value of R.

Under the above assumptions, functions $A_l(s)$ exhibit rather definite features. Functions $A_l(s)$ corresponding to one and the same step function $\Pi(r - R)$ are presented in Figure 5.8 for different values of l. One can see that the angular inhomogeneities (defined by the number of harmonic and Bessel functions) and the finiteness of the radial function manifest themselves in the definite sequence of the intervals, where the given amplitude makes the greatest contribution to the total intensity. The angular inhomogeneities are inherent in the given harmonic, no matter what the radial function, while the value of R plays the role of a scale factor along the abscissa axis. Therefore, the changes in relative contributions from different harmonics are to a large extent connected with their numbers. This fact can serve as the basis for developing an algorithm to decompose the scattering intensity into sum (5.28).

Indeed, suppose we specify $L + 1$ functions $\rho_l^{(k)}(r)$ satisfying the space restrictions and determining functions $A^{(k)}(s)$, and by equation (5.28) intensity $I^{(k)}(s)$, which is not in agreement with $I(s)$. In order to fit

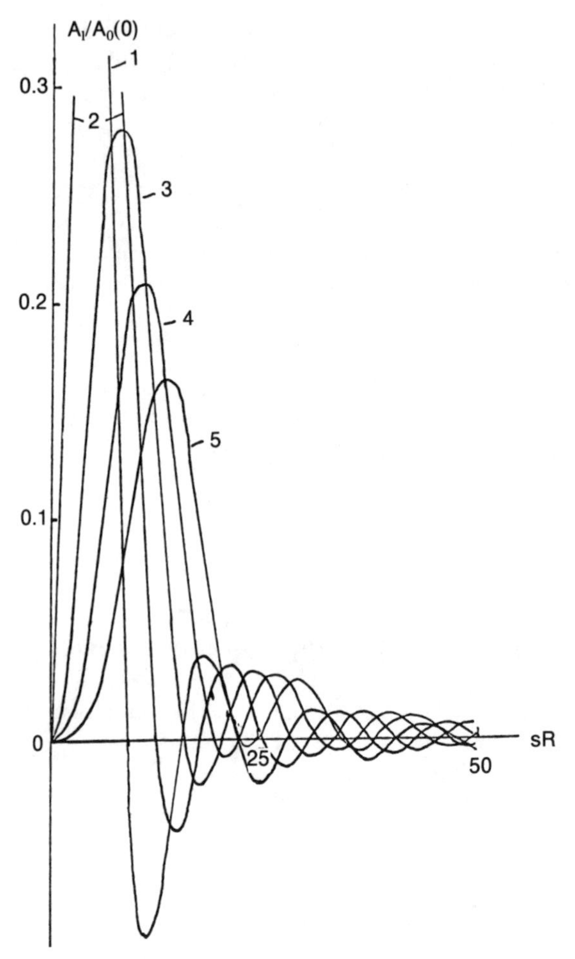

Figure 5.8. Scattering amplitudes for partial densities with $\rho_1(r) = \Pi(r - R)$. Curves (1)–(5) correspond to $l = 0$–4.

the intensity $I(s)$ with the relative contributions of different harmonics remaining unchanged, one can redistribute the amplitudes as follows:

$$\tilde{A}_l^{(k)}(s) = A_l^{(k)}(s)[I(s)/I^{(k)}(s)]^{1/2} \tag{5.29}$$

The obtained amplitudes $\tilde{A}_l^{(k)}(s)$ will obey equation (5.28), but the set $\tilde{\rho}_l^{(k)}(r)$, specified by $\tilde{A}_l^{(k)}(s)$ according to equation (5.16), will not meet the space restrictions. Therefore, we assume that

$$\rho_l^{(k+1)}(r) = \tilde{\rho}_l^{(k)}(r)\Pi(r - R) \tag{5.30}$$

The amplitudes of the next approximation $A_l^{(k+1)}(s)$ are determined by functions $\rho_l^{(k+1)}(r)$ in line with equation (5.15).

Thus, at each step of the iteration process defined by equations (5.19) and (5.20), the partial amplitudes are redistributed in accordance with the number of harmonic l, value R, and scattering intensity $I(s)$. If no *a priori* information about radial functions, except space limitations, is available, step functions (1.26) seem to be convenient for the initial approximation. The following factors R can be used to monitor the convergence of the process:

$$R_{\mathrm{I}} = \int_0^\infty s^4[I(s) - I^{(k)}(s)]^2 ds \Big/ \int_0^\infty s^4 I(s)^2 ds \qquad (5.31)$$

i.e., the discrepancy between $I(s)$ and the intensity at the kth step; and

$$R_\rho^{(l)} = \int_0^R r^2[\rho_l^{(k)}(r) - \rho_l^{(k+1)}(r)]^2 dr \Big/ \int_0^R [r\rho_l^{(k)}(r)]^2 dr \qquad (5.32)$$

i.e., the discrepancy between the results of two successive iterations. The weighting factors s^4 and r^2 owe their origin to the Parseval theorem for Hankel transforms (Sneddon, 1951):

$$\int_0^R \rho_{lm}^2(r) r^2 dr = \int_0^\infty A_{lm}^2(s) s^2 ds$$

One can see that the process (5.29)–(5.31) is somewhat a generalization of the "box-function refinement" technique; the former is reduced to the latter in the case of a spherically symmetric particle ($L = 0$). Svergun *et al.* (1982) showed that this process should converge to the true function $I(s)$. However, when dealing with several terms of series (5.28) this fact does not guarantee convergence to the true radial function. The efficiency of the method has been verified by Svergun *et al.* (1981, 1982) with model examples, as follows: An initial model density $\rho(\mathbf{r})$ was specified, then the intensity was calculated from the corresponding radial functions at some finite interval $0 \leqslant s \leqslant s_{\max}$, after which the radial functions were restored by an iteration procedure and the model structure reconstructed. The model calculations showed that algorithm (5.29)–(5.30) leads to an accumulation of errors, caused by numerical instability of transform (5.16) when functions $\tilde{A}_l^{(k)}(s)$ have discontinuities, and drastic distortions of the results even with exact input data on function $I(s)$, the set of harmonics, and the value of R. An example in which the procedure is used to restore three model radial functions is given in Figure 5.9, where the first three nonzero radial coefficients of the uniform prolate ellipsoid of rotation with $c/a = 2$ were used as model functions.

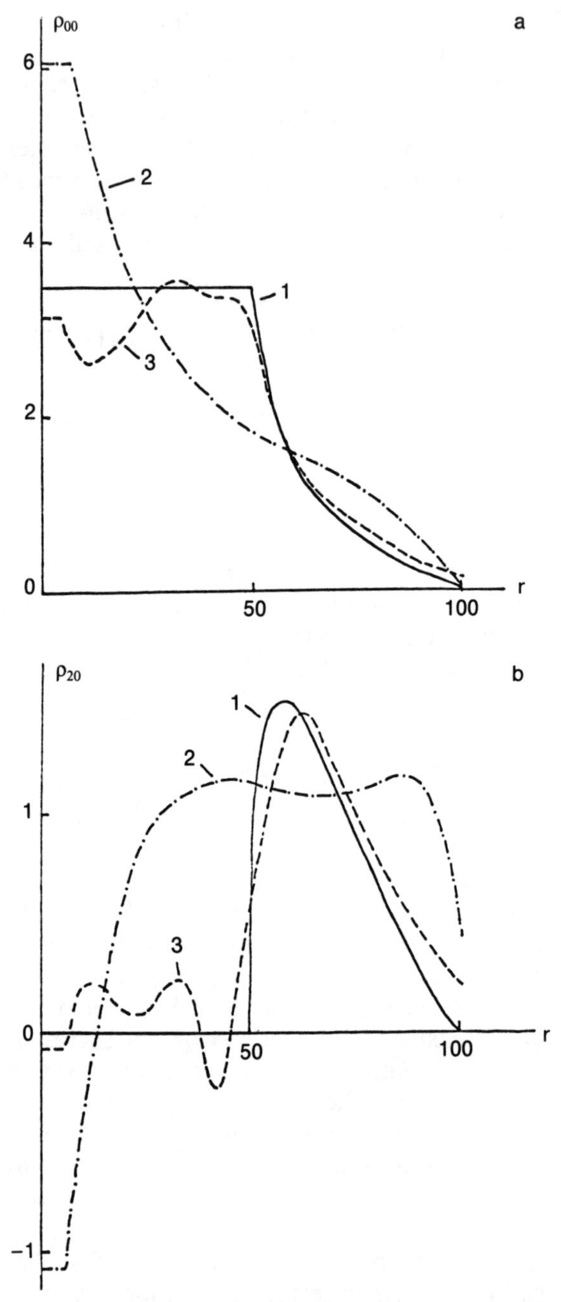

Figure 5.9. Restoration of three model radial functions from the total intensity (after Svergun *et al.*, 1982): (a), (b), (c) correspond to $l = 0, 2, 4$. (1) True functions $\rho_{l0}(r)$; (2) result of the ninth iteration, $R_{\mathrm{I}} = 1.12 \times 10^{-2}$; $R = 100$, $s_{\max} = 0.5$; (3) the stabilized solution, $R_{\mathrm{I}} = 7.8 \times 10^{-4}$.

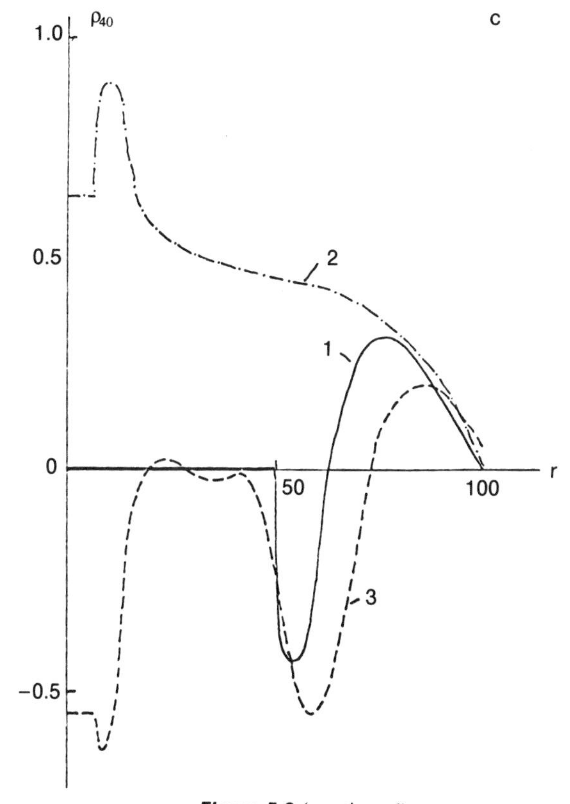

Figure 5.9 (continued)

An additional requirement was used to stabilize the solution, in that the boundedness of the density was demanded:

$$\rho_{\min} \leqslant \rho^{(k)}(\mathbf{r}) \leqslant \rho_{\max} \tag{5.33}$$

for $0 \leqslant r \leqslant R$. Here

$$\rho^{(k)}(\mathbf{r}) = \sum_{l=0}^{L} \rho_l^{(k)}(r) Y_{lk}(\omega)$$

Indeed, the distortions in radial functions leading to considerable density fluctuations mean that conditions (5.33) would not be satisfied. To meet them, it is appropriate to introduce corrections $\Delta\rho_l(r)$ in the radial functions, the factor R_{I} being minimum. It can be written as follows:

$$\rho_{\min} - \rho^{(k)}(\mathbf{r}) \leqslant \Delta\rho(\mathbf{r}) \leqslant \rho_{\max} - \rho^{(k)}(\mathbf{r}) \tag{5.34}$$

and

$$\min\left\{\int_0^\infty \left\{I(s) - 2\pi^2 \sum_{l=0}^{L} [A_l^{(k)}(s) - \Delta A_l(s)]^2\right\}^2 s^4 ds\right\} \tag{5.35}$$

where

$$\Delta\rho(\mathbf{r}) = \sum_{l=0}^{L} \Delta\rho_l(r) Y_{lk}(\omega), \qquad 0 \leqslant r \leqslant R$$

while functions $\Delta A_l(s)$ are related to $\Delta\rho_l(r)$ by transforms (5.15). The corrections can be transformed analytically to reciprocal space using equation (5.15), by representing the corrections as a superposition of Laguerre polynomials $L_n^{(a)}(r^2)$ (Abramowitz and Stegun, 1964; Stuhrmann, 1970a). Functional (5.35) can be linearized, if the obtained functions $\Delta A_l(s)$ are regarded as small enough when compared with $A_l^{(k)}(s)$. Then problem (5.34)–(5.35) can be solved by linear programming and the desired corrections found. The corresponding equations for the case of axial symmetry are as follows: The corrections are taken in the form

$$\Delta\rho_l(r) = r^l \exp(-r^2/2) \sum_{n=0}^{K} C_n L_n^{(l+1/2)}(r^2)$$

where K is the maximum order of Laguerre polynomials. Coefficients C_n are given by the solution vector $\{x\}$ of the linear programming task (Dantzig, 1963)

$$\min \sum_{\nu=1}^{\beta} b_\nu x_\nu, \quad \sum_{\nu=1}^{\beta} a_{\mu\nu} \geqslant d_\mu, \qquad 1 \leqslant \mu \leqslant \alpha$$

where

$$b_\nu = b_{ln} = (-1)^n \int_0^\infty s^4 A_l^{(k)}(s)[I^{(k)}(s) - I(s)]\exp(-s^2/2)s^l L_n^{(l+1/2)}(s^2) ds$$

and

$$x_\nu = C_{ln}, \qquad \beta = (L+1)(K+1), \qquad l = [(\nu-1)/K],$$
$$n = \nu - (K-1)l - 1$$

α is the number of restrictions, $\alpha > \beta$, while r_μ and θ_μ specify a grid of restrictions,

$$a_{\mu\nu} = r^l \exp(-r_\mu^2/2) L_n^{(l+1/2)}(r_\mu^2)(-1)^\mu Y_{l0}(\theta_\mu)$$

and

$$d_\mu = \tfrac{1}{2}[1-(-1)^\mu][\rho^{(k)}(r_\mu, \theta_\mu) - \rho_{\max}] + \tfrac{1}{2}[1+(-1)^\mu][\rho_{\min} - \rho^{(k)}(r_\mu, \theta_\mu)]$$

Usually, restrictions (5.33) are to be applied when the iteration procedure does not lead to a substantial improvement in factor R_I. As to the validity of the linearization of functional (5.35), one may judge it by changes in factor R_I following application of the restrictions. It is seen from Figure 5.9 that application of the restrictions permits one to improve the solution considerably. Svergun *et al.* (1982) showed that the structure of axially symmetric particles can be restored using the considered method. By means of model examples the latter was proved to be stable with respect to experimental errors, i.e., to statistical noise in the scattering curve $I(s)$ and to deviations in R as well as ρ_{min} and ρ_{max}, where R was varied up to $\pm$ 10% of its actual value and ρ_{min} and ρ_{max} to $\pm$ 20%. In practice, most inherent errors may be associated with the fact that any particle is represented by infinite series (5.10), while we are searching for a finite representation (5.28). However, as already noted, the multipole expansion converges rapidly for globular particles, and the first few terms of the series would describe the whole structure with sufficient accuracy. Figure 5.10 represents the restoration of a model body, comprising two touching solid spheres, from its intensity curve. A finite number of harmonics ($L = 4$) was used in the restoration, although the intensity of scattering by the body was represented by infinite series (5.10).

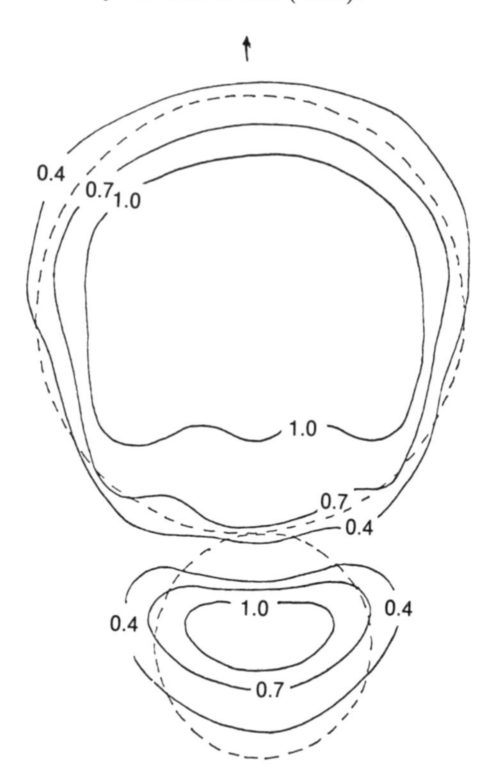

Figure 5.10. Restoration of model body comprising two touching spheres. Density of the spheres $\rho(\mathbf{r}) = 1$; radii $R_1 = 54.5$ and $R_2 = 27.3$. The cross section made by the plane through the rotation axis is shown (marked by an arrow); dashed curves are cross sections of spheres and solid curves are density levels of the restored structure.

The possibility of applying the considered method for structure determination depends, in any particular case, on *a priori* information being available, because only by employing such information can one choose the set of harmonics Y_{lk} correctly. The value of R can be estimated from small-angle data, by comparing parameters R_g, D, and the anisometry of the particle; ρ_{min} and ρ_{max} can be evaluated from the chemical composition of the particle and solvent. The resolution of the method is determined by two parameters: (1) the angular resolution, which depends on the maximum number of harmonics in expansion (5.28), namely $\pi/(l+1)$, and (2) the radial resolution, which is $2\pi/s_{max}$.

In the axially symmetric case, it is not difficult to construct the structure using the obtained radial functions. The only ambiguities are caused by changes in sign of the radial function $\rho_l(r)$ and rotation of partial structure $\rho_l(\mathbf{r})$ through angle π. The number of independent solutions is 2^{M-2}, where M is the number of harmonics taken into account. Usually, M does not exceed 4 or 5, so the solution can be found by sorting out all possible structures allowing for additional information. Therefore, axially symmetric particles can very conveniently be treated by the considered method.

In the general case, this method involves more complicated problems. First, for the iteration procedure to operate it is absolutely necessary that the above assumption about a certain harmonic Y_{lk} should be valid for each l. Furthermore, when reconstructing density $\rho(\mathbf{r})$ using the restored radial functions, all possible rotations of the partial densities must be taken into account. Consequently, application of the method in the general case, if possible, requires a lot of additional information about a particle. Thus, data on its subunit structure, symmetry axes, and features of the quaternary structure may be of help when selecting the most significant harmonics in expansion (5.10). If *a priori* information is available, a stepwise increase in resolution can be used. For example, if L is small enough, one can choose values of k for given l that result in a structure consistent with *a priori* information at this level of resolution. The independent rotations of partial densities can also be performed at each step, and one can refine the obtained structure by taking higher multipoles into consideration. Such approaches, however, depend to a great extent on the information available; moreover, they can hardly be formulated as straightforward procedures.

5.4.4. Examples of Direct Structure Determination

We now present two applications of the method described in Section 5.4.3 for determining the biomolecules, namely investigation of bacteriophage T7 and immunoglobulin M in solution.

Bacteriophage T7 is a large bacterial virus, with molecular mass

5.6×10^7 daltons and diameter about 870 Å, containing a double-stranded DNA molecule that affects *E. coli* bacteria. The phage has long been the subject of studies by various physical and chemical methods, and some general geometrical and weighting parameters of the phage as well as its shape have been thoroughly investigated. At the same time, several problems about its inner structure still remain to be solved. The main problem, dealing with interactions between the intraphage DNA and proteins, has been investigated by a number of researchers. Models of DNA packing (spherical shells or coaxial solenoid) were proposed by Richards *et al.* (1973), Earnshaw and Harrison (1977), and Kosturko *et el.* (1979). The results of Serwer (1976) and Agamalyan *et al.* (1982) provided evidence that there is a protein core in the central part of the phage head. It should be noted that the cited investigations were carried out by various methods (X-ray and neutron scattering, electron microscopy, circular and linear dichroism). Each method requires specific sample preparations and therefore involves a specific state of the particles under investigation. It is thus very difficult to bring these data together in order to construct a general picture of the inner structure.

Rolbin *et al.* (1980b) and Ronto *et al.* (1983) carried out a small-angle X-ray scattering investigation of T7 bacteriophage in solution and obtained the scattering curve up to $s = 1.35$ Å^{-1}, corresponding to a resolution of 4.7 Å (Figure 5.11). A number of geometrical and weighting

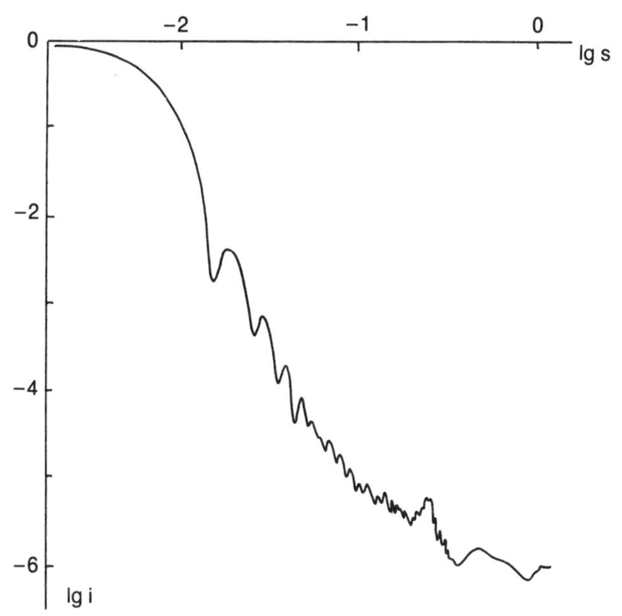

Figure 5.11. Entire small-angle scattering curve of bacteriophage T7 (after Rolbin *et al.*, 1980).

parameters of the particle were determined and its uniform model suggested. However, these data were extracted mainly from the initial part of the curve ($s < 0.1$ Å^{-1}) and most of the experimental information remained useless. Unfortunately such a situation is usual in small-angle scattering investigations.

Svergun *et al.* (1982) applied the above method to interpret the scattering curve of bacteriophage T7. According to *a priori* information, the phage can be regarded as axially symmetric with a good degree of accuracy. The initial parameters for the direct method (values of R, ρ_{max}, and the set of harmonics) were determined using both small-angle results and additional information. A portion of the scattering curve up to $s_{max} = 0.57$ Å^{-1} corresponding to radial 12-Å resolution (Figure 5.12) was treated. The angular resolution was 25°, where the harmonic with maximum number $l = 7$ was used. The obtained result — the electron map in bacteriophage T7 in a cross section containing the rotational axis — is presented in Figure 5.13. The cross section of the phage head is seen to be a smoothed hexagon. The tail appears as a circular cylinder of about 180 Å. This is a globular protein core with diameter about 240 Å in the central part of the head. Intraphage DNA forms sectorial regions of concentric packing, the distance between layers being 25 Å. There is a cylindrical area with radius of 80 Å of higher DNA concentration near

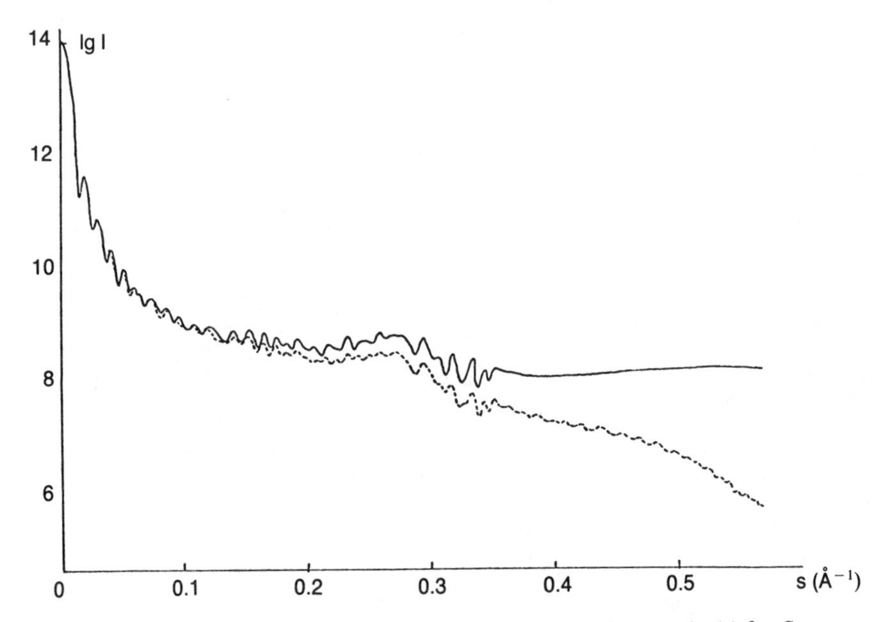

Figure 5.12. Part of the T7 scattering curve interpreted by a direct method (after Svergun *et al.*, 1982): (1) experiment; (2) scattering from recovered structure.

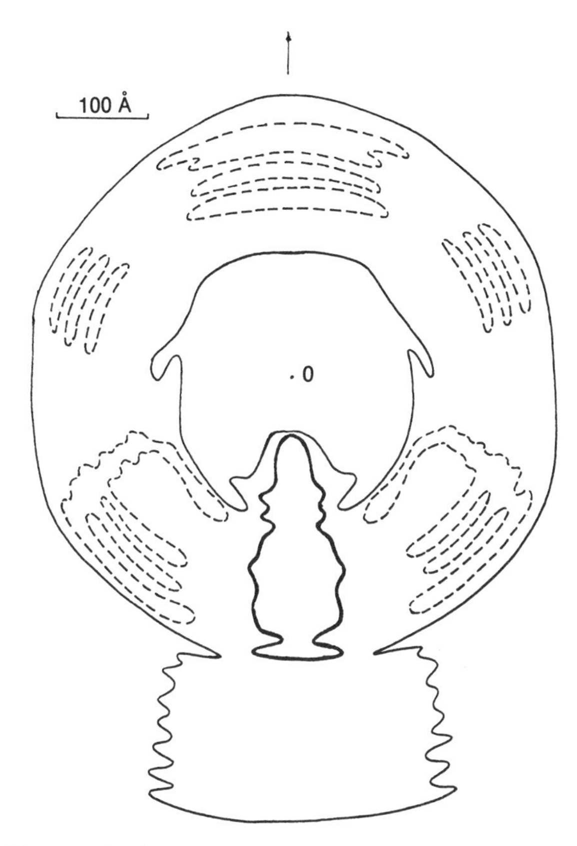

Figure 5.13. Electron density map of T7 bacteriophage in a cross section through the symmetry axis (marked by arrow); after Svergun *et al*. (1982). Solid line denotes level 0.38 $e\text{Å}^{-3}$ (hydrated protein), dashed line 0.42 $e\text{Å}^{-3}$ (strongly hydrated DNA), and thick line 0.52 $e\text{Å}^{-3}$ (slightly hydrated DNA); $R_I = 2.2 \times 10^{-2}$.

the tail, while on the whole the phage DNA is hydrated much more strongly than the protein. The scattering curve of the restored structure is shown in Figure 5.12 and is in excellent agreement with the experimental curve up to values of s corresponding to radial resolution. All T7 structural features specified are in good agreement with data provided by other methods. The obtained information about its inner structure permits a general model of DNA packing and its interactions with proteins to be proposed.

Direct determination of the structure and conformational transitions in immunoglobulin M (IgM) provides another example. According to the results of various methods the native IgM molecule represents a flat starlike structure consisting of five subnits (Glynn and Steward, 1981;

Wilhelm *et al.*, 1978). Each subunit is similar to the molecule of immunoglobulin G (IgG), which was investigated by Huber and Bennett (1983) using X-ray single-crystal analysis. Essential conformational transitions of the IgM molecule take place during the process of functioning, as well as under certain changes in the surroundings (temperature, pH); see Wilhelm *et al.* (1978), Litman and Good (1978), and Kayushina *et al.* (1982). The direct method has been used by Svergun *et al.* (1985) and Kayshina *et al.* (1986) to interpret small-angle X-ray scattering curves of IgM solutions. Experiments were carried out at two temperatures (23 °C and 58 °C) corresponding to native (IgM_{23}) and modified (IgM_{58}) states of the molecule. The scattering curves (Figure 5.14) were measured up to $s_{max} = 0.35$ Å^{-1} (radial resolution 18 Å).

The values of R were estimated using experimental data (190 ± 10 Å for IgM_{23} and 160 ± 10 Å for IgM_{58}). The value of ρ_{min} was zero, the value of ρ_{max} (the difference between the electron densities of the protein and solution) was 0.08 eÅ^{-1}. *A priori* information was used to choose a set of harmonics Y_{lk}. The flattened general shape of native IgM_{23} can be described by a set of even axial harmonics $Y_{2n,0}$ ($n = 0, 1, \ldots$), while the starlike structure (equatorial density distribution with fivefold symmetry relative to the z axis) is given by the sectorial harmonic Y_{55} (see Figure 5.8). The set Y_{00}, Y_{20}, Y_{40}, Y_{55}, Y_{60} was used for the IgM_{23} molecule, the presence of odd harmonics not changing the results. At the same time, the Y_{10} harmonic plays a significant role for IgM_{58} and provides evidence that

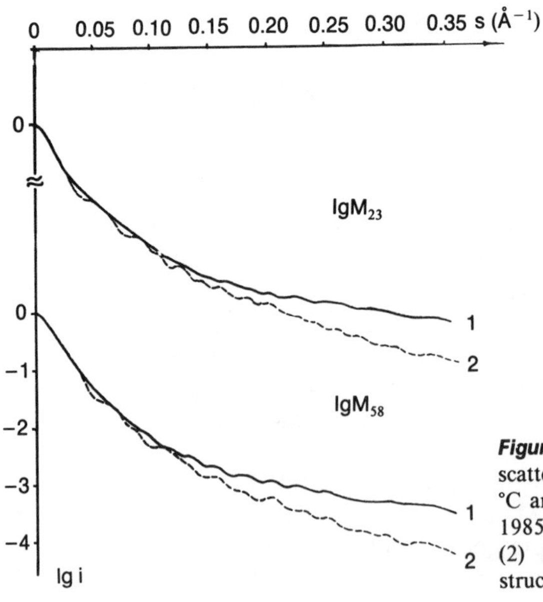

Figure 5.14. Small-angle X-ray scattering curves from IgM at $T = 23$ °C and 58 °C (after Svergun *et al.*, 1985): (1) experimental curves; (2) scattering from the restored structures.

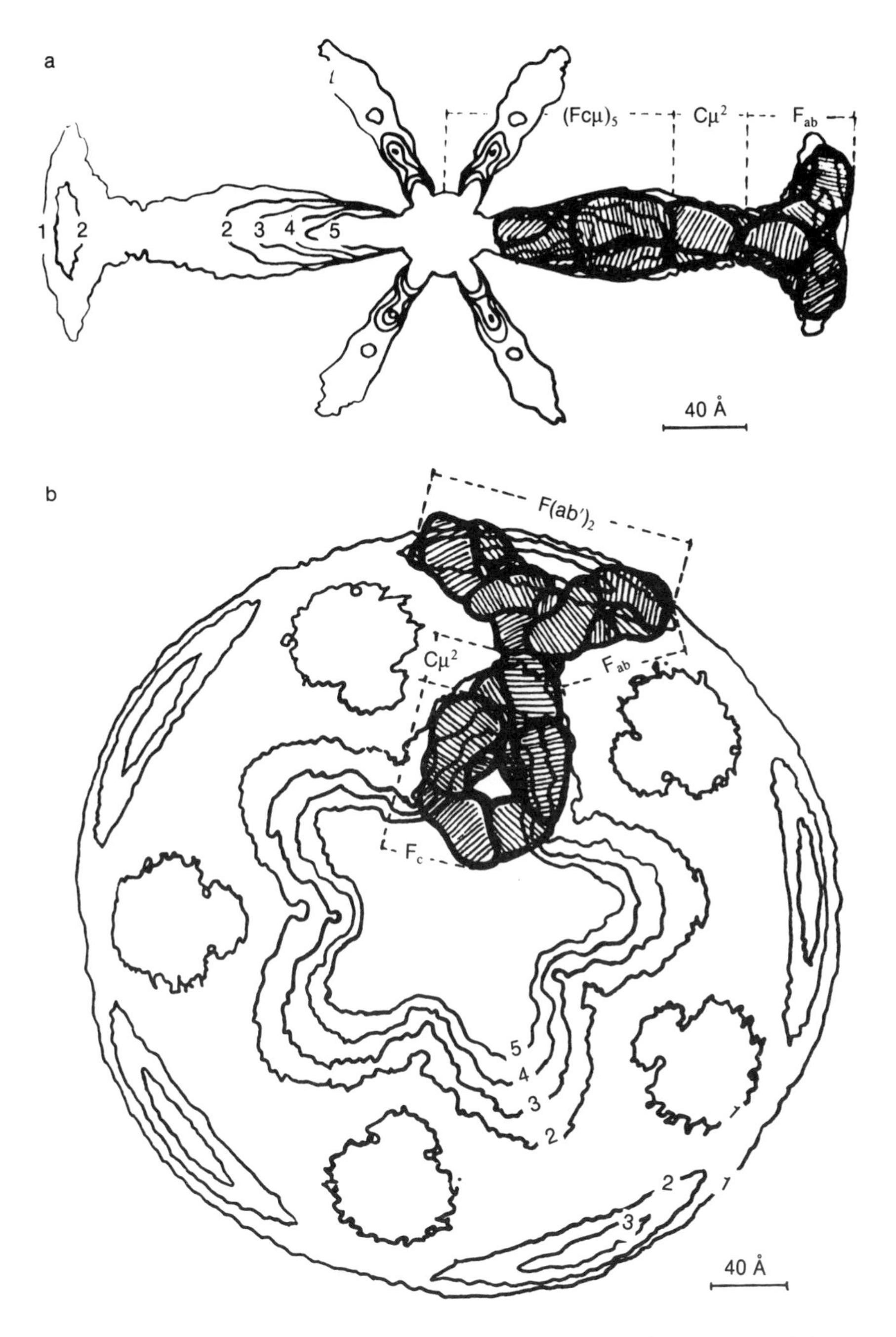

Figure 5.15. Electron density map in IgM_{23} (after Svergun *et al.*, 1985): (a) axial cross section; (b) sectorial cross section (plane $z = 0$). Isolines (1)–(5) correspond to levels 0.35, 0.36, 0.37, 0.39, 0.40 $eÅ^{-3}$.

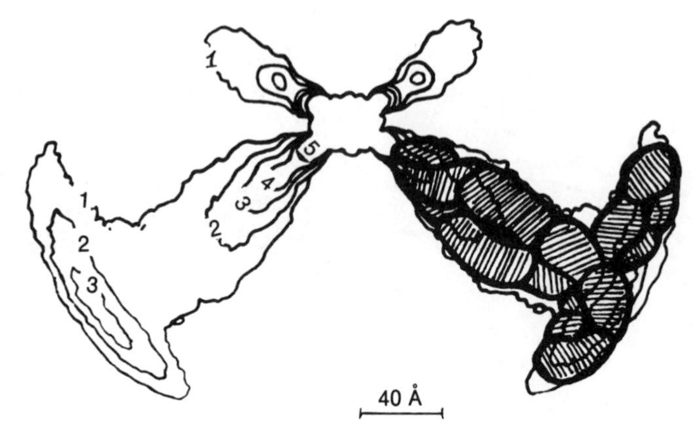

Figure 5.16. Electron density map in an axial cross-section of IgM_{58} (after Svergun *et al.*, 1985). Notations are the same as in Figure 5.15a.

the molecule is asymmetric with respect to the plane $z = 0$. In order to have a number of harmonics the same as for native IgM, the set Y_{00}, Y_{10}, Y_{20}, Y_{40}, Y_{55} was used for IgM_{58}.

After the radial functions were restored by means of the algorithm described in Section 5.4.3, the reconstruction of the structures of IgM_{23} and IgM_{58} was carried out. All possible combinations of signs of radial functions were sorted out, where the "sign" of $\rho_l(r)$ means the sign of the first maximum of $A_l(s)$. The alternative sequence of signs, typical for flattened particles, gives the best results for IgM_{23}: $0\ \bar{2}4\ \bar{6}$; it is clear that the signs of $\rho_5(r)$ as well as rotations of the fifth partial structure in the equatorial plane do not affect the reconstruction. Figure 5.15a gives the isolines of the restored structure in the cross section containing the z axis (the axial cross section; the fifth harmonic is omitted here for the sake of clarity). Figure 5.15b shows the same isolines in the plane $z = 0$. Fitting the molecule IgM_{23} to the obtained electron density map is carried out in both figures; an IgG-like particle is shown according to the results of Huber and Bennett (1983). When constructing a map of IgM_{58}, the sign combination $0\ 1\ \bar{2}\bar{4}$ seems to be most probable; the axial cross section together with a drawn IgG molecule is given in Figure 5.16. The results obtained by the direct method confirm general knowledge about the structure and functioning of the IgM macromolecule.

5.5. Conclusion

The development and application of direct methods are probably the most important directions in the field of small-angle scattering studies of monodisperse systems. It is very attractive to restore the particle structure directly, without laborious sorting out of a lot of models (as in the modeling method) and without careful sample preparations and accurate additional measurements (as in contrast variation or isomorphous replacements). One should, however, bear in mind that direct methods also require additional information about the sample. Moreover, the possibilities of their application and of determining a unique solution are to a great extent dictated by the features of the scattering density $\rho(\mathbf{r})$.

When the density is described by the one-dimensional function $\rho(\xi)$ and the problem is reduced to determination of the signs of the scattering amplitudes, unique restoration of the particle structure can be performed. In many practical applications the particle can be represented by a two-dimensional function: uniform particles by $F(\omega)$ and axially symmetric particles by $\rho(r, \theta)$. An approach using spherical harmonics proved very fruitful for such particles. The particle structure may be restored with minimum *a priori* information, using the method described in Section 5.4. Resolution up to 10–20 Å can be achieved.

The problem of restoring the structure is most difficult in the general case of a three-dimensional function $\rho(\mathbf{r})$. More or less unique recovery of the structure is hardly to be expected. The iteration procedure described in Section 5.4.3 could be applied in some cases; however, a large amount of *a priori* information is absolutely necessary for it to operate. We hope that the further development of direct methods will permit more complete utilization of the information contained in small-angle scattering data.

III

Polymers and Inorganic Materials

6

Investigations of Polymer Substances

The structure of polymers is a very important field of application for small-angle X-ray and neutron scattering methods. Small-angle scattering research can be conducted both in solutions of various concentrations and in a solid (amorphous, crystalline) state. These investigations very often provide information on the structure of polymers not obtainable by other methods.

Polymer substances as a subject of study by small-angle scattering have several specific features. Thus, as a rule, the polymer macromolecules both in solution and in an amorphous state form a fairly loose conformation (the statistical coil); crystals of polymers consist of randomly oriented aggregates of lamellar layers. The polymer samples are practically always polydisperse — they contain molecules with different molecular mass. Therefore, special interpretation methods, different from those discussed in Part II, are used for analyzing the structure of polymers.

Section 6.1 deals briefly with the basic conformation models of polymer macromolecules and the patterns of small-angle scattering by these models; the influence of polydispersity of the polymer samples is analyzed. Section 6.2 demonstrates application of the models under discussion to structural analysis of polymers in solution and in an amorphous state.

Methods for investigating lamellar structures formed by polymers crystallized from solution or from a melt are described in Section 6.3. Certain types of oriented polymer objects and possibilities for analyzing their structure using anisotropic small-angle scattering are discussed in Section 6.4.

6.1. Models of Polymer Chains

In this section we dicuss briefly the basic models used to describe chain macromolecules in solution and present expressions for the small-angle scattering intensity obtained by these models. We shall not elaborate on the conformation theory of chain macromolecules, which can be found in a number of monographs by, for example, Volkenshtein (1963), Morawetz (1963), and Flory (1969).

6.1.1. Gaussian Chains

The model in which the chain is visualized as a combination of N linked elements (statistical segments), each l long (among other things they can be represented by N monomeric units each of dimension l) is very convenient for describing the conformation of the chain macromolecule. This schematic concept is shown in Figure 6.1. There might be two types of interaction in this chain: short range, between adjacent elements of the chain; and long range, between elements distant from each other in the chain but close in space.

Here, only the presence of short-range interations can be assumed; whether or not there are long-range interactions is determined by the solvent containing the polymer chain and their presence leads to additional effects, that will be discussed in Section 6.1.4.

Polymer macromolecules are capable of assuming a multitude of conformations, so it is reasonable to talk only about their statistically averaged characteristics in solution. The average distance between the elements of the chain is one of the most important such characteristics. It can be demonstrated that for the ith and jth elements of the chain,

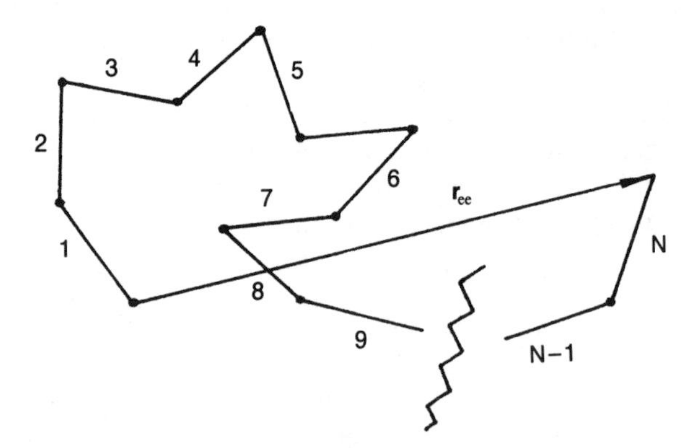

Figure 6.1. Scheme of a chain molecule. Statistical segments are marked with numbers.

sufficiently distant from each other such that $|i-j|>30$, which permits statistical laws to be applied, the probability density that the distance between the elements equals r_{ij} is given by the Gaussian distribution

$$P(r_{ij}) = \left(\frac{3}{2\pi\langle r_{ij}^2\rangle}\right)^{3/2} 4\pi r_{ij}^2 \exp\left(-\frac{3r_{ij}^2}{2\langle r_{ij}^2\rangle}\right) \tag{6.1}$$

where $\langle r_{ij}^2\rangle$ is the root-mean-square distance between the segments:

$$\langle r_{ij}^2\rangle = \int_0^\infty r_{ij}^2 P(r_{ij}) dr_{ij}$$

These are called Gaussian chains (Flory, 1969).

A free-linked chain is the simplest model of the Gaussian chain (Kuhn, 1934) and is assumed to consist of N vectors $\mathbf{l}_i$ of equal length l, but the angles between adjacent vectors can, with equal probability, assume any value.

The root-mean-square distance between the ends of the chain is

$$\langle r_N^2\rangle = \langle r_{\mathrm{ee}}^2\rangle = \left\langle\sum_{i,j}\mathbf{l}_i\mathbf{l}_j\right\rangle = Nl^2 + 2\sum_{i>j}\langle\mathbf{l}_i\mathbf{l}_j\rangle$$

Since there is no correlation between the segments, $\langle\mathbf{l}_i\mathbf{l}_j\rangle = 0$ for $i \neq j$ and

$$\langle r_{\mathrm{ee}}\rangle = Nl^2 \tag{6.2}$$

This result, that the root-mean-square distance between the ends of the chain is proportional to the number of statistical segments, is of a very general character for unperturbed chains. The mean radius of gyration of the free-linked chain is represented by the relation

$$R_{\mathrm{g}}^2 = \frac{l^2N(N+2)}{6(N+1)} \approx \frac{l^2N}{6} = \frac{\langle r_{\mathrm{ee}}\rangle^2}{6} \tag{6.3}$$

Therefore, if $L = lN$ is the contour length of the chain and M_L the molecular mass per unit weight, we have

$$R_{\mathrm{g}}^2 = lL/6 = Ml/6M_L = K_{\mathrm{r}}M \tag{6.4}$$

where the characteristic ratio $K_{\mathrm{r}} = R_{\mathrm{g}}^2/M$ depends only on the structure of the monomers of a certain chain.

Another model of a Gaussian chain is represented by the chain with given valence angle differing from the previous chain by fixing the valence angle α between the bonds (Figure 6.2). In this case (Eyring, 1932)

$$\langle r_{ee}\rangle^2 = Ll\frac{1+\cos\alpha}{1-\cos\alpha}$$

which agrees with equation (6.2), assuming that the statistical segment is equal to $l' = l(1+\cos\alpha)/(1-\cos\alpha)$. For this model equations (6.3) and (6.4) are also true with l replaced by l'. Analogous results can also be obtained when considering correlations between internal angles of rotation. In general, relations (6.2)–(6.4) are true for any linear (unbranched) Gaussian chain with appropriate choice of the value of the chain statistical element l.

In order to calculate the intensity of small-angle scattering by the chain molecule, the Debye equation (2.16) should be averaged over all conformations. It will readily be seen that

$$\left\langle\frac{\sin(sr_{ij})}{sr_{ij}}\right\rangle = \int_0^\infty P(r_{ij})\frac{\sin(sr_{ij})}{sr_{ij}}\,dr_{ij}$$

Taking into account that the weight of each term with given distance r_{ij} is

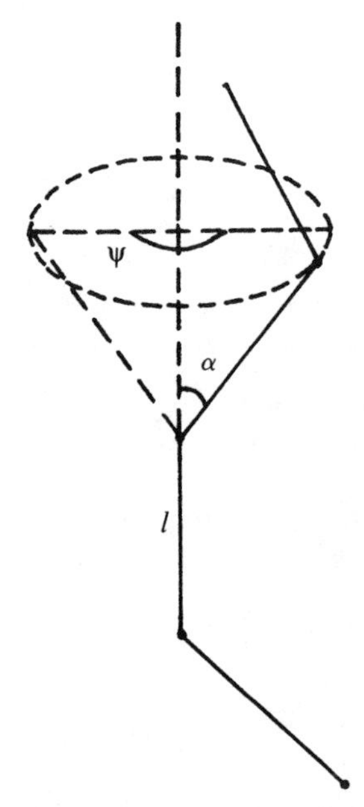

Figure 6.2. Chain with fixed angle of rotation, where l is the length of a segment, α the valence angle, and ψ the angle of rotation.

given by the value $(N - |i-j|)$ we obtain

$$i(s) = \frac{2}{N^2} \sum_{|i-j|=0}^{N} (N - |i-j|) \int_0^\infty P(r_{ij}) \frac{\sin(sr_{ij})}{sr_{ij}} = dr_{ij}$$

for the normalized scattering intensity of the chain molecule.

Substitution of the Gaussian distribution yields

$$i_G(s) = 2(e^{-x} + x - 1)/x^2 \tag{6.5}$$

where $x = s^2R_g^2$ (Debye, 1947). Thus, the normalized small-angle scattering intensity by the ideal Gaussian chain is determined by the only parameter, namely R_g. Guinier's approximation (see Section 3.1.1) holds for the initial part of curve (6.5). However, the asymptotic value of scattering by the Gaussian chain at large s does not obey Porod's law for compact particles (Section 2.4.4); it is easy to see that, as s tends to infinity, $i_G(s) \sim s^{-2}$ and

$$\lim_{s \to \infty} [s^2 i_G(s)] = \frac{2[1 - 1/(s^2R_g^2)]}{R_g^2} \tag{6.6}$$

If the scattering intensity is known in absolute units $[I_G(s) = i_G(s)M]$ then along with R_g, the molecular mass and the characteristic ratio K_r for a given chain are determined from this curve.

All the above-mentioned principles refer to linear macromolecules. In the case of branched macromolecules with the length of the side chain comparable to that of the main chain, the concept of the "distance between the ends of the chain" does not make sense. The geometry of the branched chain can be characterized only by the value of its radius of gyration R_{gb}. The more the chain is branched, the more compact it becomes; therefore the ratio

$$g = R_{gb}^2/R_g^2$$

where R_g is the radius of gyration of the linear macromolecule of the same degree of polymerization, serves as a measure of the extent of branching (see, for example, Zimm and Stockmayer, 1949).

6.1.2. Persistent Chain

The Gaussian coil serves as a very good approximation for describing the behavior of a number of polymers in solution in the absence of long-range interactions, even with a degree of polymerization down to several dozens (the molecular weight amounts to several thousands). At the same time, there are certain polymers for which this model is not applicable even with a molecular weight of hundreds of thousands. This situation is governed by the high rigidity of these chains (for example, some derivatives of cellulose), because of which the number of independent segments in the chain becomes too small for the statistical laws to be used. (In this sense such chains are similar to short flexible chains.) The persistent (wormlike) chain is convenient for the description of such macromolecules (Porod, 1949; Kratky and Porod, 1949).

The model is a limiting case of the chain with fixed valence angles, discussed in the previous section: with fixed chain length L and distance between the ends r_{ee}, the length of the links approaches zero. Such a chain has continuous curvature, which can be characterized by the mean cosine of the angle between the tangents to the chain at two selected points:

$$\langle \cos \psi \rangle = \exp(-l_p/a)$$

where l_p is the contour length of the chain between these two points. The value of a, the persistent length of the chain, serves as a measure of chain rigidity and is defined as the chain piece with the mean cosine of the angle between the ends equal to $1/a$.

The root-mean-square distance between the ends of the chain and the radius of gyration are given by (Benoit and Doty, 1953)

$$\langle r_{ee}^2 \rangle = 2a^2 \left[\frac{L}{a} - 1 + \exp\left(-\frac{L}{a} \right) \right]$$

and

$$R_g^2 = a^2 \left\{ \frac{L}{3a} - 1 + \frac{2a}{L} \left\{ 1 - \frac{a}{L} \left[1 - \exp\left(-\frac{L}{a} \right) \right] \right\} \right\}$$

As a tends to infinity, $\langle r_{ee}^2 \rangle \approx L^2$ and $R_g^2 \approx L^2/12$ (rigid rod), while for $a \ll L$ one has $\langle r_{ee}^2 \rangle = 2aL$ and $R_g^2 = aL/3$ (Gaussian chain with $l = 2a$). Hence, with the help of the persistent chain, it is possible to represent both the most rigid and most flexible macromolecular chains.

Small-angle scattering intensity by the persistent chain can be split qualitatively into three sections. As s tends to zero as well as for the other particles, it is determined by the Gaussian chain. As s tends to infinity, large-scale (compared with the thickness of the chain) characteristic features of its winding cease to affect the scattering, which will be determined by the scattering asymptotic values for a thin rod of length L, namely $i_p(s) \sim \pi/sL$ (see Section 2.4.5). There will be an interval in the intermediate region of angles within which scattering by the persistent chain will conform to scattering by the Gaussian chain with $l = 2a$.

Figure 6.3 presents schematically scattering curves from Gaussian and persistent chains on a Kratky plot [$s^2 i(s)$ versus s]. The transition point s^* from the asymptotic values for the Gaussian coil to the asymptotic values for the rod can be determined by the condition of intersection:

$$\frac{2}{R_g^2} = \frac{12}{Nl^2} = \frac{\pi s^{*2}}{s^* Nl}$$

from which $s^* l = 12/\pi$, i.e., $a = 1.91/s^*$. Thus, even qualitative analysis of scattering by the persistent chain permits one (if the transition point can be localized with a certain degree of reliability) to evaluate its most important parameter, namely, the persistent length.

The Monte Carlo method is used for quantitative calculations of the scattering intensity by persistent chains: computers assist in generating a large number of random chain configurations at a preset ratio L/a (where l and a are finite, but small) and the calculated scattering intensities are averaged (Peterlin, 1960). In the ensuing years analytical methods of calculating $i_p(s)$ were also developed. The equation (Sharp and Bloomfield, 1968)

$$i_p(s) = \frac{2[\exp(-x) + x - 1]}{x^2} + \left[\frac{4}{15} + \frac{7}{15x} - \left(\frac{11}{15} + \frac{7}{15x}\right) e^{-x}\right] \frac{2a}{L}$$

where $x = Las^2/3$, was derived and yields good results when $L > 20a$ and $s < 2/a$.

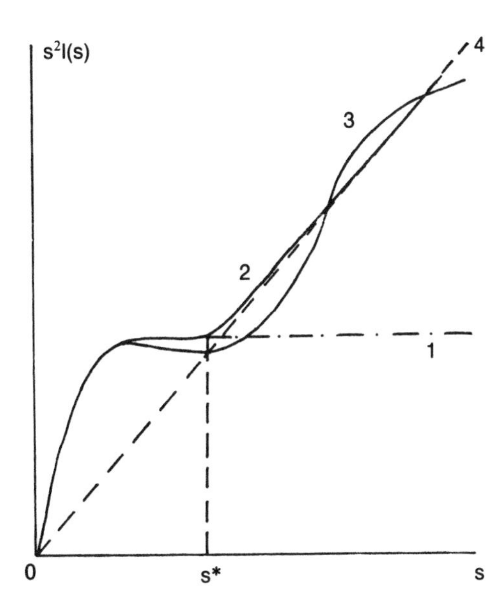

Figure 6.3. Scattering by different schematic models of chain molecules: (1) Gaussian chain; (2) persistent chain; (3) persistent chain with persistent curvature; (4) asymptotic behavior of scattering by an infinitely thin rod.

The model of the persistent chain was futher developed in subsequent work (Kirste, 1967a,b) using, as an additional condition, the limitations for the rotation angle ψ (Figure 6.2) characterized by the "preferable radius of curvature" of the chain, R_{curv}. This procedure leads to the appearance of properties subject to the curtailment of the chain and, in turn, to oscillations in the scattering curve, the period of which diminishes with growth of the ratio R_{curv}/a (Figure 6.3, curve 3).

6.1.3. Perturbed Chains

The models of polymer chains examined above are based on the assumption that there are no long-range interactions in the macromolecule, which assumption is not always valid. Actually, the conformation of the macromolecules depends substantially on the solvent.

The osmotic pressure of the polymer solution Π can be expressed in the form

$$\Pi = (RT/M)c + A_2c^2 + A_3c^3 + \cdots \tag{6.7}$$

where c is the concentration of the solution and M is the molecular mass of the polymer. The virial coefficients (for lower concentrations, basically A_2) govern the character of the interaction between of the polymer and the solvent; when $A_2 < 0$ (a good solvent) the segments repel one another and the chain "unwinds," while when $A_2 > 0$ (a poor solvent) attraction predominates and the chain contracts. In an ideal solvent (the θ solvent) $A_2 = 0$ and distant interactions in the chains are practically nonexistent, so the models we have described are applicable. That is why, as a rule, the experiments are attempted at the θ point.

It happens, however, that for various reasons measurement in the θ solvent are found to be inadvisable or impossible. One of the simplest methods of taking interactions into account is to consider the model of the chain with "self-avoiding walk," which obeys all the laws governing the formation of unperturbed chains, except that a given statistical element cannot remain in the volume of space previously occupied by another element. Calculations using this model give, for asymptotic values as $|i - j| \to \infty$, the probability distribution

$$P(r_{ij}) = Cr_{ij}^h \exp[-(r_{ij}/\sigma)^t] \tag{6.8}$$

where

$$C = \frac{t}{\sigma^{h+1}\Gamma((h+1)/t)} \quad \text{and} \quad \sigma^2 = \frac{\langle r_{ij}^2 \rangle \Gamma((h+1)/t)}{\Gamma((h+3)/t)}$$

$\Gamma(x)$ being the Gamma function (Fisher, 1966). Distribution (6.1) is a special case of equation (6.8) for $h = t = 2$. This model gives the radius of gyration as

$$R_g^2 = K_{r\varepsilon} M^{1+\varepsilon} \tag{6.9}$$

where $\varepsilon = (t-2)/t$ and $K_{r\varepsilon}$ are constants.

Thus, the nonlinear relation between the squared radius of gyration and the molecular mass is a characteristic feature of the perturbed chain. Parameters ε, h, and t are, to a certain extent, of an empirical nature. The sign of ε agrees with the sign of A_2 (naturally, $\varepsilon = 0$ when $A_2 = 0$). For sufficiently long molecules in good solvents, $\varepsilon = 0.2$.

Efforts at calculating small-angle scattering curves from perturbed chains are mainly associated with the use of distributions of type (6.8) and conditions (6.9). The asymptotic function $i(s)$ can generally be expressed as

$$\lim_{s \to \infty} [M i(s)] = \frac{f(\varepsilon, h, t)}{s^{2/(1+\varepsilon)} K_{r\varepsilon}^{1/(1+\varepsilon)}} \tag{6.10}$$

where f is a constant that depends on ε, h, and t (Kirste and Oberthür, 1982). The value of ε can therefore be assessed by analyzing the decrease in $i(s)$ at higher s. By and large, it is not possible to derive a general equation for small-angle scattering by the perturbed chains, similar to equation (6.5), since the behavior of the scattering curves will depend on the molecular weight of the polymer dissolved and on the solvent type and temperature.

6.1.4. Molecular Mass Distribution

Basic models of the structures of polymer chains in solution and small-angle scattering curves derived from these models were discussed in the previous section. When the polymer solutions are ideal monodisperse systems comprising polymers of identical length (equal molecular mass), the total intensity of the scattering is proportional to the intensity of scattering by one particle, as discussed in Part II. In practice, however, the solutions of the same polymer always contain particles with different molecular mass — the polymer solutions are, to a certain extent, polydisperse. The solution polydispersity can be characterized by the numerical distribution $N(M)$, which yields the relative number of molecules with a given molecular mass M, or by the weight (molecular mass) distribution $W(M)$, proportional to the weight fraction of the molecules with a given mass:

$$W(M)dM = MN(M) \Big/ \int_0^\infty MN(M)dM$$

Any parameter X of the polymer molecule determined while studying this system is obtained by appropriate averaging over $N(M)$ or $W(M)$. The numerical average

$$\langle X\rangle_N = \int_0^\infty X(M)N(M)dM \Big/ \int_0^\infty N(M)dM$$

the weighting average

$$\langle X\rangle_W = \int_0^\infty X(M)W(M)dM$$
$$= \int_0^\infty X(M)MN(M)dM \Big/ \int_0^\infty MN(M)dM$$

and the z-average

$$\langle X\rangle_z = \int_0^\infty X(M)N(M)M^2dM \Big/ \int_0^\infty N(M)M^2dM$$

are of the greatest practical importance. The parameter $\langle X\rangle_W/\langle X\rangle_N = U_X$ can serve as an estimate of the distribution width.

The small-angle scattering curve for a polydisperse system (see Section 2.3.3) is

$$I(s) = \int_0^\infty i(s, M)W(M)dM = \langle i(s)\rangle_z\langle M\rangle_W \tag{6.11}$$

which therefore makes possible direct determination of the average weight value of the molecular mass and the z-average of the other structural parameters, particularly $\langle R_g^2\rangle_z$. We stated in Section 2.3.3 that the aim of investigations of polydisperse samples involves finding either the form factor or the distribution $W(M)$ [or $D(R)$]. For polymers, the kind of distribution $W(M)$ can often be predicted from the polymerization conditions. Some distributions are known (see, for example, Bowen, 1971); the Shultz distribution

$$W(M) = \frac{M^k}{\Gamma(k+1)}\left(\frac{k+1}{\langle M\rangle_W}\right)^{k+1}\exp\left[-(k+1)\frac{M}{\langle M\rangle_W}\right] \tag{6.12}$$

is one of the most common, where $k = 1/U_M$ and $U = U_M - 1$. When the latter distribution is employed to calculate the scattering intensity by the polydisperse system of Gaussian coils, we obtain (Greschner, 1973)

$$\langle i(s)\rangle_z = \frac{2[(1+U\xi)^{-1/U} + \xi - 1]}{(1+U)\xi} \tag{6.13}$$

where $\xi = \langle R_g^2 \rangle_z s^2/(1+2U)$. For the asymptotic value one has

$$\lim_{s\to\infty} s^2 \langle M \rangle_W \langle i(s) \rangle_z = \frac{12M_L}{l}\left(1 - \frac{6M_L}{s^2\langle M \rangle_N l}\right) \tag{6.14}$$

If the value of $\langle M \rangle_W$ is known, then $\langle M \rangle_N$ can in principle be estimated from this asymptote. The value of l is assumed constant for the given polymer and is therefore determined in the same manner as for the monodisperse system. Quantity $\langle R_g^2 \rangle_W$ can be determined for Gaussian chains by the equation (Altgelt and Schulz, 1960)

$$\langle R_g^2 \rangle_W = \frac{1+U}{1+2U} \langle R_g^2 \rangle_z$$

In the cases of the other chains, with proportionality not implemented [see equation (6.4)], the relation between $\langle R_g^2 \rangle_W$ and $\langle R_g^2 \rangle_z$ is more complex (Oberthür, 1978).

6.2. Polymers in Solution and in the Amorphous State

The main possibilities of determining the aforementioned structural characteristics of polymers will now be discussed briefly. Attention here will chiefly be paid to the methodical aspects. Examples of specific investigations can be found in original papers or reviews (for instance, Kirste and Oberthür, 1982; Higgins and Stein, 1978; Kratky, 1982).

6.2.1. Polymers in Solution

6.2.1.1. General Parameters

Certain structural characteristics of polymers in solution can be determined from the initial part of the small-angle scattering curve without any assumptions as to the specific features of the polymer chain and the shape of the molecular-mass distribution $W(M)$. Such characteristics include the average weight mass $\langle M \rangle_W$, the z-average radius of gyration, and the second virial coefficient. Double extrapolation of the scattering curve to $s = 0$ and $c = 0$, suggested by Zimm, is generally used for their determination. It can be seen from equation (6.11) that the initial part of the infinitely diluted solution can be expressed as

$$I^{-1}(s) = (\langle M \rangle_W \langle i(s) \rangle_z)^{-1} \approx \langle M \rangle_W^{-1} [1 + (s^2/3) \langle R_g^2 \rangle_z]$$

where $I(s)$ is assumed to be measured in absolute units; for absolute measurements see Section 3.3.2. Interference effects due to measuring polymer solutions of finite concentration can be taken into account through the use of a well-known Einstein equality for the intensity of light scattering by liquids and molecular-dispersed solutions:

$$I(s) \sim c\left[\frac{\partial}{\partial c}\left(\frac{\Pi}{RT}\right)\right]^{-1}$$

On introducing equation (6.7) it can be stated that, for small concentrations,

$$cI^{-1}(s) \approx (\langle M \rangle_W \langle i(s) \rangle_z)^{-1} + 2cA_2Q(s)$$

where $Q(s)$ tends to 1 as s tends to zero. The relationship between $c/I(s)$ for various concentrations and the argument $s^2 + c_0c$ (with constant c chosen such that the terms are of the same order) is plotted on a Zimm plot. Simultaneous extrapolation to zero values of s and c (Figure 6.4) is

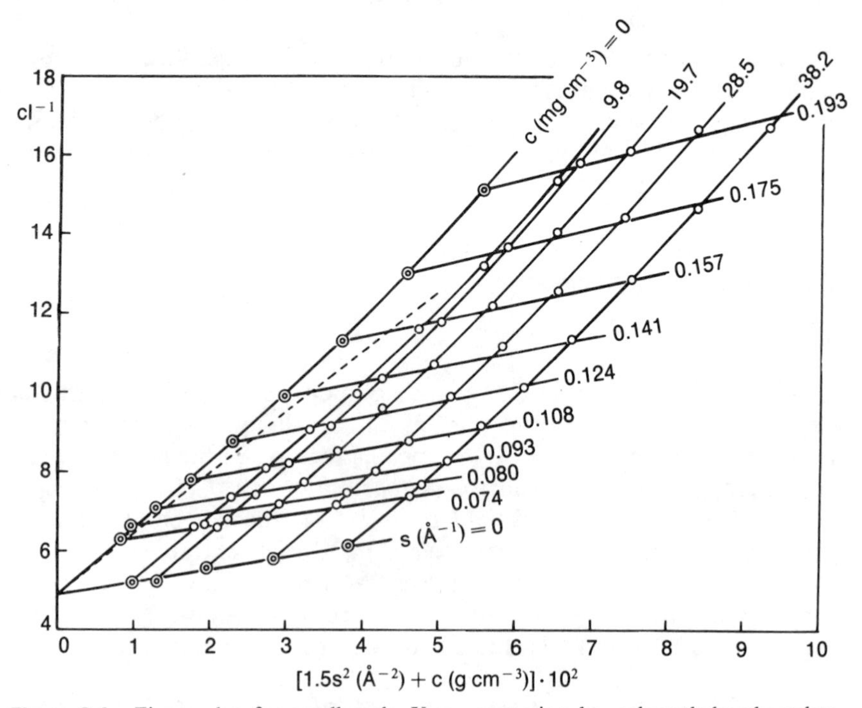

Figure 6.4. Zimm plot for small-angle X-ray scattering by polymethylmethacrylate solution (after Kirste *et al.*, 1968).

conducted on the basis of the data obtained. The point $s = c = 0$ then gives $\langle M \rangle_W$, while $\langle R_g^2 \rangle_z$ and A_2 are determined by the slopes of the straight lines corresponding to $c = 0$ and $s = 0$, respectively. Such extrapolation is usually used for polymer samples instead of extrapolation to zero concentration according to the Guinier plot for particle solutions (Section 3.3.1). In principle, these two extrapolations are similar. The Zimm plot generally yields a higher value for $\langle R_g^2 \rangle_z$ compared with the Guinier plot, and therefore comparison of both methods very often assists in improving the result (Kirste and Wunderlich, 1968).

Determination of common parameters permits one to reach a number of conclusions in relation to the structure of the polymer chain in a given solvent. In practice, the parameters are usually measured for various fractions of the same polymer in an effort to obtain relationships between $\langle R_g^2 \rangle_z$, A_2, and $\langle M \rangle_W$ that make it possible to judge whether or not the given chain is Gaussian, to study the effect of the solvent on molecules of different lengths, and so on. Besides, determination of $\langle R_g^2 \rangle_z / \langle M \rangle_W$ for some value of K_r known for a given polymer permits parameter g to be evaluated, hence characterizing the degree of branching (Section 6.1.1).

6.2.1.2. Polydispersity

In principle, an effort can be made to determine the molecular-mass distribution $W(M)$ from the scattering curve $\langle i(s) \rangle_z$ with known form factor $i(s)M$. (The methods of calculating poydispersity in the general case, discussed further in Section 7.2, are applicable.) However, two-parameter distributions of type (6.12) are frequently used in investigations of polymers. The availability of such distribution is taken account of when calculating model intensities by polymer chains, while its parameters, together with those of the hypothetical model, can be improved, for example, by fitting the model intensity to the experimental data by the least-squares method. However, in a number of cases determination of quantities $\langle M \rangle_W$ and U is quite sufficient for an assessment of the degree of polydispersity of the sample, taking into account the possibility of independent determination of $\langle M \rangle_N$ by osmometric methods.

6.2.1.3. Use of the Asymptote

Analysis of the scattering curve from the polymer solution at large angles permits one to clarify additionally various specific features of its structure. Thus, in the case of a perturbed Gaussian chain, the value of ε, determining the effect of the excluded volume, can be assessed by asymptote (6.10). The degree of branching of the Gaussian chain can be estimated by analyzing the asymptote of $\langle i(s) \rangle$ (see, for example, Kirste and Oberthür, 1982). Besides, important characteristics of the structure of a given polymer chain can be obtained by modeling. The Kratky plot (Figure 6.3) usually helps one to make an immediate choice between the (Gaussian or persistent) models for describing the structure of a given chain. Then, the chain parameters are sought in order to provide the best correlation with experimental data. The polydispersity of the sample is also taken into account (as stated above) in the course of modeling. Furthermore, the factor of the chain thickness must quite often be taken into account in the persistent models. Actually, in the course of calculations in Section 6.1.2 the chain was regarded as infinitely thin. With finite thickness

$$i_{pc}(s) \approx i_p(s) \exp(-R_c^2 s^2/2)$$

(compare with Section 2.4.5). If the condition $R_c \ll a$ is fulfilled, the second factor can be neglected.

The above-mentioned asymptotes can also be used to calculate the value of the statistical segment l. It is noteworthy, however, that the point of inflection for the persistent chain (Figure 6.3) very often cannot be localized simply, and therefore modeling with fitting of parameters U, R_c, and l should be applied.

The above sections have dealt with studies of low-concentration solutions. For semidilute ($c \sim 15\%$) and concentrated solutions, interference effects can lead to very complicated relationships between the parameters obtained and the concentrations. A review of approaches used to analyze such systems can be found in Higgins and Stein (1978).

In conclusion, a few words can be said about the possibility of studying heterophase polymer objects including, first, solutions of block copolymers with chains consisting of two or more alternating groups (blocks) of various monomers. In this case, the methods of analysis used for conventional polymers are not applicable. However, in the case of a not-too-large number of blocks in the copolymer (for example, diblock copolymers) the object can be treated as a particle with several components differing from one another in scattering density, and various techniques can be adopted in order to study the structure of heterogeneous particles, discussed in Chapter 4. Thus, relations (4.23) and (4.24) hold for diblock copolymers and one can determine the values of $\langle R_g^2 \rangle_z$ for the two blocks and the distance $\langle L \rangle_z$ between their centers of mass by varying the contrast. Light, X-ray, and neutron techniques (Section 4.4.1) can be used jointly to vary the contrast, or isotropic substitution (in neutron scattering) can be employed (Section 4.2.3; see, for example, Cotton and Benoit, 1975).

6.2.2. Amorphous Polymers

There are two different approaches to studying the structure of amorphous polymers. The first is associated with the interpretation of the scattering, either by density fluctuations or by the cluster structure of the object. In this case one aims at determining thermodynamic characteristics of amorphous polymers, as well as the presence of the domain structure in the latter, through the use of small-angle scattering. The second approach is aimed at determining the structure of individual chains that comprise the polymer. It involves sample preparation, in which an insignificant fraction of macromolecules is modified in order to change their scattering capacity. In this case the object can be regarded as a dilute solution of the modified polymer in the solvent while remaining an ordinary polymer. This permits one to use the methods discussed in Section 6.1 for the data analysis and to seek the structural features of individual polymer chains in an amorphous state.

Analysis of the thermodynamic characteristics and separation of the phase in amorphous polymers is conducted on the basis of the general relations presented in Section 2.5. Numerous experiments with a number of amorphous polymers (see Wendorff and Fisher, 1973; Renninger *et al.*, 1975; Wiegand and Ruland, 1979) showed that the small-angle scattering curves were, as a rule, of the specific shape presented in Figure 6.5, curve 1. Here, the zero-intensity peak is governed by the sample heterogeneity, while extrapolation conforming to curve 2 produces the scattering intensity governed by the "fluid scattering" of the sample. Therefore, the extrapolated value of $I(0)$ should be proportional to the coefficient of isothermic compressibility and to the absolute temperature

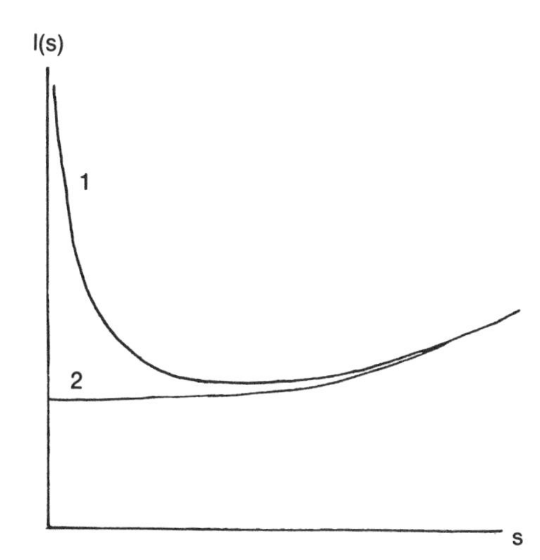

Figure 6.5. Typical small-angle scattering curves for amorphous polymers: (1) experimental curve; (2) curve corresponding to the liquid state.

in accordance with equation (2.43). Experiments have proved that this ratio is very well maintained for $T > T_g$ (glass transition temperature). When $T < T_g$, $I(0)$ does not depend practically on T and is governed by the sample compressibility at $T = T_g$. Thus, small-angle scattering measurements permit one to localize the phase transition points in amorphous polymers.

In the case of crystalline polymers, apart from analysis of their structure by the lameller model (discussed further in Section 6.3), small-angle scattering permits one to evaluate the degree of sample crystallinity (ratio between the volumes of the crystalline and amorphous phases) by examining the values of Porod's invariant (Section 2.5.2). When studying scattering with the aid of densely packed cellulose samples (Kratky, 1966), a model of a ternary system was utilized (crystalline and amorphous cellulose and air) and it was demonstrated that the volume share of the air pores in the sample did not exceed 0.7%. Investigations of loose cellulose samples, in which some samples were produced by swelling cellulose in an organic solvent with subsequent drying, serve as an example of the molecular structure analysis of the natural polymer and its cluster structure in the amorphous state. A detailed description of this investigation and references can be found in the papers of Kratky (1966, 1982).

Methods of isomorphous (isotopic) replacement are used for studying conformations of individual polymer chains in amorphous polymers by both X-ray and neutron scattering. In X-ray scattering this is achieved by preparing samples in which a small percent of molecules is modified by labels containing heavy atoms. In neutron scattering, the sample is prepared by dissolving a small number of deuterated molecules in an ordinary polymer. Isotopic exchange in neutron scattering is more convenient, since it introduces smaller distortions into the structure of modified molecules. Experiments with polymethylmethacrylate (Kirste *et al*., 1972) and polystyrene (Benoit *et al*., 1973) demonstrated that the dimensions of macromolecules were close to their dimensions in θ solvent, while the relationship between $\langle R_g^2 \rangle_z$ and $\langle M \rangle_W$ was linear. This is evidence in favor of the Flory theory (Flory, 1969), which describes the structure of amorphous polymers as aggregates of a randomly packed unperturbed Gaussian chain. Subsequent neutron experiments with various polymers have also confirmed this model; a zero value of A_2 was determined for glassy polymethylmethacrylate by Kirste *et al*. (1975). Similar results were also obtained using X-ray scattering. Thus, a number of experiments were conducted with polystyrene in a concentrated solution and in an amorphous state by dissolving copolymers of styrene and P-iodostyrene in polystyrene solution with toluene at different concentrations (Hayashi *et al*., 1976, 1977; Hamada *et al*., 1978). In order to eliminate the effect of iodine labeling, the data were extrapolated to zero iodine concentration. Data pertaining to the amorphous state (Hayashi *et al*., 1976) were correlated with results obtained by small-angle neutron scattering, and Hayashi *et al*. (1977) traced the transition from the unperturbed state of the chain in an amorphous polymer to the perturbed state in toluene solution by analyzing results for various concentrations of polystyrene solutions.

6.3. Crystalline Polymers

Crystallization from the melt or from the solution normally causes polymer substances to form structures in which crystalline regions are observed to alternate with parallel stacking of the macromolecules and amorphous structures, maintaining irregular packing of the chains. Electron microscope studies have indicated that the crystalline regions have the shape of lamellae with thickness about 10^2 Å. Crystalline lamellae with alternating amorphous layers form randomly oriented stacks within which they are packed in parallel. Data from small-angle scattering by these samples permit one to evaluate the general characteristics (such as thickness of layers and degree of crystallinity) and to model (simulate) specific features of the crystalline stacking of individual polymer chains.

6.3.1. Lamellar Model

The structure of an ideal crystalline polymer can be presented schematically as illustrated in Figure 6.6, where the cross section in the plane perpendicular to the lamellar plane is shown; C, A, and E are the thicknesses of the crystalline, amorphous, and transition layers, respectively. The density distribution in the stack can be described with the help of the one-dimensional function $\rho(z)$. For sharp boundaries, $E = 0$ and $\rho(z)$ assumes two values: density of crystalline (ρ_c) and amorphous (ρ_a) polymers, respectively, where $\rho_c > \rho_a$. Random orientation of the stacks leads to isotropic scattering by the object. If the stack dimensions are sufficiently large, scattering on their form is unobservable (see Section 2.2.1), and in this case we are faced with the averaged scattering on a two-phase lamellar structure.

There are two possible approaches to the interpretation of scattering data. The simplest consists of the utilization of general equations for a binary system. If ψ denotes crystallinity of the substance (ratio of the volume of crystalline regions to the irradiated volume of the sample) then, in the case of sharp boundaries ($E = 0$), equations (2.45) and (2.46) enable us to write

$$Q = 2\pi^2\psi(1-\psi)(\rho_c-\rho_a)^2V$$

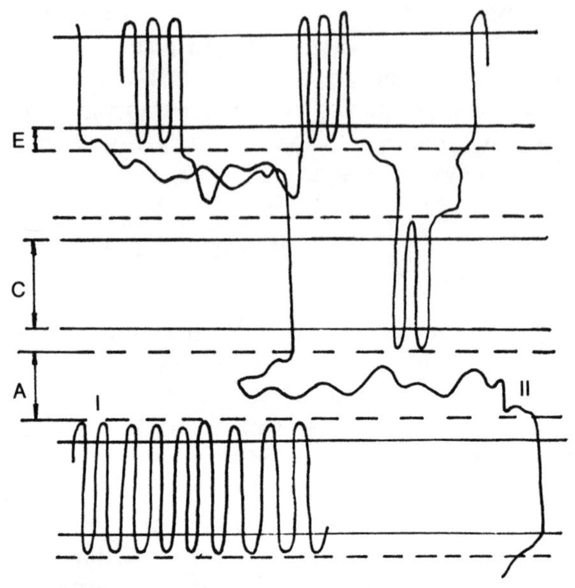

Figure 6.6. Structure of a crystalline polymer, where C, A, and E refer to crystal, amorphous, and intermediate states, respectively, and chains I and II illustrate different molecular arrangements.

and

$$S/V = \frac{\pi\psi(1-\psi)}{Q} \lim_{s \to \infty} [s^4 I(s)]$$

If $E \neq 0$ but the scattering density in the transition layers causes linear variations then (Vonk, 1982)

$$Q \approx 2\pi^2 \left[\psi(1-\psi) - \frac{E}{6}\frac{S}{V}\right] (\rho_c - \rho_a)^2 V$$

while the expression

$$I(s) \approx \frac{B}{s^4}\left(1 - \frac{2E^2 s^2}{3}\right)$$

is valid for the asymptote at large scattering angles (Ruland, 1971) and the analysis, for example, of the plot giving the relationship between $s^4 I(s)$ and s^2 allows the value of E to be assessed. It should be noted, however, that the estimates obtained with this interpretation cannot be correlated with the parameters of the lamellar model with a high degree of confidence. The value of ψ happens to be lower, as a rule, owing to the presence of the amorphous phase in the gaps between the stacks.

The second approach consists of the separation and analysis of the one-dimensional scattering intensity by an individual stack. Really, since scattering intensities by the stacks are distributed uniformly over the surface of a sphere in the reciprocal space, owing to their random orientation in the sample, the function

$$I_1(s) = 4\pi s^2 I(s)$$

complies with the averaged one-dimensional scattering intensity by one stack, namely, function $\{\mathscr{F}[\rho(z)]\}^2$. If the thickness of the crystalline and amorphous layers were not changed, there would be a strictly periodic system while $I_1(s)$ would contain a set of maxima in accordance with Bragg's law: $ns = 2\pi/(A + C) = 2\pi/D$ with n an integer. As a matter of fact, because of the finite dimensions of the stacks and the variation in the thickness of the layers, the maxima of function $I_1(s)$ are desmeared and shifted. No more than two maxima can be obtained for the isotropic samples, and the periods determined by their positions according to Bragg's law can differ substantially. Therefore, the problem of finding the parameters of the lamellar model will be discussed in a more general way.

6.3.2. Scattering by the Lamellar Stacks

It is clear that only average values can be determined experimentally when deviating from the ideally periodic lamellar structure. The lamellar-model parameters which one hopes to obtain from the small-angle scattering data include mean periodicity, mean thickness of the crystalline (C) and amorphous (A) layers, the widths of the distribution of layer thickness around the mean values ΔA and ΔC, as well as the mean crystallinity C/D. We first discuss the model containing N crystalline layers of equal thickness, while the position of their centers along the z axis can vary. If, for the sake of simplicity, we take $\rho_a = 0$, then designating $\rho_c(z)$ as the distribution of the scattering density in the crystalline layer and $p(z)$ as the distribution of the centers of the layers [$p(z)dz$ gives the average number of centers of the layers in the gap between z and $z + dz$ from a given center] (see, for example, Vonk, 1982) yields

$$I_1(s) = [NF^2(s)Z(s)] * V^2(s)$$

where $F(s) = \mathscr{F}[\rho(z)]$, $Z(s) = \mathscr{F}[p(z)]$, and $V(s) = \mathscr{F}[\Pi(z, L)]$; $\Pi(z, L)$ is the step function determining the stack dimensions. Function $p(z)$ can be expressed in the form

$$p(z) = \delta(0) + \sum_{i=-\infty}^{\infty} p_i(z)$$

where $p_i(z)dz$ is the probability that the center of the ith layer lies in the interval $[z, z+dz]$ from the given one. Function $p_i(z)$ is given by the i-fold self-convolution of function $p_1(z)$; therefore $I_1(s)$ can be expressed through the distribution function of the adjacent layers $p_1(z) = p_D(z) = \delta(z-D) * p_0(z)$, where $p_0(z)$ gives the distribution of distances between the layers in relation to the mean value D. Hence one can write

$$Z(s) = \mathrm{Re}\{[1 + G(s)]/[1 - G(s)]\}$$

where

$$G(s) = \mathscr{F}[p_0(z)]\exp(-iDs)$$

The relations obtained are readily generalized to the case in which the layer thickness can change. If $\langle F(s)\rangle$ is the average factor of layer thickness, then (Vonk, 1982)

$$I_1(s) = N[\langle F^2(s)\rangle - \langle F(s)\rangle^2] * V^2(s) + [N\langle F(s)\rangle^2 Z(s)] * V^2(s) \qquad (6.15)$$

These relations permit one to analyze the profile of the curve $I_1(s)$, which depends on the width and kind of distribution $p_0(z)$, as well as on the variations in thickness of the crystalline layers and sizes of the stacks.

Function $V^2(s) = [\sin(Ls/2)/(Ls/2)]^2$ determines the influence of the finite thickness of the stack. At higher values of L it is close to a δ-function, and does not give rise to substantial errors in the region of maxima; if $D \ll L$, then for $s \sim 2\pi/D$ we have $Ls/2 \sim \pi L/D \gg 1$. The finite sizes of the stacks can normally lead to a peak in the region of the smallest angles that should be disregarded.

Practical application of equations like (6.15) is associated with model calculations which, assuming the type of distribution to be $p_0(z)$, analyze the relationship between the parameters of the lamellar model and the position and width of the maxima on the curve $I_1(s)$. This approach was employed to develop a method by which the structural parameters could be evaluated directly from the scattering curve (Tsvankin, 1964). It is assumed that the thicknesses of the crystalline layer are distributed uniformly within the limits $[C - \Delta C, C + \Delta C]$, $p_c(z) = \mathrm{const}$, while the density transition from the crystalline to the amorphous layer is implemented linearly, i.e., the density distribution is trapezoidal. Then, equation (6.15) is replaced by

$$I_1(s) = N[\langle F^2(s, \varepsilon)\rangle - \langle F(s, \varepsilon)\rangle^2] + \langle F(s, \varepsilon)\rangle^2 I_2(s, \alpha, \beta)$$

where $\alpha = C/A$, $\beta = \Delta C/A$, and $\varepsilon = E/C$. Tsvankin (1964) set the first term constant and derived explicit expressions for $\langle F(s, \varepsilon)\rangle^2$ and $I_2(s, \alpha, \beta)$. Curves $I_1(s)$ were calculated on the basis of these equations for various values of the parameters; the relationships between these values were noted, as well as the position and width of the maxima on the curve $I_1(s)$. Specifically, calibration curves presenting values of $\eta = qd$ (where d is the position of the maximum and q its width), $k = C/(A + C)$, and X_m (calculated value of the maximum) as functions of p (calculated width of the maximum) are plotted for various values of $\beta/\alpha = \Delta C/C$ governing the variation in the thickness of layers around the mean value. It is possible for known values p and q to determine values of p, X_m, and k as well as

$$D = X_m d, \quad C = kD, \quad A = D - C$$

guided by the given value of β/α.

Subsequent development of this approach was suggested by Buchanan (1971), who modified the equations for $\langle F^2(s)\rangle$ and $\langle F(s)\rangle^2$ and took into account the first term in equation (6.15). He presented tables of calibration curves $\eta(p)$, $k(p)$, and $X_m(p)$ for various ratios β/α.

This method can be employed to determine the parameters of a lamellar model on the basis of experimental data, guided by the given value of the distribution width β/α; normally, the first maximum of function $I_1(s)$ is used. It should be noted, however, that this approach does not always produce reliable results for wide distributions (see, for example, Crist, 1973). In such cases one could use more complex models that take into account both the finite size of the stacks and the correlation between the distribution functions of the crystalline and amorphous layers (see, for example, Brämer, 1972). These approaches often yield better results with the experimental data.

6.3.3. Correlation Functions

Generally the position and width of the maxima are used in the above studies of function $I_1(s)$, but the remaining information on the function gets lost. An alternative approach is to study the corresponding correlation function. The expression

$$\gamma_1(u) = \gamma_1\left(\frac{z}{D}\right) = \frac{\psi}{1-\psi}\left[\frac{1}{\psi^2}\int_u^\infty (t-u)P_C(t)\,dt + P_{CAC} + P_{CACAC} + \cdots - 1\right]$$

was obtained (Vonk and Kortleve, 1967) for the function

$$\gamma_1(z) = \int_0^\infty I_1(s)\cos(sz)\,dz \Big/ \int_0^\infty I_1(s)\,ds$$

(analogous to the characteristic function of the thickness for flat particles;

see Section 2.4.5), where $P_{CACA\cdots C} = Q_C * P_A * \cdots * Q_C$,

$$Q_C(z) = Q^{-1} \int_z^{\infty} P_C(t)dt$$

$P_C(z)$ and $P_A(z)$ are distribution functions along the thickness of the crystalline and amorphous layers, respectively. The calculations were carried out for functions $\gamma_1(u)$ with respect to various widths of these distributions. An approximate relationship between $\gamma_1(u)$ and the distribution width is presented in Figure 6.7. It is clear that $\gamma_1(u)$ has maxima at values of z that are D-fold; sample crystallinity can be evaluated from the first minimum of this function. Kortleve and Vonk (1968) proposed a procedure for the determination of the structural parameters of a lamellar model by $\gamma_1(z)$ using $P_A(z)$ and $P_C(z)$ as two-parameter lognormal distributions of equal width. This procedure was compared with the method of Buchanan (1971) and the agreement was found to be quite satisfactory. The effects caused by various deviations from the ideal model were analyzed in subsequent works which involved function $\gamma_1(z)$ (see, for example, Vonk, 1978).

Another possibility is associated with a function similar to the chord distribution $G(l)$ (Section 2.4.2), namely, the interface distribution function $g_1(z) \sim \gamma_1''(z)$ (Ruland, 1977). The ideal form of the function is presented in Figure 6.8. It is evident that if the peaks in $g_1(z)$ do not overlap, the function can be used to determine directly the value of ψ as

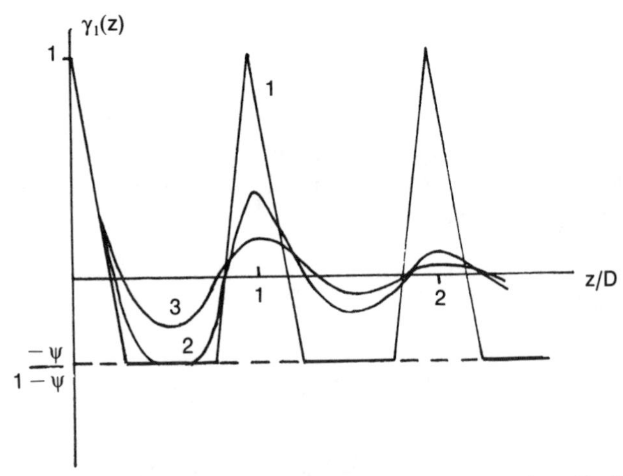

Figure 6.7. Correlation function of an ideal lamellar model for various distributions with P_A and P_C ($\psi < 0.5$): (1) corresponds to infinitely narrow distributions, (2) and (3) to gradual widening of the distribution function.

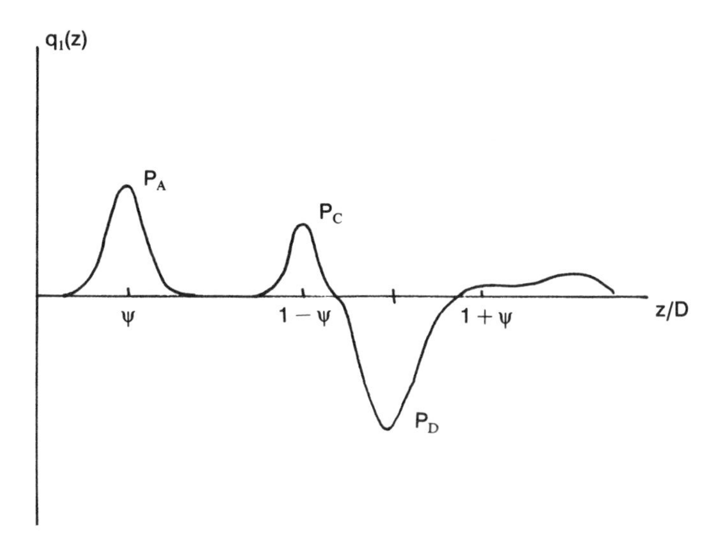

Figure 6.8. Intercept function for an ideal lamellar model.

well as the parameters of distributions P_A and P_C. Function $g_1(z)$ can be calculated from experimental data by the relation

$$g_1(z) = 4\pi^2 \int_0^\infty [s^4 I(s) - c_4]\cos(zs)ds \Big/ \int_0^\infty s^2 I(s)ds$$

Various model calculations were carried out for specific samples (Stribeck and Ruland, 1978). It was shown that with appropriate processing of the scattering curve (taking into account possible heterogeneities inside the phases and the thickness of the transition layer, accurate determination of c_4) the corresponding functions $g_1(z)$ are very convenient for determining structural parameters of the lamellar model.

6.3.4. Determination of Chain Folding

The formation of crystalline lamellae is of interest in studying crystalline polymers. For example, the question arises whether they are formed from the chains basically belonging to the same lamella (adjacent reentry, chain I in Figure 6.6) or, say, the same chain is built randomly into different lamellae (random folding, chain II in Figure 6.6). The individual properties of the chains in a crystalline polymer can be studied by small-angle neutron scattering, similar to studies of amorphous polymers. A small fraction of deuterated (or protonated, with deuterated matrix) chains, selected randomly during crystallization, is added to the solution or the melt from which the polymer is being crystallized. Determination of small-angle scattering for such crystals permits one to separate the scattering from the individual chains.

Samples of crystalline polyethylene were used as the main object in such investigations. The conformation is determined by comparison with experiment of model scattering curves for various selected methods of stacking the chains. Yoon and Flory (1977) have shown that agreement cannot be reached with experiment for polyethylene crystallized from the melt within the model framework with the adjacent reentry model. Investigations of samples crystallized from solutions (Sadler and Keller, 1977) have exhibited somewhat better agreement between this model and experiment; however, substantial differences were noted in the region of small s.

Various models were examined comprehensively by Yoon (1978) for samples crystallized from both the solution and the melt. The models were calculated using the Monte Carlo method for various periodicities (100–250 Å) and various models (adjacent reentry, random walk, adjacent return to the same lamellae with a certain degree of probability). The results permit one to state that the best agreement can be reached for the solution-crystallized sample model in which the chain returns to the same lamella with 0.7 probability; however, the site of the reentry is very remote from the site of emergence. In the case of crystallization from the melt the model of nearby (but not regularly folded) reentry gives good agreement with experimental data.

6.4. Anisotropic Systems

So far we have discussed isotropic samples. However, when studying polymer objects the samples may very frequently contain a certain predominant arrangement of molecules. Then, of course, the scattering pattern also becomes anisotropic. Such a pattern in principle contains more information related to the structure, but the methods of its analysis may be more complicated. In this section we discuss briefly the main types of anisotropic polymer studied by small-angle scattering as well as methods of structural data interpretation in these cases.

6.4.1. Oriented Amorphous Polymers

Orientation of polymer molecules in an amorphous state can be achieved by stretching the sample in a certain direction (z). Such investigation methods are, for example, of great importance for vulcanized rubber and caoutchouc. When deformation during stretching is regarded as possessing an affine character, the radius of gyration of a molecule in directions parallel and perpendicular to the stretching can be expressed in the form

$$(R_g^2)_z = (R_g^2)_\parallel = (R_g^2)_0 \lambda_\parallel^2, \qquad (R_g^2)_\perp = (R_g^2)_0 \lambda_\perp^2$$

where $(R_g)_0$ is the radius of gyration in an unstressed state and $\lambda_\parallel$ is the macroscopic coefficient of linear expansion along the axis. The condition $\lambda_x \lambda_y \lambda_z = 1$ yields $\lambda_\perp = 1/\sqrt{\lambda_\parallel}$. The deformation affinity is checked through small-angle neutron scattering, similarly to investigations of conformations of the chain in amorphous and crystalline states. The prepared sample contains a certain quantity of deuterated macromolecules. The heated sample is then stretched and cooled, and the anisotropic pattern of scattering recorded. Measurements on polystyrene (Picot *et al.*, 1977) indicated that the deformation was affine in a perpendicular direction and somewhat nonaffine in a parallel direction. Measurements of large λ are impaired due to the formation of fractures and voids in the sample and the appearance of scattering caused by them.

Measurements of oriented polymers are also possible in the solution. A method of polymer orientation in the shear gradient is described by Linder and Oberthür (1984). The polymer solution is placed in the gap between coaxial quartz cylinders. Then, by rotating one of them a shear gradient of the solution flow velocity is formed, which causes the polymer chain to alter its conformation. At not too high rotation velocities the macromolecules acquire a new stable conformation the analysis of which yields conclusions about the flexibility and mobility of the chains.

6.4.2. Fibrillar Systems

Fibrillar systems are formed by long polymer molecules that aggregate in parallel (or nearly in parallel) to a chosen direction (z-axis). The fibrils formed during aggregation of the molecules are also oriented predominantly along this axis. These structures are formed by a number of polymers of biological origin (collagen, keratin, muscles) as well as natural polymers (cellulose) and appropriately treated samples of synthetic polymers. Owing to arbitrary azimuthal orientation of the molecules, the samples usually possess cylindrical symmetry in relation to the z axis.

Small-angle scattering by these objects contains, as a rule, a set of meridional reflexes conforming to periodicity along the fibril axis. Analysis of this pattern usually includes finding the periodicity and, if possible, the function $\rho(z)$ which gives the projection of the scattering density on the axis.

Study of the structure of collagen (fibrillar protein contained in tendons, bone, skin, etc.) is a classic example of the investigation of fibrillar structure. Diffraction patterns of scattering by collagens permit one to record up to one hundred reflection orders of large periods D (ranging from 640 to 700 Å for different collagens), indicative of higher ordering along the z axis. This periodicity does not agree with the length of the collagen molecule (about 3000 Å) and is explained by regular shifting of the molecules by the value of the period when forming the fibrils (Figure 6.9). Determination of the signs of the meridional amplitudes (for example, by a trial-and-error method) permits a density projection profile to be built up inside one period attesting to the fact that a large period is formed by combination of the region of tropocollagen fibrils (triple helix of the collagen molecules) and "the gap." Neutron studies of mineralized collagen samples with the help of contrast variation shows that salts penetrate into the gaps between the fibrils (White *et al.*, 1977; see Figure 6.10). Known primary and secondary structures of the collagen permit one to examine fine components of the fibrils (see, e.g., the review of Stuhrmann and Miller, 1978).

The oriented samples of many biopolymers and their complexes have ordering not only in the direction of the fibril axis, but also in the direction of the plane perpendicular to it. Thus, the fibers of muscles are made of microfibrils, each consisting of regular combinations of thick threads (myosin) and thin threads (actin); their packing is hexagonal. Therefore the diffraction pattern, apart from being meridional, has a number of other reflexes as well. Analysis of scattering by these structures permits the basic model of muscular contraction to be improved. At present it is possible to carry out time-resolved measurements at various stages of muscle contractions using synchrotron radiation (Huxley *et al.*, 1982; Gadjiev and Vazina, 1984). It is not our purpose to present in

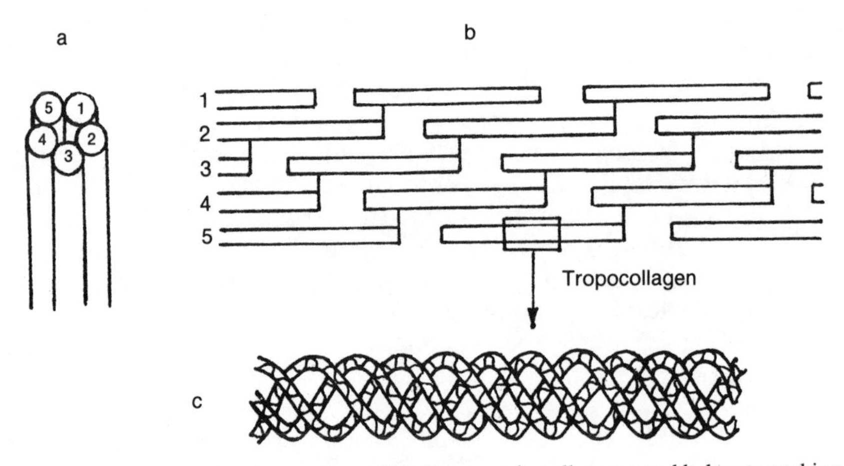

Figure 6.9. Scheme for the structure of fibrillar protein collagen; a and b denote packing of tropocollagen fibrils and c their ternary helix.

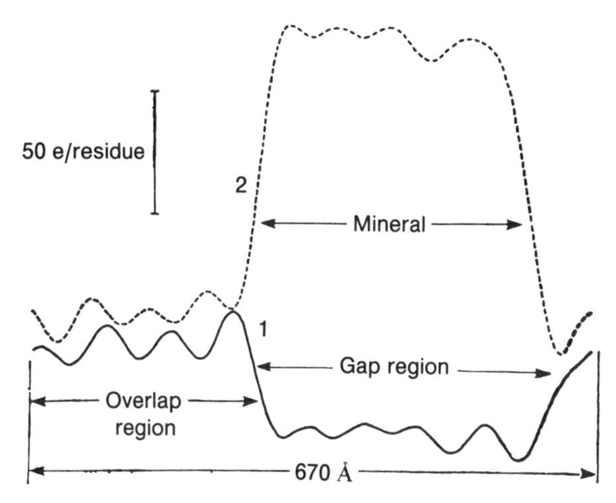

Figure 6.10. Scattering density within a period of turkey leg tendon (after Stuhrmann and Müller, 1978): (1) native state; (2) calcified state.

detail methods for interpreting such scattering patterns rich in reflexes. [Small-angle scattering aspects of the analysis of muscle structure can be found in Farugi and Huxley (1978).] Features of the helical macromolecule structure (such as DNA) will also not be considered here. In this case the small-angle scattering region alone is insufficient for an interpretation (Vainshtein, 1966).

Oriented fibrillar polymers of nonbiological origin are obtained mainly in the course of deforming isotropic crystalline polymers. Changes in the structure of crystalline polymers were studied during their monoaxial stretching or contraction, including reverse transitions with shrinkage of oriented samples (Gerasimov and Tsvankin, 1969; Genin *et al.*, 1973; Gerasimov *et al.*, 1974). The shape of small-angle scattering reflexes (Figure 6.11) was shown to depend substantially on the form and orientation of separate crystalline regions (crystallites). Typical structural transition of isotropic lamellar structures into oriented fibrillar structures was shown to occur during both stretching and contracting by comparing experimental data with X-ray images calculated for various structural modes. The reverse transition, studied for thermal shrinkage of films, contains an intermediate stage of the oriented lamellar structure, which then transforms to the isotropic structure. It should be noted that the value of a large fibrillar period mainly depends upon the deformation conditions instead of the period D of the lamellar model (Section 6.1.3). Lamella–fibril transitions are evidently associated with the planes sliding relative to each other, along which folding of the polymer chains occurs in the crystalline layers.

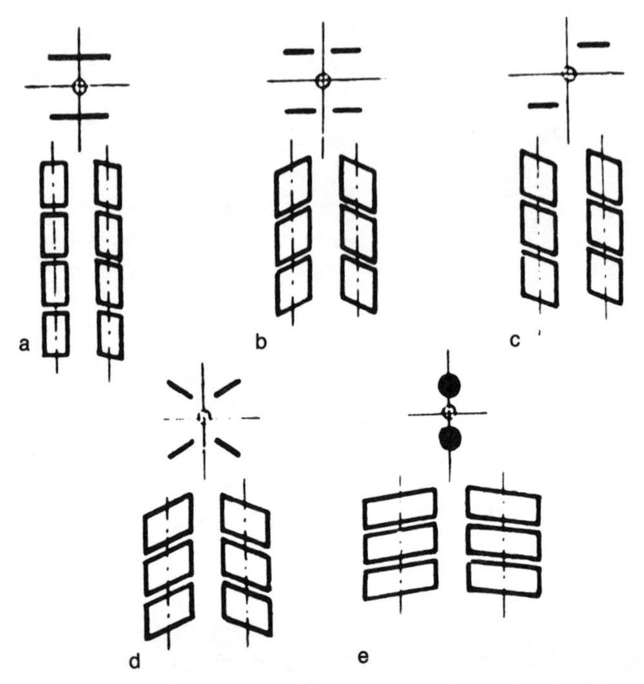

Figure 6.11. Structure models of crystallites and shape of small-angle reflections. Structures with axial texture: (a) stoke; (b) four-point reflection; (c) two-point raflection; (d) inclined four-point reflection; (e) spherical reflection (after Gerasimov *et al*., 1974).

6.4.3. Lamellar Systems

A number of substances (crystalline polymers, membranes, various liquid crystals) have a lamellar structure. The possibilities for structural investigation of such substances through the use of small-angle scattering depend, to a great extent, on the ordering of the lamellae positions: the higher the degree of orientation in relation to a certain direction, the greater the amount of information that can be obtained regarding the lamellar structure. Thus, substantially more information (as compared with the information obtained when studying isotropic crystalline polymers) can be obtained in the course of studying polymer films produced through consistent stretching, rolling, and annealing of crystalline polymers at various temperatures (see, for example, Pope and Keller, 1975). Such films consist of crystalline layers oriented parallel (or almost parallel, depending on the temperature) to each other. Their stretching leads to a change in the distances between the lamellae and the angles of their deflections, which are easily examined on a diffraction pattern.

Nevertheless, it is not possible to formulate the problem of restoring the density profile $\rho(z)$ along the lamellar thickness when studying even completely oriented crystalline polymers, since the system is not strictly periodic. For oriented lamellar systems with equal motives, the diffraction pattern contains a number of meridional reflexes (see Section 1.3.5). Solving the phase problem for this set permits the density distribution to be restored within the limits of the lamellar-system period. Studies of photoreceptor cells of vertebrates consisting of stacks of disks formed by photoreceptor membranes are examples of such investigations. The photoreceptor rod contains up to 1000 such disks, with the membranes containing about 50% protein (mainly rodopsin; Figure 6.12). The ordering of disk positions permits one to record up to 10–20 reflection orders. The periodic function $\rho(z)$ of the scattering density distribution is symmetric with respect to the center of the disk and can be

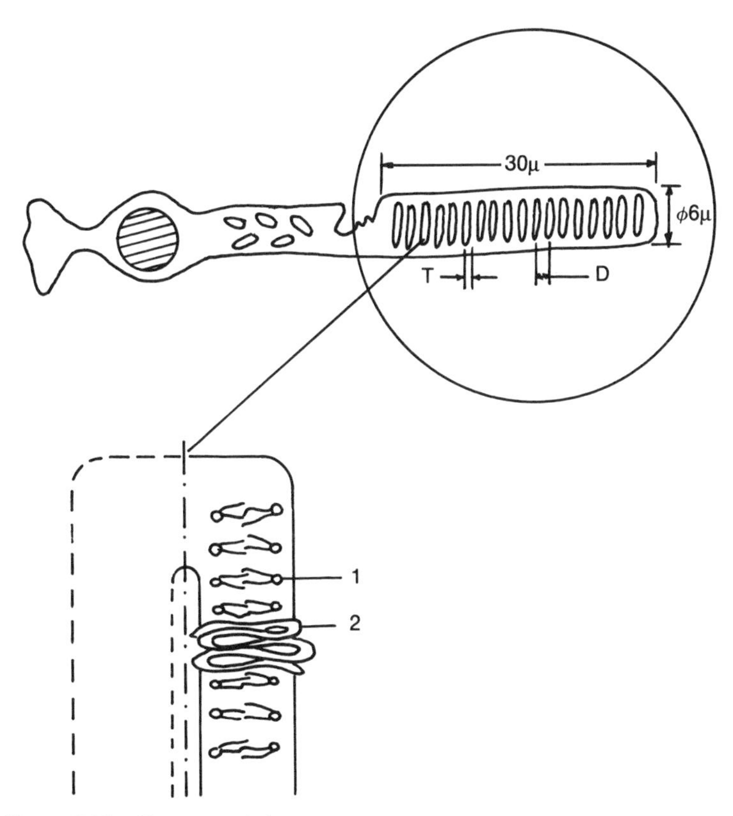

Figure 6.12. Structure of photoreceptor rod and disk: (1) lipids; (2) rodopsin.

determined from the experimental data by the expression

$$\rho(z) = \frac{2}{D} \sum_{h=1}^{N} \pm (I_h L_h)^{1/2} \cos\left(2\pi z \frac{h}{D}\right)$$

where D is the period (for photoreceptor rods $D \approx 300$ Å), h is the number of the reflection and I_h its integral intensity, while L_h is the Lorentz factor different from unity with disordering of the rods in the sample (the origin of the coordinates is placed at the center of the disk). The problem here consists of determining the signs of the corresponding structural amplitudes. The solution of this problem very often necessitates simply sorting out possible sign combinations while taking into account that $\rho(z)$ should assume a permanent value in the interdisk space (filled with the solvent). Other approaches to determining the signs include the use of contrast variation in neutron scattering (Chabre *et al.*, 1975) as well as X-ray experiments using a sample with various degrees of swelling (i.e., with different periods D), thus permitting one to construct a continuous function $\pm [I(s)]^{1/2}$ and determine with more confidence the points at which the sign changes (Franks and Lieb, 1979). Another approach to the calculation of $\rho(z)$ is associated with the one-dimensional Patterson function, either for direct phase determination by the recursive method (Worthington, 1973) or for restoration of the continuous intensity distribution and recording the "zero" reflex value (Worthington, 1981).

Determination of differences in membrane density profiles for dark-adapted and decolorized (illuminated) states is the basic problem in research on photoreceptor disks. With this aim experiments were carried out using both separated retinas and the retinas of live vertebrates (frogs). However, a number of investigations (such as Blaurock and Wilkins, 1969, 1972) show no substantial changes in the profile $\rho(z)$; an insignificant increase (about 5 Å) of a large period and a small shear of the density occur, as a rule, in the external part of the disk toward the interdisk space. A light-damaged state at high light irradiation was also investigated by Krivandin *et al.* (1981). The density profile in such a state contains substantial changes on the external surface of the membrane (Figure 6.13) and can be explained by the irreversible aggregation of rodopsin molecules. It should be noted that the investigation of the photoreceptor membrane structure shows a number of methodical features. Although various, previously discussed methods of phase determination usually lead to identical sets of signs, the necessity of recording very weak effects with light irradiation requires strict accounting of various errors. Thus, the theoretical value of the Lorentz factor for strong deorientation of rods $L_h = h$ is very often found to be unsuitable (see, for example, Blaurock and Wilkins, 1969). More exact determination of $\rho(z)$ can be achieved by taking into account possible deviations from strict periodicity in the

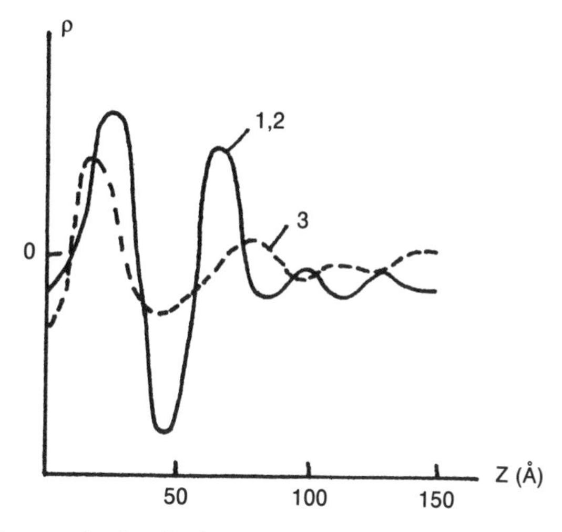

Figure 6.13. Electron density distribution of a photoreceptor disk of *Rana temporaria* frog: (1) dark-adapted; (2) decolorized; (3) light-damaged retina (after Krivandin *et al.*, 1981).

packing of disks (Schwartz *et al.*, 1975). By and large, work involving the investigation of details of the photoreceptor membrane can serve as a good illustration of structural studies of ordered lamellar systems.

6.5. Conclusion

A large number of various physical and physicochemical methods (nuclear magnetic resonance, light scattering, electron diffraction, UV spectroscopy, fluorescent labels, osmometry, ultracentrifugation, differential thermal analysis, and others) are used for studying the structure of polymers. It was not the aim of this chapter to compare these methods with the method of small-angle scattering of X rays and neutrons or to present the theoretical bases of polymer physics and chemistry. We mainly tried to answer questions regarding the potential uses of small-angle scattering to determine structural characteristics of polymers in various states and to outline the possibilities and limitations of the method as applicable to polymer substances.

The light-scattering method is closest to the small-angle scattering techniques and involves the same relations for visible light diffraction with the exception of a change in the scattering density concept (see Section 4.4.1). Historically, this method was employed first for polymers, and therefore many ideas (such as the Zimm plot) were taken from it. It was originally thought that light scattering is applicable only to sufficiently large ($R_g \sim 10^3$ Å) macromolecules, but the possibility of its application to particles with $R_g \sim 10^2$ Å was demonstrated some years ago (see, for example, Serdyuk and Grenader, 1974). Therefore, it is often possible to combine light scattering results with the use of small-angle X-ray and neutron scattering (see Section 6.2.1 and also Section 4.4.1). A comparative survey of work pertaining to these methods can be found, for example, in the review of Higgins and Stein (1978).

In conclusion, we note that a more profound analysis of polymer structures very often necessitates a special investigation of these features. Thus, in the study of aggregates of helical macromolecules of biological origin (nucleus acids, β-structures of proteins) a separate analysis is required for specific scattering features from such molecules and their aggregates (Vainshtein, 1966). A paracrystal conception (Hosemann and Bagchi, 1962) is found to be very useful when studying the disruption of the packing of molecules in fibrillar and lamellar structures. In this chapter we did not elaborate on these special approaches to the interpretation of scattering patterns by partially ordered systems since, as a rule, they require information on reflexes in the regions of large-angle scattering.

7

Structural Studies of Inorganic Materials

The structure of inorganic materials is a very important field for small-angle scattering application from both fundamental and applied aspects. Inhomogeneities of submicron sizes (10^1–10^4 Å), which can be identified by the small-angle scattering technique, exist or develop in many classes of inorganic substances in response to various impacts. The inhomogeneities are represented by, for example, defect agglomerates in single crystals, phase separations in crystalline and amorphous systems, particles and pores in powders and ceramics, and high-density clusters and statistical fluctuations in glasses and liquids. The size and distribution pattern of inhomogeneities exercise noticeable influence on the physical properties of inorganic materials.

A great variety of structurally different objects can be studied by small-angle scattering, and this is responsible for the fact that no general methods for interpreting data on small-angle scattering from inorganic objects are available at present. Although in many cases it is still possible to adopt the approximation of particle scattering or, for example, of the binary system, the structural interpretation of small-angle scattering data requires, as a rule, careful consideration of individual features of the object under study. Moreover, sometimes it is difficult to formulate even the direct problem of small-angle scattering (how to calculate the scattering intensity), since the fundamental properties of a specific condensed phase must be selected from a wide range of equally significant physical parameters. For this reason, efforts in the field of small-angle scattering studies of inorganic condensed phases have actually been confined to a series of experimental applications to different classes of objects.

At the same time, the phenomenon of small-angle scattering from a sample is highly informative for the study of inorganic substances. This phenomenon indicates the presence of inhomogeneities of colloidal size in the specimen. It allows evaluation of their distribution pattern (typical sizes) and magnitude (contrast) regardless of the actual structure of a sample. Therefore, the origin and disappearance of small-angle scattering by a sample subjected to different physical impacts is a highly promising approach to the study of inorganic materials.

The present chapter covers the main aspects of small-angle scattering applications to structural studies of various inorganic substances. Section 7.1 offers a review of methods for studying the superatomic structure found in crystalline substances. Section 7.2 considers one of the most important (mainly in terms of applied studies) trends in small-angle scattering of highly dispersed systems, that is, the calculation of the particle-size distribution function. This technique is essential in small-angle scattering studies of porous bodies, powders, and colloidal solutions. The methods for studying amorphous substances and liquids are practically identical and are discussed in Section 7.3.

7.1. Crystalline Materials

Small-angle scattering studies of crystalline substances show altogether different aspects of these materials as compared to X-ray and neutron structure analysis. Small-angle scattering experiments are aimed at measuring the scattering intensity near the origin of the reciprocal lattice (see Section 2.1.1). In contrast to the determination of atomic crystal structure, these experiments are focused on the identification of disturbances in this structure, the characteristic sizes of which considerably exceed the interatomic distances in the crystal (e.g., clusters, dislocations, phase separation in alloys). The present section covers the basic methodical aspects of these studies. (For more detailed reviews of small-angle scattering investigations of crystalline objects, see Gerold and Kostorz, 1978; Kostorz, 1982.) A serious problem in small-angle scattering studies is presented by the occurrence of strong double Bragg scattering, which suppresses the "structural" small-angle scattering. In the case of perfect crystals, the double scattering can be avoided by means of appropriate orientation of the sample. In studies of polycrystalline objects, the basic solution lies in the use of radiation with a wavelength more than twice the lattice spacing. This is possible only in the case of cold-source neutron beams, and sometimes in synchrotron radiation studies (see Chapter 8).

7.1.1. Defects in Single Crystals

The presence of individual point defects in the crystal matrix produces only scattering which does not depend on the angle in the small-angle scattering range. However, high concentration of defects, achieved, for example, by irradiation of single crystals with fast neutrons or other particles, can cause vacancy clustering. Vacancy clusters exhibit a lower scattering density than the crystal matrix as well as a distinct small-angle scattering. Analysis of the latter gives information on the size and, sometimes, the shape of these clusters.

Haubold and Martinsen (1978) described small-angle X-ray scattering research on copper single crystals serving as an example of detection of vacancy clusters. The samples were exposed to 3-MeV irradiation with electrons at a temperature of 4 K with subsequent annealing at different temperatures. In this experiment, at temperatures below 200 K scattering was caused by point defects alone. Further increase in temperature brought about scattering from the vacancy clusters. Figure 7.1 shows Guinier plots for samples annealed at 200 K, 260 K, and 300 K, and the corresponding radii of gyration of clusters. At $T = 300$ K the clusters contain some 50 vacancies and the small-angle scattering curve shows distinct secondary peaks, indicative of their more or less uniform size (about 7 Å).

Larger voids are observed in single crystals of aluminum irradiated with fast neutrons. Although the voids have been detected by small-angle X-ray scattering and transmission electron microscopy (Epperson *et al.*, 1974), the most complete data were obtained by means of small-angle neutron scattering (Hendricks *et al.*, 1974) with a wavelength of 8 Å greatly exceeding the limit for the occurrence of Bragg reflections (4.6 Å). In this experiment three linear position-sensitive detectors were placed in parallel. This allowed one to obtain an anisometric pattern and to propose a model of the void shape — a truncated octahedron with a radius of gyration of 215 Å. Further experiments contributed to the improvement of the model and gave rise to an assumption that the structure and size of the voids depend greatly on the transformation of a small (about 0.1 at%) amount of aluminum atoms into silicon atoms during exposure. Figure 7.2 reflects the dependence of the void radius of gyration and swelling (relative increase in their volume) on annealing temperature obtained from small-angle X-ray scattering data (Hendricks *et al.*, 1977; Lindberg *et al.*, 1977). The diagram shows that a sharp growth of voids or an alternation in their shape starts at $T \sim 306$ °C close to the melting point of Al–0.1 at% Si alloy. These data imply that at temperatures below 306 °C the growth of voids is stabilized, owing to the precipitation of silicon atoms on their surfaces.

Dislocations are another type of defect in single crystals studied by small-angle scattering. Small-angle neutron scattering techniques with long wavelengths should be used, since the scattering intensity by dislocations is very low and the effect of double Bragg scattering ought to be completely eliminated. The scattering intensity of unpolarized neutrons

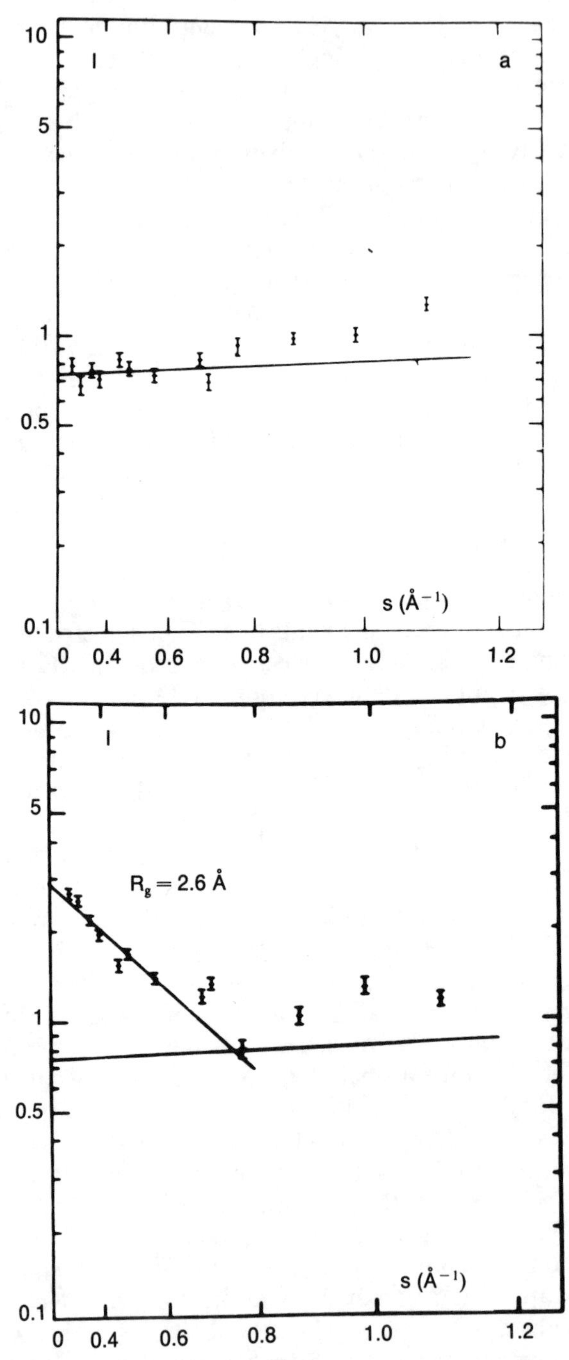

Figure 7.1. Guinier plot of small-angle X-ray scattering by a single crystal after electron irradiation and annealing at various temperatures: 200 K (a); 260 K (b); 300 K (c) (after Haubold *et al.*, 1978).

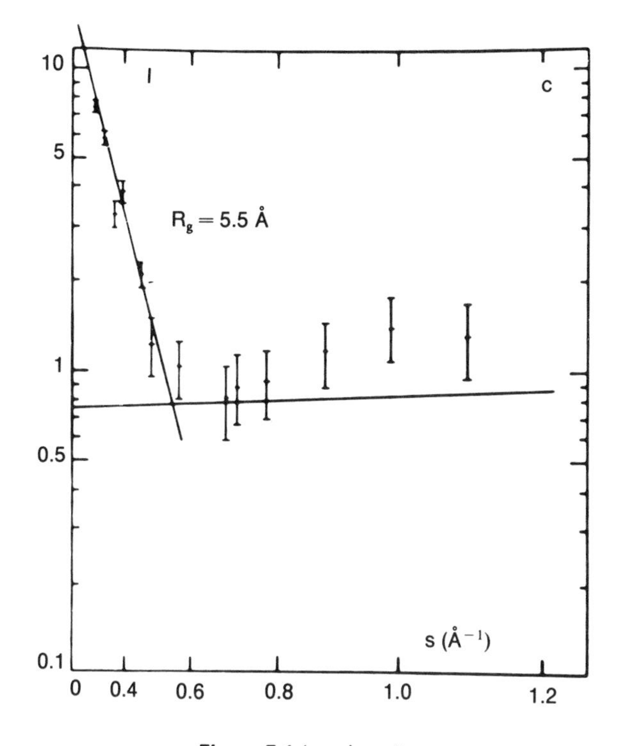

Figure 7.1 (continued)

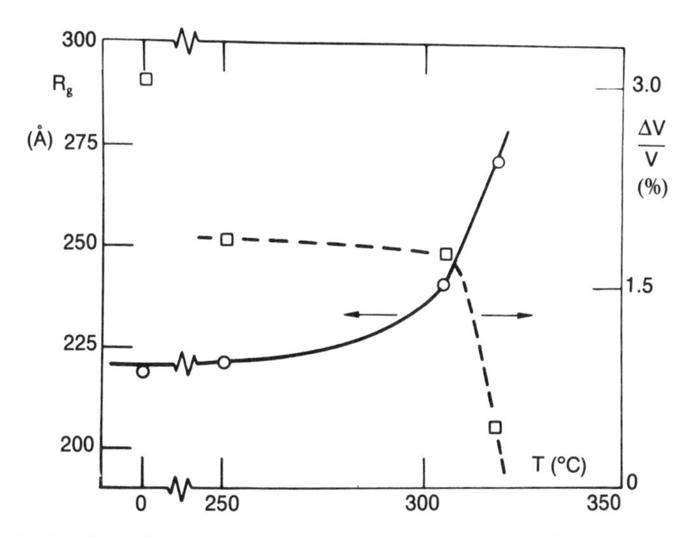

Figure 7.2. Radius of gyration of voids (R_g) and swelling ($\Delta V/V$) as a function of annealing temperature for an Al single crystal irradiated with fast neutrons at 55 °C (after Lindberg *et al.*, 1977).

by dislocations in a ferromagnetic metal sample with N atoms may be expressed as (Schmatz, 1975)

$$I(\mathbf{s}) = \frac{1}{N}[b_n^2|A_n(\mathbf{s})|^2 + b_m|A_m^{\perp}(\mathbf{s})|^2]$$

where b_n and b_m are the nuclear and magnetic scattering lengths (see Section 1.5), and $A_n(\mathbf{s})$ and $A_m(\mathbf{s})$ are the Fourier transforms of nuclear and magnetic density fluctuations caused by dislocations; $A_m^{\perp}(\mathbf{s})$ is the component perpendicular to the vector $\mathbf{s}$.

In an elastically isotropic crystal (Seeger and Kröner, 1959)

$$A_n(\mathbf{s}) = i\frac{1-2\nu}{1-\nu}\frac{[\mathbf{bs}]}{V_a s^2}\int_L \exp(i\mathbf{sr})d\mathbf{r}$$

where V_a is the atomic volume, $\mathbf{b}$ is the Burgers vector, and L is the dislocation line. This expression allows the intensity of nuclear scattering to be calculated for random arrangement of dislocations in the sample. Moreover, it shows that screw dislocations do not contribute to the scattering pattern, while the contribution of edge dislocations is highest when $\mathbf{b}$ is perpendicular to s. Experiments with small-angle nuclear scattering by dislocations are designed mainly to measure the polar scattering diagram (correlation between the intensity and the scattering vector orientation at fixed $\mathbf{s}$) obtained from samples deformed to different shear stresses. Changes in polar diagrams present a qualitative evaluation of the dislocation arrangement with respect to the slip plane.

Magnetic neutron scattering by dislocations in ferromagnetics exceeds considerably the nuclear scattering (screw dislocations contribute to small-angle magnetic scattering). Nuclear and magnetic scattering effects can be distinguished by parallel measurements of the scattering curves for magnetic fields near the saturation point and well above the saturation field (where the magnetic scattering vanishes). Monitoring of polar diagrams in certain cases provides sufficient data with which to compile models of the prevailing dislocation arrangement in the sample.

It should be noted, however, that scattering effects from dislocations are weak and very difficult to detect, even in the case of perfect single crystals; it is even more difficult in the case of polycrystals. For this reason small-angle scattering experiments generally admit of a certain qualitative evaluation only. For a more detailed review on the theory and practice of dislocation studies see, for example, Schmatz (1975).

7.1.2. Phase Separation in Alloys

The study of alloys is one of the most significant trends in small-angle scattering applications. The very first small-angle scattering study of Guinier (1938) was focused on the study of the structure of alloys (Al–Cu and Al–Ag). This work provided evidence for the existence of areas of coherent precipitates (Guinier–Preston zones). Small-angle scattering studies of alloys are of great practical value, since submicron phase separation occurring in alloys as a result of aging, annealing, and so on, to a great extent affects their physical properties (yield stress, resistivity, magnetic susceptibility).

Two principal mechanisms have been proposed for coherent precipitation in alloys, i.e., phase separation within the crystal lattice with only elastic distortions. One of them is the nucleation and growth of a new phase, while the other, which occurs at temperatures below the spinodal temperature T_s, is spinodal decomposition, namely, the growth of the amplitude of concentration fluctuations with wavelengths exceeding some critical value λ_c. These mechanisms exhibit different kinetics of scattering in time. Phase separation in the metastable solid solution at $T > T_s$ starts with nucleation and is followed by the growth of the nuclei (that is, the nucleation and growth of particles within the matrix). At $T < T_s$ the solution is unstable and spinodal decomposition starts immediately, accompanied by changes in the density but not in the size of concentration inhomogeneities. According to Cahn's linear theory of spinodal decomposition (Cahn, 1962), the variation in the small-angle scattering intensity with time is given by

$$I(s, t) = I(s, 0)\exp[2R(s)t]$$

where $R(s)$ is the amplification factor (factor of the growth rate of the concentration wave with wavelength $2\pi/s$), which is related to the thermodynamic properties of the alloy.

In a number of cases small-angle scattering studies allow one to evaluate mechanisms of coherent scattering in alloys both quantitatively and qualitatively. Figure 7.3 shows an example of small-angle X-ray scattering research on Al–22 at% Zn alloy (Agarwal and Herman, 1974). The samples were melted and quenched; aging was studied at 65 °C. Figure 7.3a presents the resulting small-angle scattering curves, which agree very well with spinodal decomposition theory (the crossover point corresponds to the critical wavelength λ_c); Figure 7.3b shows the calculated function $R(s)$.

Two separation mechanisms were compared in the small-angle neutron study of manganese–copper alloy (Vintaikin *et al.*, 1978). In this experiment, samples with different manganese content were analyzed

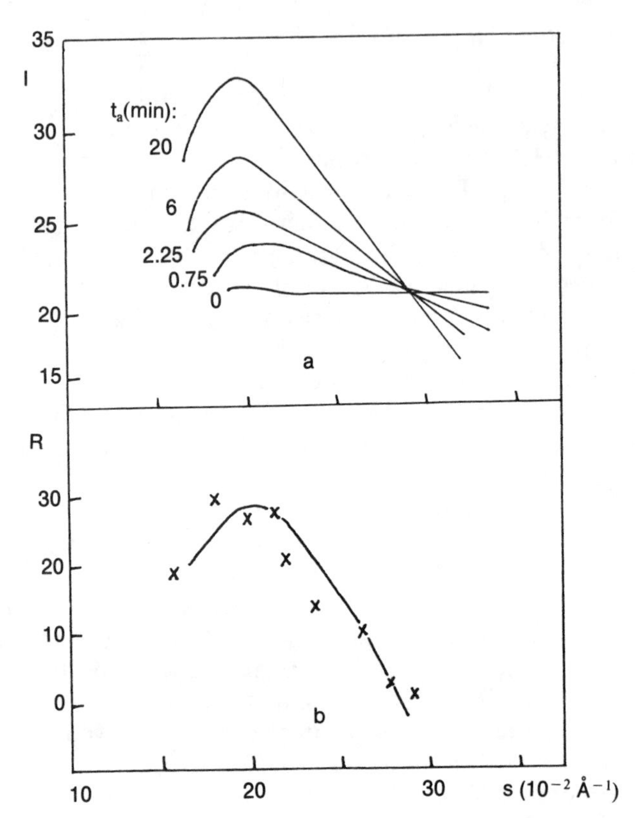

Figure 7.3. Small-angle X-ray scattering analysis of an Al–22 at% Zn alloy from the liquid state aged at 65 °C (after Agarwal and Herman, 1974): (a) evaluation of intensity profile $I(s)$; (b) amplification factor $R(s)$ derived from (a).

after quenching and annealing at 380 °C. In the case of the first sample (48.5 at% Mn) this temperature was higher than the theoretical value of T_s, but in the other samples (59.7, 66.8, 75.1 at% Mn) it was lower. Figure 7.4 shows small-angle scattering curves for the four samples after quenching and annealing over different periods of time. The first sample (Figure 7.4a) exhibits a shift in the scattering peak toward smaller angles at the initial stages of annealing, which is evidence of the growth in size of the inhomogeneities. At the same time, none of the other samples produce a distinct shift in the scattering peaks, in accordance with Cahn's linear theory. Besides, the asymptotic behavior of the scattering intensity of the first sample at high s is proportional to s^{-4} even at the initial stages of annealing, and this indicates the presence of particles with sharp boundaries. Other samples after annealing times less than 6 h show a growth in

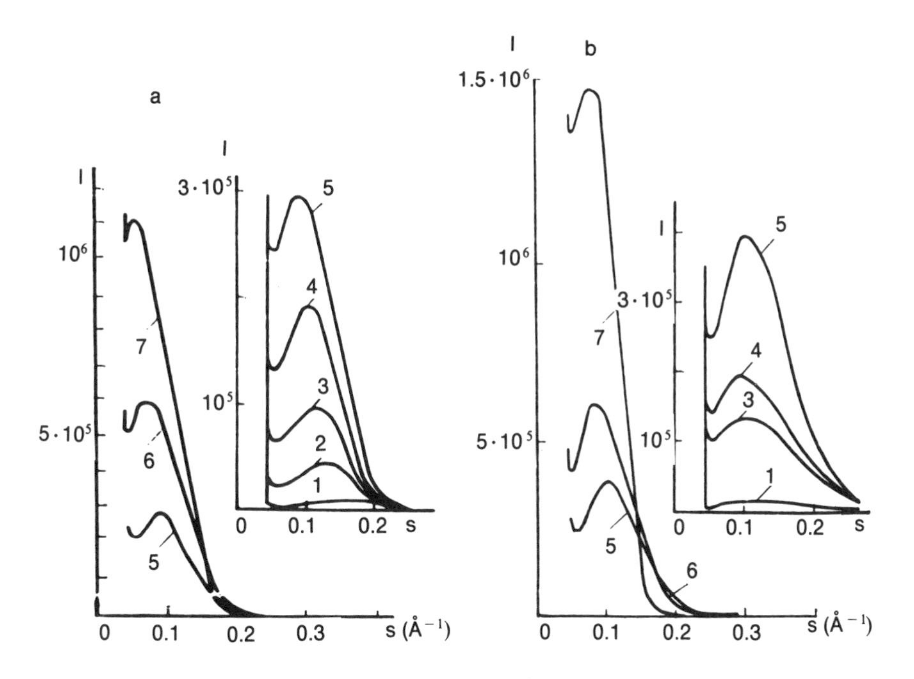

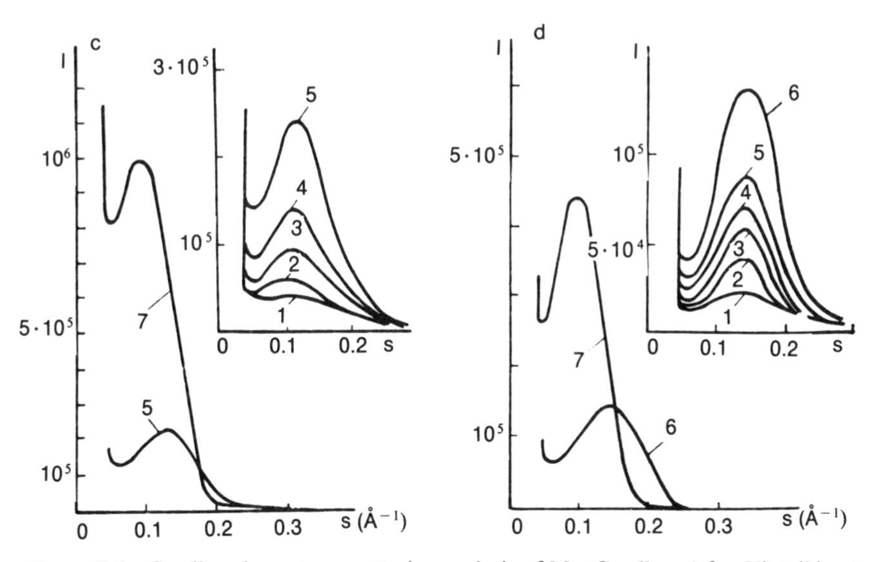

Figure 7.4. Small-angle neutron scattering analysis of Mn–Cu alloys (after Vintaikin *et al.*, 1978): (1) scattering after quenching; (2)–(7) correspond to annealing at 380 °C over 0.5, 1, 2, 4, 6, and 10 h, respectively. Alloy composition at% Mn: (a) 48; (b) 59.7; (c) 66.8; (d) 75.1.

function $s^4I(s)$ with growth in s, which means that phase interfaces in these samples are somewhat smeared. On the basis of the results obtained the general conclusion that the first alloy decomposes in accordance with the mechanism of nucleation and growth of a new phase, while the others exhibit spinodal decomposition, was drawn.

A number of small-angle scattering studies have aimed at determining the separation mechanism in alloys (see, for example, Gerold and Kostorz, 1978). It should be noted, however, that it is often difficult to interpret clearly small-angle scattering data by means of a single mechanism of decomposition, in particular, because of the limitations of Cahn's theory. (For a more general expression of the time dependence of the scattering intensity, see Cook, 1970.)

Small-angle neutron scattering also enables the study of magnetic properties of alloys during phase separation. This problem has been tackled by Vintaikin *et al.* (1977), who investigated the impact of thermomagnetic treatment (TMT) on the phase-separation features of a ternary solid solution iron–chromium–cobalt. The samples were quenched in water from a temperature of 1300 °C and annealed in a magnetic field of 6 kE. Figure 7.5 shows small-angle scattering curves for demagnetized samples of Fe–31.0 wt% Cr–22.5 wt% Co alloy, annealed at 580 °C (close to the Curie point of this alloy). One can see that variations in the scattering intensity in directions perpendicular and

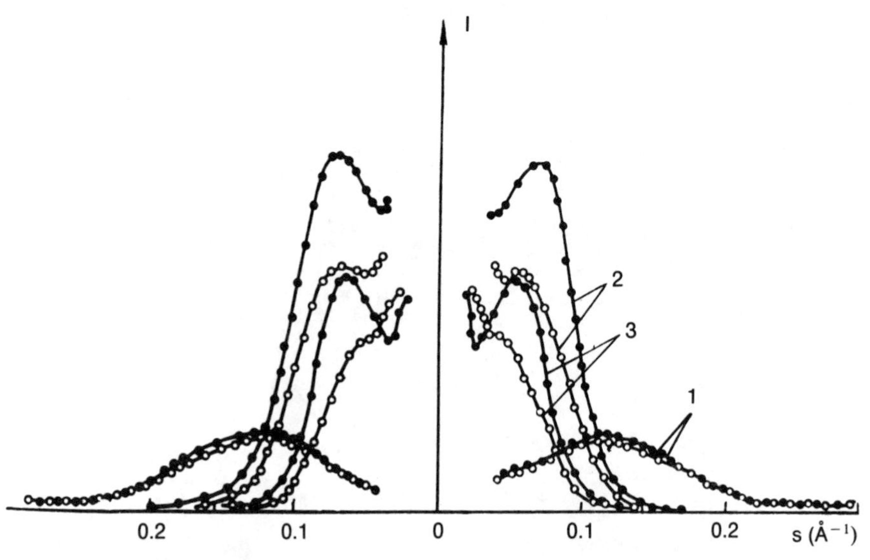

Figure 7.5. Small-angle neutron scattering on Fe–Cr–Co alloys after TMT at 580 °C (after Vintaikin *et al.*, 1977): (1)–(3) correspond to 0.5, 2.5, and 6h tempering; full circles refer to $\mathbf{s} \perp \mathbf{H}_{TMT}$, open circles to $\mathbf{s} \parallel \mathbf{H}_{TMT}$.

parallel to the magnetic field used in TMT grow with an increase in annealing time (untreated samples show isotropic scattering).

Segregation of nuclear and magnetic contributions to scattering was performed by measuring the annealed samples, magnetized in a 9-kE field, in directions perpendicular and parallel to the field (in conformity with maxmium and zero contributions of magnetic scattering). The scattering curves along and perpendicular to $\mathbf{H}_{TMT}$ coincided. This points to the fact that the amplitudes of concentration waves parallel and perpendicular to $\mathbf{H}_{TMT}$ exhibit a practically similar increase. Proceeding from these data, the authors interpret the scattering anisometry (Figure 7.5) caused by preferential magnetization of domains in the direction of $\mathbf{H}_{TMT}$. Annealing combined with thermomagnetic treatment at high temperatures also resulted in phase separation of anisometric shape. These data agree with those of electron microscopy and are confirmed by results obtained by Vintaikin *et al.* (1977) from measurements of diffuse X-ray scattering near knots 006, 500, and 600 of the reciprocal lattice.

Another approach to the study of phase separation in alloys is the analysis of invariants of small-angle scattering curves; this allows for quantitative evaluation of the size and magnitude of separation (for example, by determining the radii of gyration and forward scattering). In addition, it has been shown by Gerold (1967) that the determination of Porod's invariant for alloys with different amounts of each component enables one to find the miscibility gap limits. Indeed, we now consider a binary alloy after phase separation. If the concentration of the second component is X_b, then in conformity with equations (2.44) and (2.45) we can write

$$Q = 2\pi^2 V_0(\rho_b - \bar{\rho})(\bar{\rho} - \rho_a) \tag{7.2}$$

where $\bar{\rho} = \rho_a(1 - X_b) + \rho_b X_b$, ρ_a and ρ_b being the densities of the atomic scattering length of the first and second components. If phase separation occurs under such conditions at concentrations of $X_{b1} \leqslant X_b \leqslant X_{b2}$ (concentrations of $X_b < X_{b1}$ and $X_b > X_{b2}$ result in the formation of solid solution), then equation (7.2) must hold for all X_b from this range and, in particular, for X_{b1} and X_{b2}. Proceeding from this we obtain

$$Q_c(X_b) = (X_{b1} + X_{b2})X_b - X_{b1}X_{b2} \tag{7.3}$$

where

$$Q_c(X_b) = X_b^2 + Q(X_b)/[2\pi^2 V_0(\rho_a - \rho_b)^2]$$

Hence determination of $Q(X_b)$ for various X_b (in absolute units) allows us to find the linear dependence (7.3) and, further, the miscibility gap determined by X_{b1} and X_{b2}. Figure 7.6 (Gerold, 1977) illustrates this

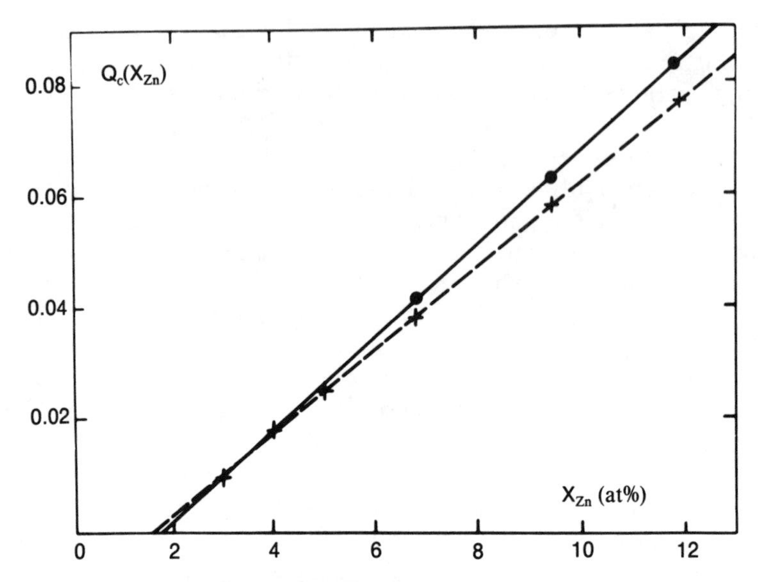

Figure 7.6. Estimation of the miscibility gap in Al–Zn alloy (after Gerold, 1977): (●) relationship (7.3) is used; (×) with correction for atomic volume changes.

approach adapted to evaluation of the miscibility gap in aluminum–zinc alloy (solid line), where X_{b1} and X_{b2} are 1.7 and 81 at% Zn, respectively. The dashed line represents the corrected calculations, which indicate a change in the atomic volume caused by replacement of the aluminum atom by Zn (in this case $X_{b1} = 1.8$ and $X_{b2} = 71$ at% Zn). Gerold (1977) has shown that this approach is valid for ternary systems as well.

7.2. Polydisperse Objects. Calculation of Size Distribution

In studies of a good number of highly dispersed inorganic materials the sample can be interpreted as a matrix with inclusions ("particles") of colloidal size. This model relates to such substances as porous materials (where particles are represented by pores), liquids with cluster structure (clusters), alloys which tend to decompose following nucleation of the new phase (phase separation), and colloidal solutions. In these cases the phase interfaces are sharp and, as a rule, the particles are randomly distributed. Such objects may therefore be viewed as isotropic polydisperse systems (see Section 2.3.1).

One of the most important tasks when investigating such systems is to determine the particle-size distribution $D_N(R)$, which is of great practical value as the particle sizes affect the properties of highly dispersed objects. Determination of the particle-size distribution function for different classes of objects can be aided by, for example, electron microscopy, sedimentation techniques, and chromatography. However, small-angle scattering methods are much more advantageous for the analysis of sample polydispersity, as the measurements do not depend on the physical nature of the particles insofar as they differ when compared to the matrix scattering density. For this reason small-angle scattering techniques are widely used in research on highly dispersed objects.

In line with Section 2.3, scattering from a dilute polydisperse system can be expressed, to within a constant multiple, by the equation

$$I(s) = \int_0^\infty i_0(sR) D_N(R) m^2(R) dR \tag{7.4}$$

where R is an effective particle size, $i_0(sR)$ is the mean particle form factor, and function $m(R)$ relates R to the particle volume. In this section we examine the principal methods of calculating function $D_N(R)$ from equation (7.4) with the given form factor. [In many cases regarding the particles as spherical in shape is a good approximation, in which case $i_0(sR)$ represents the form factor for a uniform sphere, $\phi^2(sR)$; see equation (1.27).]

It should be noted that for systems with a high concentration of particles within the matrix, equation (7.4) can sometimes be violated by the effects of interparticle interference (Section 2.3.3). In general, extrapolation to zero concentration is required (see Section 3.3.1), but several methods described below do not respond noticeably to these effects.

7.2.1. Analytical Methods

The simplest approach to an analysis of polydispersity is to describe the distribution $D_N(R)$ by an analytical function with few parameters, and then determine these parameters from experimental data. In this case various two-parameter distributions are normally used. For example, if we assume that $D_N(R)$ fits the Maxwell distribution

$$D_N(R) = (R/R_0)^n \exp[-(R/R_0^2)] K(n)/R_0 \tag{7.5}$$

involving the two parameters R_0 and n, where $K(n) = 2/[(n-1)/2]!$, $n \geqslant -1$, then the initial portion of the curve for scattering from spheres in such a polydisperse system satisfies the equation

$$\ln[I(s)] = \text{const} - \{(n+4)\ln[1 - s^2 R_0^2/5]\}/2$$

The correlation curves for $\ln[I(s)]$ versus $\ln(s^2 - \sigma^2)$ for different values of σ can be plotted. For $\sigma = \sqrt{5}/R_0$ one has a straight line inclined at an angle of $(n+4)/2$. This method, suggested by Shull and Roess (1947), yields the distribution parameters (7.5) in the initial portion of the scattering curve.

Similar approaches have been suggested by Mittelbach and Porod (1965), Sjöberg (1974), Plestil and Baldrian (1976), and others. Mittelbach and Porod (1965), for example, carried out simultaneous approximations of the particle shape (an ellipsoid of rotation) and polydispersity (a single-parameter distribution). The particle anisometry and distribution parameters $D_N(R)$ can be calculated by determining the mean invariants from the scattering curve. Hence, in conformity with equation (6.11), we obtain for the experimental radius of gyration of globular particles $[m(R) \sim R^3]$ that

$$\langle R_g^2 \rangle_z = \int_0^\infty R_g^2(R) D_N(R) R^6 dR \Big/ \int_0^\infty D_N(R) R^6 dR$$

In particular, for a system of spheres $\langle R_g^2 \rangle_z = 3\langle R^8 \rangle / 5\langle R^6 \rangle$, and for ellipsoids of rotation with $c/a = v$, one has $\langle R_g^2 \rangle_z = (2+v^2)\langle R^8 \rangle / 5\langle R^6 \rangle$, where

$$\langle R^k \rangle = \int_0^\infty D_N(R) R^k dR$$

A similar dependence on particle shape and size distribution can be written for other experimental invariants. By following the above approach, we find that comparison of these relationships provides the anisotropy v and function $D_N(R)$.

These methods are simple and, in a number of cases, yield adequate results. However, it should be noted that the distribution pattern of $D_N(R)$ can hardly be forecasted in studies of inorganic objects. For this reason these methods are less efficient than similar parametrization of the molecular mass distribution for polymers (Section 6.1.4).

Another aspect of the analytical approach is aimed at the explicit solution of equation (7.4), possible for certain form factors. Some work has been done on the derivation of an explicit expression for $D_N(R)$. For example, the following equation of Letcher and Schmidt (1966) is widely used to calculate the distribution of a system of uniform spheres along the radii:

$$D_N(R) = \frac{\text{const}}{R^2} \int_0^\infty [s^4 I(s) - c_4] \left\{ \cos(2sR)\left[1 - \frac{2}{(sR)^2}\right] - \frac{2\sin(sR)}{sR}\left[1 - \frac{1}{2(sR)^2}\right] \right\} ds \tag{7.6}$$

A generalized solution of equation (7.4) has been suggested by Fedorova and Schmidt (1978), which is valid when the form factor can be expressed as

$$i_0(sR) = J_\nu^2(sR)/(sR)^\beta, \qquad m^2(R) \sim R^{2\alpha} \tag{7.7}$$

where $J_\nu(x)$ is the Bessel function of the first kind and order ν, β is an integer, and α is the effective number of dimensions ($\alpha = 3$ for globular particles, 2 for platelets, 1 for thin rods). Hence equation (7.4) can be rewritten in the form

$$s^n I(s) = \int_0^\infty sJ_\nu^2(sR)RD_N(R)R^m dR \tag{7.8}$$

where $n = \beta + 1$ and $m = 2\alpha - \beta - 1$. The integral on the right-hand side of equation (7.8) is the Titchmarsh transform, for which the inverse transform is

$$R^m D_N(R) = -2\pi \int_0^\infty J_\nu(sR)N_\nu(sR)s\frac{d}{ds}[s^n I(s) - c_n]ds \tag{7.9}$$

where $N_\nu(x)$ is the Bessel function of the second kind and order ν, and $c_n = \text{const}$. By selecting

$$c_n = \lim_{s\to\infty} [s^n I(s)]$$

and integrating by parts, we obtain

$$R^m D_n(R) = \int_0^\infty [s^n I(s) - c_n]\varphi(sR)ds \tag{7.10}$$

where

$$\varphi(x) = 2\pi\frac{d}{dx}[xJ_\nu(x)N_\nu(x)]$$

$$= 4\left\{1 + \frac{\pi}{2}N_\nu(x)[J_\nu(x)(1 - 2\nu) + 2xJ_{\nu-1}(x)]\right\} \tag{7.11}$$

Relationships (7.10) and (7.11) are adequate for calculating function $D_N(R)$ whenever the form factor is given by equation (7.7). Table 7.1, previously published by Fedorova and Schmidt (1978), presents explicit expressions for $i_0(sR)$, ν, m, and $\varphi(x)$ and allows one to evaluate $D_N(R)$ from the experimental curve for a system with a specific particle shape. In cases (1)–(6), (8), (10), and (11) R is the radius of a sphere or of a cylinder

Table 7.1. Explicit Expressions with Which to Calculate $D_N(R)$ for Particles with Several Form Factors

Shape	$i_0(sR)$	n	m	$\varphi(x)/4$
1. Uniform sphere	$J^2_{3/2}(sR)/(sR)^3$	4	2	$\cos(2x)(1 - 2/x^2)$ $- 2\sin(2x)(1 - 1/2x^2)/x$
2. Long cylinder perpendicular to plane of scattering				
3. Thin disk in plane of scattering	$J^2_1(sR)/(sR)^2$	3	1	$1 - (J/2)N_1(x)*[J_1(x) - 2xJ_0(x)]$
4. Sphere (Fraunhöfer diffraction)				
5. Long, randomly oriented cylinder	$J^2_1(sR)/s(sR)^2$	4	1	
6. Spherical shell	$J^2_{1/2}(sR)/(sR)$	2	2	
7. Thin, randomly oriented platelet	$J^2_{1/2}(sR)/s^2(sR)$	4	0	
8. Long cylinder (Fraunhöfer diffraction)	$J^2_{1/2}(sR)/s(sR)$	3	0	$-\cos(2x)$
9. Thin rod perpendicular to incident beam and in plane of scattering	$J_{1/2}(sR)/(sR)$	2	0	
10. Hollow cylinder perpendicular to beam	$J^2_0(sR)$	1	1	$1 + \frac{1}{2}\pi N_0(x)[J_0(x) - 2xJ_1(x)]$
11. Randomly oriented hollow cylinders	$J^2_0(sR)/s$	2	1	

cross section, in (7) it is half a lamellar thickness, and in (9) it is half the length of a rod.

This approach permits function $D_N(R)$ to be computed for certain particle shapes without assumptions about this function. (We mentioned earlier that this approach is most frequently used for spheres.) However, it must be noted that c_n is to be determined as precisely as possible; an incorrect estimate of this quantity brings about strong termination effects in integral (7.10), which always has to be calculated within finite limits $(0, s_{max})$. One way of correctly evaluating c_n is to examine the dependence of function $s^nI(s)$ on s^n (see Section 3.3.3). Besides, equation (7.10) implies that

$$\int_0^\infty [s^nI(s) - c_n]ds = 0$$

which may also be helpful in the evaluation of c_n. Polynomial extrapolation of $I(s)$ when $s \to \infty$ (see Brill *et al.*, 1968) is also possible.

When the quantity c_n cannot be determined to a sufficient degree of accuracy (for example, when the reliability of measurements in the wide-angle scattering range is inadequate) relationship (7.9) can be used directly. However, this introduces the problem of differentiating the experimental data and the result is very much affected by random errors. Therefore in this case care must be taken in the statistical processing of the data (such as smoothing; see Section 9.3).

It is sometimes useful to describe polydispersity by means of a correlation function. Letcher and Schmidt (1966) have shown that function $\gamma(r)$ for a polydisperse system [correlated with $I(s)$ by general transform (2.22)] can be written in the form

$$\gamma(r) = \int_r^{\infty} V_R \bar{V}^{-1} \gamma_R(r) D_N(R) dR \tag{7.12}$$

where $\gamma_R(r)$ is the characteristic function with a particle of given shape and size R, $V_R = m(R)$, and

$$\bar{V} = \int_0^{\infty} m(R) D_N(R) dR$$

is the normalizing constant. In particular, the following expression can be derived from equation (7.12) for spherical particles:

$$D_N(R) = -\frac{d}{dr}\left[\frac{\gamma''(r)}{r}\right]_{r=2R} \tag{7.13}$$

where R is the radius of a sphere.

The practical application of equation (7.13) for the calculation of $D_N(R)$ is fairly difficult, since in this case it is necessary to differentiate the function $\gamma(R)$ obtained from the experimental data three times. It has been shown by Walter *et al.* (1983) that when function $I(s)$ is given at equally spaced angular points, the expression for $\gamma(R)$ is written in the form of a Fourier series and differentiation is performed analytically, while termination effects are reduced by extrapolation in accordance with Porod's law, in a way similar to equation (3.33). This allowed the authors to write an explicit expression for $D_N(R)$ using $I(s)$ and which (as anticipated) coincides with equation (7.6) if integration in the latter is replaced by summation using the trapezoidal rule. It should be noted, however, that as in the calculation of small-angle scattering invariants (see Section 3.3), the use of equation (7.12) is more convenient than equation (7.4).

7.2.2. Numerical Methods

Methods for the numerical solution of equation (7.4) comprise another group of techniques for determining size distributions. These techniques do not require either *a priori* assumptions on the distribution pattern or extrapolation of function $I(s)$. However, their use does entail a certain instability of the solution.

One way of solving equation (7.4) is to reduce it to a set of linear equations by numerical integration, an approach used by Hendricks *et al.* (1974), Plavnik *et al.* (1976), and Vonk (1976). For each experimental value of $I(s_i)$ we may write

$$I(s_i) \approx \sum_{j=1}^{n} D_N(R_j) m^2(R_j) i_0(s_i R_j) W_j \Delta R_j$$

where R_j are the knots where function $D_N(R)$ is to be determined in the interval $(R_{\min}, R_{\max})$, $j = 1, 2, \ldots, n$; ΔR_j and W_j are quantities depending on the step in this interval and on the quadrature that is applied. For example, when the trapezoidal rule is applied for uniform spacing, $\Delta R_j = (R_{\max} - R_{\min})/(n-1)$, $W_1 = W_n = \frac{1}{2}$, $W_k = 1$, $k = 2, 3, \ldots, n-1$. The number of these equations is N in accordance with the number of knots. In matrix form this can be presented as

$$\mathbf{AD} = \mathbf{I} \tag{7.14}$$

where

$$A_{ij} = m^2(R_j) W_j \Delta R_j i_0(s_i R_j)$$

and

$$I_i = I(s_i), \quad D_j = D_N(R_j); \qquad i = 1, 2, \ldots, N, \quad j = 1, 2, \ldots, n$$

When $N > n$, the system can be solved and the distribution determined. However, vector $\mathbf{I}$ as well as matrix $\mathbf{A}$ incorporate errors, involving both measurements and numerical integration. Thus, the solution of equation (7.14) represents a typical incorrectly stated problem (Tikhonov and Arsenin, 1977) and requires regularization by the introduction of physically valid restrictions. The above methods differ in the way these limitations are introduced. Hendricks *et al.* (1974) proceed from the condition of nonnegative $D_N(R)$; Pavnik *et al.* (1976) and Vonk (1976) use Tikhonov's classical regularization (Tikhonov and Arsenin, 1977), that is, solving the equation

$$(\mathbf{A} + \alpha\mathbf{\Omega})\mathbf{D} = \mathbf{I} \tag{7.15}$$

instead of equation (7.14) where $\alpha > 0$ is the regularization parameter and Ω is the stabilizer. Plavnik *et al*. (1976) use the integral norm of the second derivative of the solution as stabilizer, while Vonk (1976) combines this with the sum of the squares of the $D_N(R)$ values at the ends of the interval (R_{min}, R_{max}). The optimum regularization parameter in these cases is selected by trial and error (comparing the results obtained for different values of α).

A more general approach is based on the method of indirect transformation (Glatter, 1980a). This method implies that function $D_N(R)$ is represented in the interval (R_{min}, R_{max}) as the superposition of orthogonal functions, i.e., cubic B-splines, in the form

$$D_N(R) = \sum_{\nu=1}^{m} c_\nu B_\nu(R) \tag{7.16}$$

The method of indirect transformation is discussed in Section 9.7.2, with function $D_N(R)$ replaced by $\gamma(r)$ and transformation (7.4) by the Fourier transform, which, however, does not make any fundamental difference. It should be noted that this method admits simultaneous smoothing and introduction of instrumental corrections in the experimental data. The determination of the coefficients of decomposition (7.16) is reduced to the solution of a set of linear equations with the assistance of a regularization technique; the procedure by which the optimum value of α is selected is supported by a graphical method (see Figure 9.9). The method described by Glatter (1980a), which permits one to calculate $D_N(R)$ simultaneously with the introduction of instrument corrections, has recently acquired wide application.

Regularization methods applied to the solution of integral equation (7.4) are typical for solving incorrectly stated problems and will be discussed in Chapter 9, which is devoted to the treatment of experimental data. For example, the reduction of the problem to the solution of system (7.14) is quite similar to the matrix-inversion technique employed in slit-height collimation desmearing (see Section 9.4.3).

In the above methods, the form factor $i_0(sR)$ is usually assumed to be a given function, i.e., some definite shape (and internal structure) of the particle is postulated. In practice, however, the sample may contain particles of different shapes. In this case equation (7.4) involves a certain mean form factor, which describes, for example, the mean anisometry of particles. Plavnik (1979, 1984) suggested a simple method for an approximate evaluation of polydispersity for systems of globular particles. If the variable R is taken as the radius of gyration, we can write

$$I(s) = \int_0^\infty D_V(R) i_0(sR) R^3 dR \tag{7.17}$$

For globular particles the function $R^3 i_0(sR)$ attains the highest peak in the region, where $sR \approx 2$, and the main contribution to integral (7.17) is made by particles with $R_m \approx 2/s$. Therefore, we may write, approximately,

$$D_V(R_m) \approx s^4 I(s) \left[\int_0^\infty t^3 i_0(t) dt \right]^{-1}$$

Both papers discuss different types of form-factor approximations. Plavnik (1984) suggests the form factor $t^3 i_0(t) \sim t^3/(1+t^4)$ for particles with anisometry satisfying $1 \leqslant \nu \leqslant 2.5$; the upper integration limit is $t_{max} = sR_{max}$, where R_{max} is the radius of gyration of the largest particle and is determined empirically from the scattering curve. The final equation, presented by Plavnik (1984) for particle distribution along the radii of gyration, is

$$D_V(R_m) = s^4 I(s)/\ln(2s/s_{01}); \qquad R_m = 2/s, \quad s > s_{01} \tag{7.18}$$

where s_{01} is the scattering angle for which $s^4 I(s)$ is one order of magnitude lower than its maximum value ($s_{01} < 2/R_m$). This relationship enables easy computation of the particle-size distribution function for polydisperse systems of globular particles. It is clear, however, that the obtained function $D_V(R)$ characterizes the particle size in the object only approximately and requires further refinement.

Numerical methods enable the particle-size distribution function to be determined with a high degree of precision subject to minimum *a priori* assumptions. [The range of particle sizes (R_{min}, R_{max}) within the system is usually known.] However, in contrast to analytical methods and with the exception of the approximate method described by Plavnik (1984), numerial methods require a considerable amount of computation and can be realized efficiently by means of computerized data processing. Comparison of different methods for standard models has been dealt with in a couple of papers (Walter *et al.*, 1983; Glatter, 1980a). It has been shown that analytical methods (Brill *et al.*, 1968; Walter *et al.*, 1983), for careful extrapolation when $s \to \infty$, yield approximately the same results as numerical algorithms (Vonk, 1976; Glatter, 1980a) with the optimum regularization parameter. Figure 7.7 shows a comparison of particle-size distributions obtained by different numerical methods applied to experimental small-angle X-ray scattering data for arabinogalactan in solution (compact copolymer particles of nearly spherical shape). Figure 7.7a presents the calculated asymptotic behavior of the scattering intensity [see equation (3.28)] and Figure 7.7b shows the obtained distributions $D_V(R)$. One can see that the methods suggested by Brill *et al.* (1968) and Walter *et al.* (1983) yield the same results; the algorithm of Glatter (1980a) offers a bimodal distribution very close to the former one, while the approximate method (Plavnik, 1984) yields a less distinct monomodal function.

Thus we can see that various methods have been developed for the calculation of particle-size distributions which, to a different extent, employ *a priori* assumptions on the distribution pattern, form-factor type, and particle anisometry. Therefore, selection of the most efficient method for calculating function $D_N(R)$ greatly depends on the amount and nature of available information about the object under study.

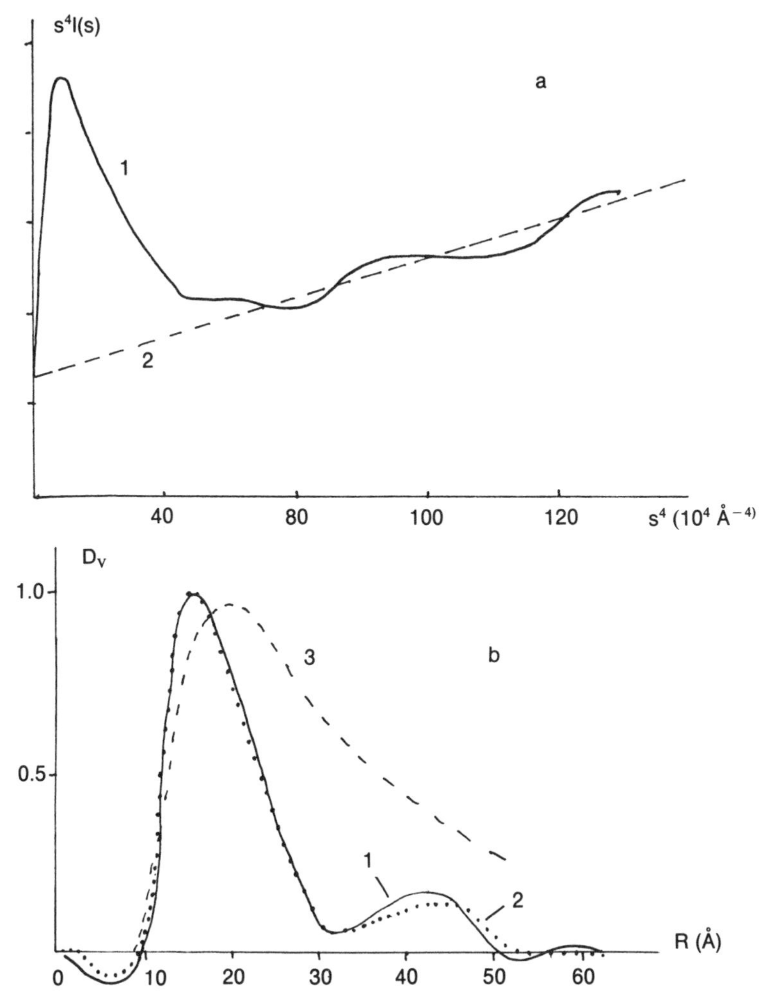

Figure 7.7. Particle-size distribution for arabinogalactan in solution. (a) Determination of asymptotic behavior and the value of c_4: (1) experimental curve; (2) asymptotic expression (3.28). (b) Restored distributions: (1) after Brill *et al.*, 1968, and Walter *et al.*, 1983; (2) after Glatter, 1980a; (3) after Plavnik, 1984.

7.3. Amorphous Solids and Liquids

The structure of amorphous solids and liquids is fairly similar. Both exhibit no anisometry and are characterized by the short-range order of the molecular distribution. Therefore, the methods of diffraction studies of amorphous solids and liquids are practically identical.

In the absence of inhomogeneities of submicron size, the scattering intensity in these objects at small angles is practically independent of the angle (see Section 2.5.1), while in the wide-angle range there is an extended halo caused by the existence of the short-range order. Small-angle scattering observed in multicomponent amorphous solids and liquids is normally indicative of phase separation and the presence of concentration inhomogeneities in the object.

7.3.1. Study of the Structure of Glasses

The classical field of small-angle scattering applications for the study of inhomogeneities of superatomic range and for their impact on the physical properties of amorphous bodies consists of vitreous substances. The small-angle scattering technique is probably the only method capable of direct evaluation of the presence of submicroscopic inhomogeneities, together with direct determination of their dimensions in single- and multicomponent glasses.

In single-component glasses, when the sample is of sufficiently high quality, the small-angle scattering intensity is practically independent of the angle. Figure 7.8a shows small-angle X-ray scattering curves obtained by Golubkov and Porai-Koshits (1981) that exhibit no structural inhomogeneities of submicron size. In the opinion of the authors, the inhomogeneities that have been observed in quartz glasses have been caused by surface effects and technological defects and the zero-angle scattering intensity is a function only of the thermal density fluctuations. Proceeding from relationship (2.43), we may write

$$I(0)=\rho^2 k_B T \beta V_0 \tag{7.19}$$

where $\rho = F(0)/v_1$ is the mean scattering density in the object. Figure 7.8b presents temperature dependences for $I(0)$. Each sample is seen to have its own temperature T_0 (close to the vitrification point) at which "freezing" of the structure takes place, i.e., no change in fluctuation level occurs with further decrease in T. On the other hand, extrapolation of straight lines (7.19) for $T < T_0$ into the zone of smaller values of T does not result in $I(0)=0$ if $T\rightarrow 0$ K; this is explained by the dependence of the isothermal compressibility coefficient on temperature.

The study of liquation processes (phase separation) in multicomponent glasses has great practical significance. Probable mechanisms of phase separation have already been discussed in Section 7.1.2. The study of 80% PbO–15% B_2O_3–5% A_2O_3 (wt%) glasses by Acuña and Craievich (1979), whose samples were obtained from solution and quickly cooled, showed phase separation in conformity with the mechanism of spinodal decomposition. On the other hand, a study of borosilicate glass B_2O_3 + SiO_2 (Vasilevskaya *et al.*, 1980) indicated no changes in the small-angle X-ray scattering curves during annealing. Moreover, the scattering inten-

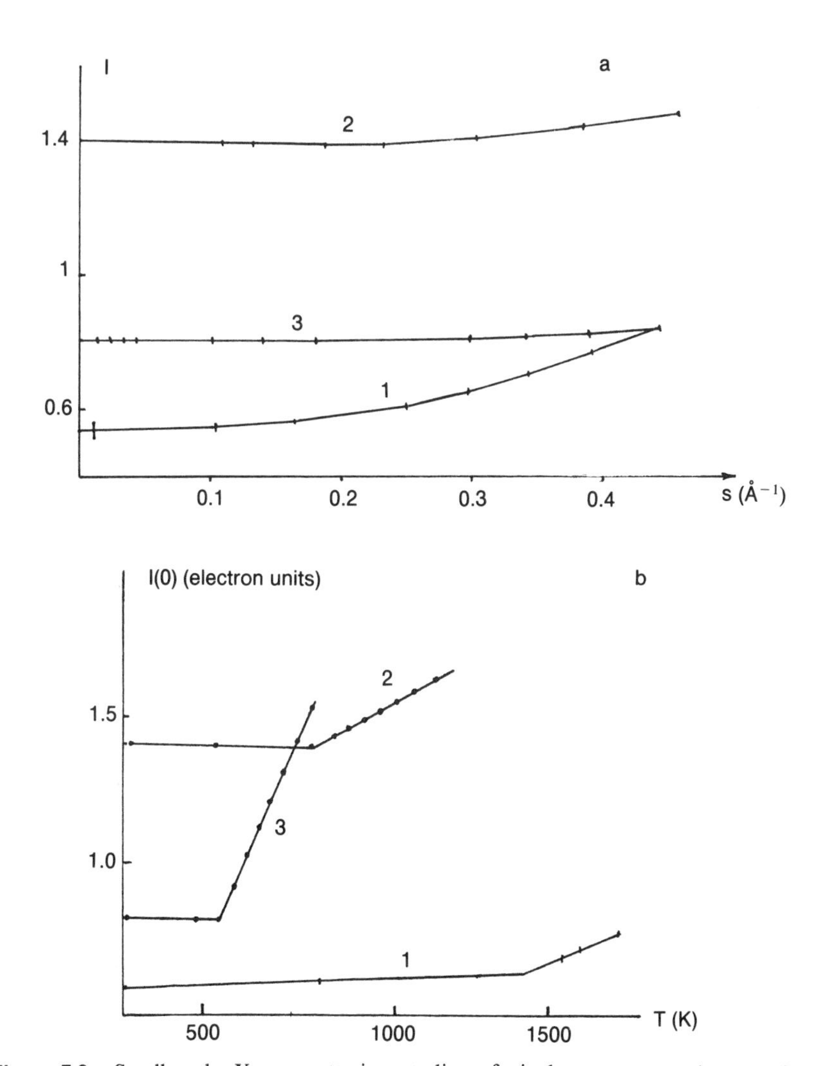

Figure 7.8. Small-angle X-ray scattering studies of single-component glasses (after Golubkov and Porai-Koshits, 1981): (a) intensity function; (b) zero-angle scattering as a function of temperature: (1) SiO_2; (2) GeO_2; (3) B_2O_3.

sity was two orders of magnitude lower than the intensity that would be obtained if phase separation took place. [The latter was evaluated proceeding from the expression for the density fluctuations of binary system (2.44).] The results were obtained for samples with different SiO_2 content (from 4 to 50 mol%), and this allowed the authors to conclude that these glasses exhibit no phase separation. However, samples with SiO_2 content exceeding 10 mol% showed, apart from thermal fluctuation scattering, certain small-angle scattering from zones approximately 10–15 Å in size (Figure 7.9). Vasilevskaya *et al.* (1980) have interpreted this phenomenon as the formation of a "pseudophase structure," that is, of zones with high silica content (similar to the concentration fluctuation; see Section 7.3.2). This explanation also correlates with temperature dependences for $I(0)$ calculated in this paper. Similar results were later obtained for borogermanite glasses B_2O_3–GeO_2 (Vasilevskaya *et al.*, 1981), the only exception being that the formation of the pseudophase structure was observed even at minimum GeO_2 concentrations.

Other examples of analyzing the structure of single- and multicomponent glasses have been discussed by Porai-Koshits *et al.* (1982). The small-angle scattering technique has proved very useful in the study of porous glasses (see, for example, Kranold *et al.*, 1983). It should be noted

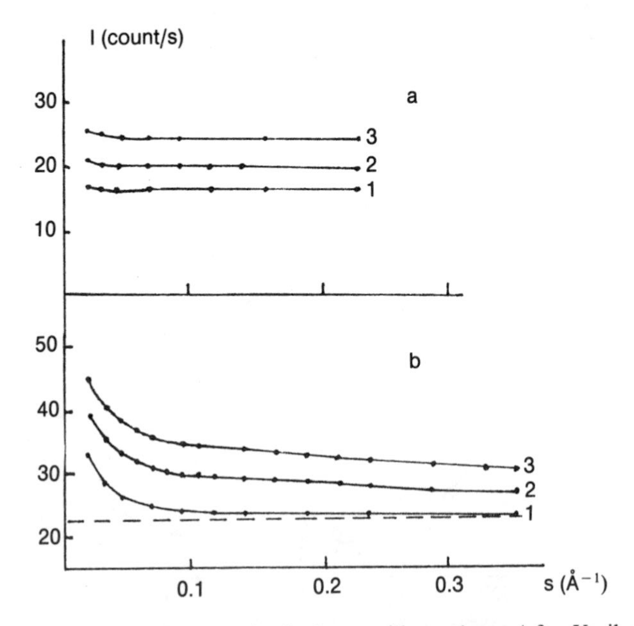

Figure 7.9. Small-angle X-ray scattering by borate silicate glasses (after Vasilevskaya *et al.*, 1980): (a) SiO_2 content 4 mol%, curves (1)–(3) correspond to 200 °C, 300 °C, and 375 °C; (b) SiO_2 content 25.4 mol%, curves (1)–(3) correspond to 250 °C, 350 °C, and 425 °C.

that, until now, most of the studies were conducted on "ordinary" (silicate, borate, etc.) glasses transparent to visible light. Small-angle scattering methods, however, can also be applied effectively to new types of glass, i.e., to metallic glasses, and to semiconductors which, until recently, were beyond the scope of materials being studied.

7.3.2. Concentration Fluctuations and Clusters

It has been shown (Section 2.5.1) that scattering from amorphous and liquid objects can be correlated with their thermodynamic features (thermal fluctuations). Multicomponent systems can also exhibit fluctuations in component concentration and bring about small-angle scattering. In line with Bhatia and Thornton (1970), for binary systems

$$I(s) = (\Delta b - \delta\bar{b})^2 S_{CC}(s) + \bar{b}S_{NN}(s) \tag{7.20}$$

where $\Delta b = b_2 - b_1$ is the difference between amplitudes of atomic scattering from the components, $\bar{b}$ is a mean scattering length, and $\delta = (\partial V/\partial c)/V$ describes the change in atomic size, c being the concentration of the second component and V the volume of the given number of atoms. In this case, the first term on the right-hand side of equation (7.20) represents scattering by concentration fluctuations, and

$$S_{CC}(0) = k_B T(\partial^2 g/\partial c^2)^{-1}$$

where g is the specific Gibbs free enthalpy. The second term corresponds to scattering by thermal fluctuations and its value at $s = 0$ is expressed by equation (7.19). If the system has a critical point, then the impact of concentration fluctuation near that point becomes highly pronounced. In accordance with the Ornstein–Zernike theory (see, for example, Krivoglaz, 1969)

$$S_{CC}(s) = S_{CC}(0)/(1 + \xi^2 s^2) \tag{7.21}$$

where ξ is the correlation length that characterizes the wavelengths of concentration fluctuations. In the case of large concentration fluctuations the second term in equation (7.20) can be neglected, and equation (7.21) presents the expression for the scattering intensity.

Small-angle scattering experiments involving relationships (7.20) and (7.21) have been applied extensively to the study of various alloys in a liquid state near the critical point. In particular, small-angle neutron scattering studies of Al–Zn alloy (with critical composition of 39.5 at% Zn) at temperatures of 10–100 °C above the critical point ($T_c = 322$ °C) exhibited a change in correlation length from 22 to 5 Å (Schwann and

Schmatz, 1978). This technique has recently been adopted for the analysis of amorphous alloys. Work on this subject, including methods of analysis, is reviewed in Steeb and Lamparter (1984).

Another mechanism governing the formation of submicron inhomogeneities in liquids and amorphous solids is the development of clusters comparing aggregates of atoms (molecules) of one and the same phase. Such a structure can be determined by means of conventional techniques. The occurrence of small-angle scattering testifies to the presence of clusters, the sizes of which are evaluated by, for example, the Guinier approximation. This can be illustrated by the paper of Zaiss *et al.* (1976) in which copper–bismuth melts have been studied; the presence of a cluster structure has been observed within a broad range of bismuth concentration. Figure 7.10 presents the Guinier and Ornstein–Zernike approximations for Cu–50 at% Bi melt. The Guinier approximation shows better correlation with experimental data. The fact that the estimates from equation (7.21) for the interval of 30–70 at% Bi account for too low a value of the correlation length ($\xi = 1.4$ Å) testifies to the presence of the cluster structure rather than concentration fluctuations, and the cluster diameter calculated by the Guinier approximation amounts to quite a reasonable value of 6–8 Å.

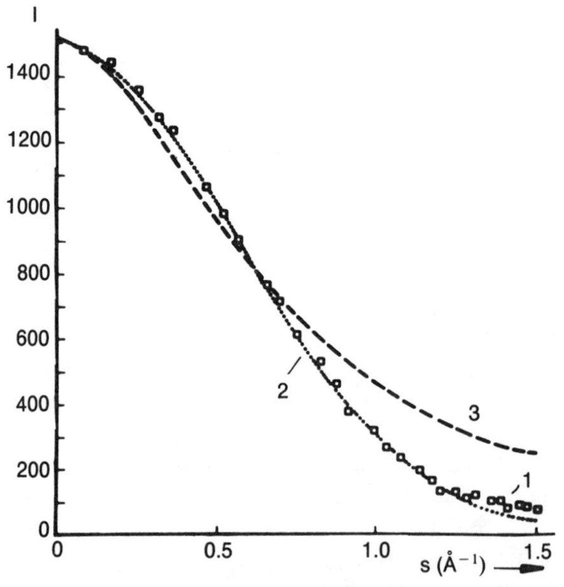

Figure 7.10. Small-angle neutron scattering data of Cu–50 at %Bi alloy (after Zaiss *et al.*, 1976): (1) experimental data; (2) Guinier approximation; (3) approximation according to equation (7.21). The specimen temperature is 5 °C above the melting point.

7.4. Conclusion

The present chapter illustrates the possibilities of small-angle scattering application to the structure analysis of various inorganic materials. This technique is instrumental for determining the general features of the superatomic structure of substances, and in a number of cases provides evidence for or against different theoretical models of superatomic structure of inorganic objects. At the same time, it should be noted that the efficiency of small-angle scattering application, as in the study of biopolymers in solution, greatly depends on the possibility of selecting an adequate model. Small-angle scattering experiments usually constitute part of a wide-ranging program of studies that involves the analysis of different physical parameters and properties of an object.

The study of inorganic materials often entails certain experimental difficulties. For example, inorganic compounds often contain heavy atoms which strongly absorb X rays; in such cases the small-angle X-ray scattering experiments should be conducted on very thin samples. In the analysis of crystalline substances care must be taken with respect to the impact of double Bragg scattering, while in the study of amorphous objects surface effects must be taken into account. The quality of samples and the technology of their preparation must never be underestimated. The study of inorganic materials requires, as a rule, careful examination of experimental conditions so as to ensure that the small-angle scattering picture obtained really reflects the structure of the sample under study and is not distorted by side effects.

This chapter shows only briefly the potentialities of small-angle scattering applications to the study of inorganic substances. Further efforts in this field will involve time-resolved experiments and research on anomalous scattering in synchrotron beams and analysis of the magnetic structure of metals and superconductors by means of neutron scattering, and will ensure reliable recording and interpretation of small-angle scattering data in anisotropic systems. More information related to various aspects of small-angle scattering applications is available in the articles and reviews cited herein.

IV

Instrumentation and Data Analysis

8

X-Ray and Neutron Instrumentation

The basic procedure for any analysis involving structure determination, as outlined in Chapter 3–7, is to accumulate reliable experimental small-angle scattering data. The experimental technique for X-ray and neutron structure analysis involves small-angle scattering instrumentation, which is reviewed elsewhere (Glatter and Kratky, 1982; Schelten and Hendricks, 1978). Here we deal only with specific features of the small-angle scattering experimental technique, basic principles underlying the design and construction of small-angle scattering facilities, and some aspects of the experimental procedure. The main difficulty of small-angle scattering intensity measurements is that the powerful primary beam does not permit one "to approach" the sufficiently small angles very closely.

The general design principles of small-angle scattering facilities, their main elements, and specific parameters are considered in Section 8.1. A small-angle instrument consists of the following main units: radiation source; collimator of the narrow beam, usually with monochromator; specimen block; and detector of the scattering radiation. Each of these units may be implemented in a number of variants. There is no doubt that the entire instrument construction and conditions for intensity measurements are determined primarily by the type of radiation source. Thus, instruments are classified in accordance with three main radiation sources: the characteristic radiation of the X-ray tubes (Section 8.2), synchrotron radiation (Section 8.3), and beams of thermal neutrons (Section 8.4).

8.1. Basic Designs of Instrumentation

We consider the main prerequisite for measuring the intensity of small-angle scattering and the basic characteristics of the corresponding facilities. High-angle resolution necessary in the study of disperse systems is the most important condition and this feature distinguishes the small-angle scattering instrument from other diffraction apparatus.

8.1.1. Angular Resolution

Any small-angle scattering experiment aims at measuring the scattering intensity for small values of the scattering vector $s = 4\pi(\sin\theta)/\lambda$, essentially smaller than a^{-1}, where a is an interatomic distance (Section 2.2). If we wish to study particles or other density inhomogeneities at a scale of 10–1000 Å, the scattering intensity in the range of s from 10^{-4} Å^{-1} up to 10^{-2} Å^{-1} must be measured. It is necessary to use even lower values of s when studying larger particles. Today, the upper limit for such particles is close to 1 μm. The unique way of working with such small values of s is to carry out the intensity measurements at small scattering angles, because the wavelengths used in experimentation cannot much exceed the interatomic distances. Nonetheless, even a small increase in wavelength facilitates the small-angle scattering measurements.

We shall now consider the general scheme of small-angle scattering experimentation. Formation of a narrow beam of incident radiation is provided by a collimation system (Figure 8.1). A set of two circular diaphragms permits us to approach plane-wave conditions (Section 1.1) with a precision up to r/R, where r and R are respectively the diameter of the diaphragms and the distance between them. The value of r/R determines the size of the projection of the incident beam on the detector plane. This value, together with the sample–detector distance R_D, determines the angle $2\theta_{min}$ and correspondingly s_{min}. This angle is the smallest permissible angle for measurements and represents the upper limit of the size of inhomogeneities that may be investigated with the given instrument.

In reality, however, the value of s_{min} is essentially higher, because the uncertainty Δs of the scattering vector s is also given by the finite size of the specimen, the finite spatial resolution of the detector, and the finite spectral interval of the radiation beam.

We now consider the contribution of all these effects to the uncertainty Δs (Schelten, 1981). In the general case

$$\Delta s \approx \frac{\partial s}{\partial(2\theta)}\Delta(2\theta) + \frac{\partial s}{\partial\lambda}\Delta\lambda = \frac{2\pi}{\lambda}\Delta(2\theta) + s\frac{\Delta\lambda}{\lambda}$$

where $\Delta(2\theta)$ is the uncertainty in the scattering angle and $\Delta\lambda$ is the wavelength bandwidth of the monochomator. The value of $\Delta(2\theta)$ is given approximately by (Figure 8.1b)

$$\Delta(2\theta) = \Delta\Omega_1^{1/2} + \Delta\Omega_2^{1/2}$$

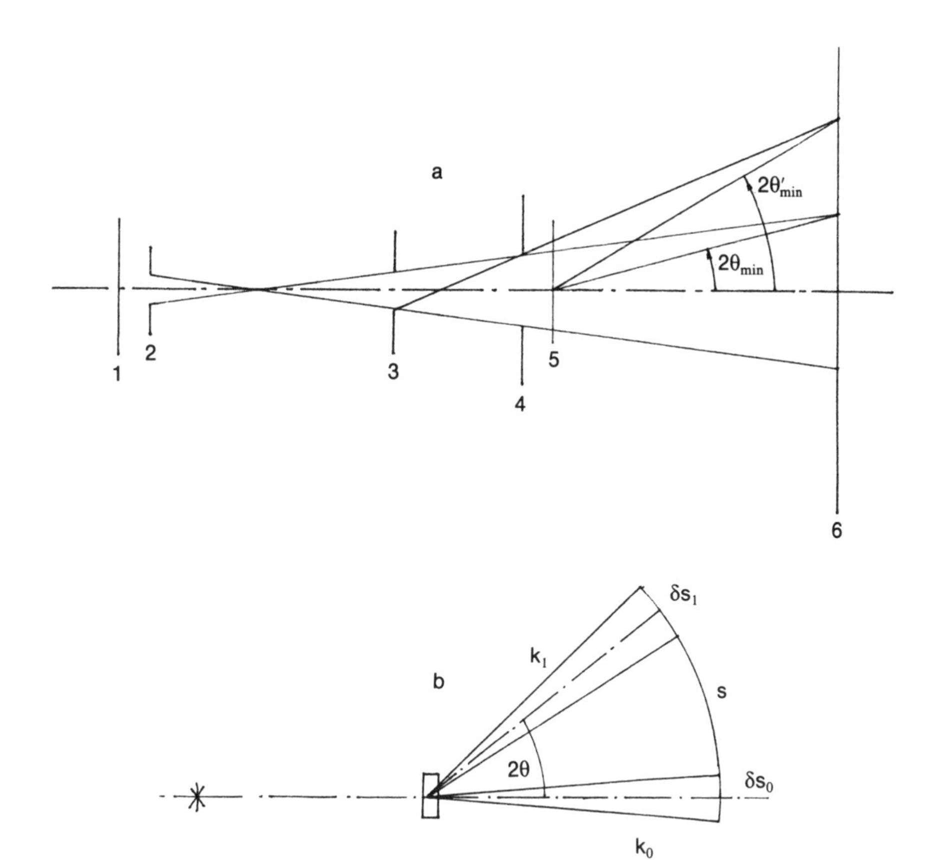

Figure 8.1. Formation of primary and scattered radiation beams in real (a) and reciprocal (b) space: (1) radiation source; (2), (3), (4) circular diaphragms; (5) sample; (6) recording plane.

where $\Delta\Omega_1$ is the solid angle of the primary beam collimation system, $\Delta\Omega_2$ is the solid angle subtended by the cross sections of the illuminated sample and detector element. Then quantity Δs may be expressed simply as a sum of three terms:

$$\Delta s = \frac{2\pi}{\lambda}\Delta\Omega_1^{1/2} + \frac{2\pi}{\lambda}\Delta\Omega_2^{1/2} + s\frac{\Delta\lambda}{\lambda} \tag{8.1}$$

with the relative spreading given by

$$\frac{\Delta s}{s} = \frac{\Delta\Omega_1^{1/2}}{2\theta} + \frac{\Delta\Omega_2^{1/2}}{2\theta} + \frac{\Delta\lambda}{\lambda} \tag{8.2}$$

Equations (8.1) and (8.2) determine the angular resolution of the small-angle scattering instrument.

Apart from s_{min} and Δs, there is a third parameter s_{max}, the maximum value of s, for which intensity measurements are performed. This parameter gives the smallest sizes of spatial inhomogeneities that can be fixed in an experiment.

8.1.2. Main Characteristics of Instruments

We have just calculated the general features of the angular resolution for the ideal collimation system. In any real collimation system one always observes "parasitic," or background, scattering that originates from scattering by the edges of the diaphragms, the sample holder, the atmosphere along the beam path, the various windows in the device, and so on. Thus, an actual experiment may be started not from angle $2\theta_{min}$ but from the much greater angle $2\theta'_{min}$ (Figure 8.1a). The latter is to be determined somewhat indefinitely, as scattering by the object must exceed the background scattering over the given range of small angles. The decrease in $2\theta'_{min}$ and the progressive approach of the value of $2\theta_{min}$ is one of the most delicate aspects in small-angle scattering experimentation.

The small-angle scattering intensity may be expressed as

$$\frac{dW}{d\Omega} = P_0 \Delta\Omega_1 \frac{\Delta\lambda}{\lambda} \Delta\Omega_2 S d \exp(-\mu d) \frac{d\sigma}{d\Omega} \tag{8.3}$$

where P_0 is the density of the incident radiation per unit area, time, solid angle, and relative wavelength band; S, d, and μ are the sample cross section, thickness, and attenuation coefficient, respectively; $\Delta\Omega_1$, $\Delta\Omega_2$, and $\Delta\lambda/\lambda$ have the same meaning as in equations (8.1) and (8.2); and $d\sigma/d\Omega$ is the differential cross section of the sample per unit volume. If the maximum value of $dW(s)/d\Omega$ is provided for given parameters of angle resolution determined by $\Delta\Omega_1$, $\Delta\Omega_2$, $\Delta\lambda/\lambda$, and the background scattering, the construction of the camera will be of optimum quality.

The contemporary small-angle scattering camera is a device attached to a given radiation source and utilizing a given detector system. We shall now enter into some details connected with the parameters of the small-angle scattering instrument as a whole. First, we examine the intensity P_0 of the primary beam falling on the object. For increasing P_0 one should use a high-power source, suitable geometry of the collimation system, and correct choice of the monochromatization conditions.

Another instrument parameter, in a purely geometric sense, is the sample scattering volume V_0. It is advantageous to deal with a larger value of V_0 because the scattering intensity is proportional to V_0; however, the values of d and S cannot be increased arbitrarily. The choice of optimum sample thickness (Section 1.6) is naturally determined by the absorption of the radiation. The increase in S leads to a considerable rise in length of the small-angle scattering diffractometer in order to preserve the identical resolution.

Thus, the basic features of the primary beam, the collimation system, and the examined object are: angular divergence, area and intensity of background scattering, utilized wavelength and corresponding bandwidth, primary beam intensity, and sample size.

A system for detecting the scattering intensity is another important feature of the small-angle scattering device. Measurement of the scattering intensity may be organized in line with two principally different patterns (Figure 8.2), namely, sequential (one-channel) and parallel

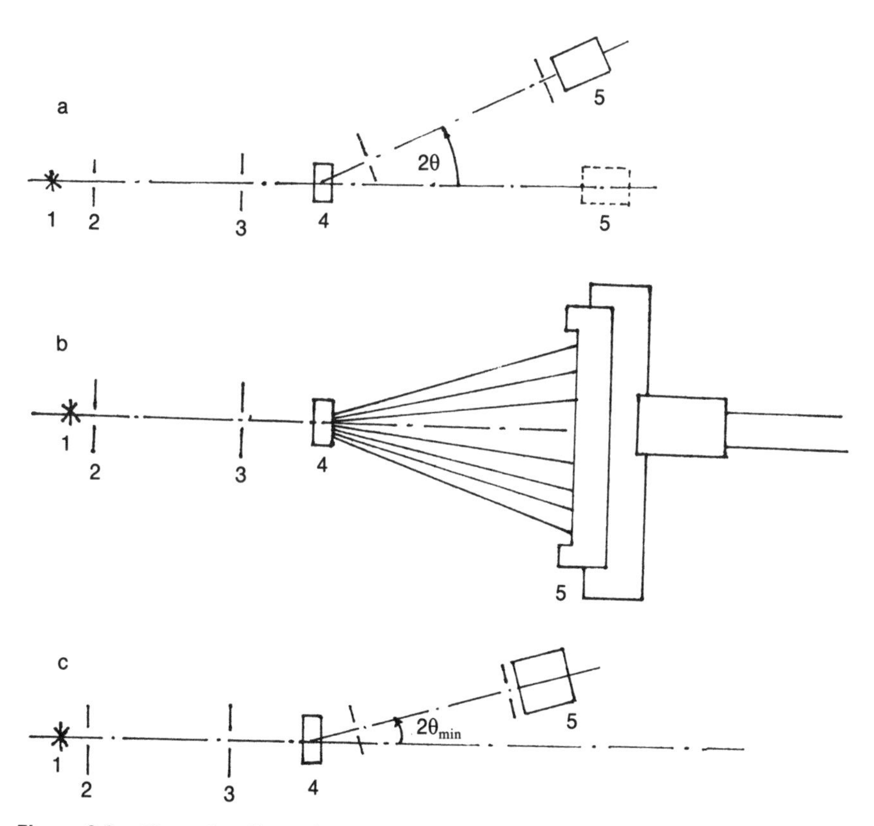

Figure 8.2. General outlines of small-angle scattering measurements for various types of detector: (a) one-channel; (b) position-sensitive; (c) energy dispersive detector. (1) Radiation source, (2) and (3) collimation system, (4) sample, and (5) detector.

(multichannel) mode of data collection. The second pattern is known in two variants. Nowadays, the first variant utilizing the position-sensitive detector is widely used. The second variant with energy-dispersive detector is only used in some specific cases. The parallel collection of scattering information permits simultaneous measurement of the scattering intensity over a large range of values of s. At times, this approach enables one to obtain immediately the whole scattering curve of an object.

Now we shall consider some of the detector-system parameters. One of the basic characteristics of a detector is its counting efficiency A_D, i.e., the fraction of the number of particles fixed by the detector. Usually the value of A_D ranges from 1 to 0.1; the lower values are of no practical significance. Each detector is characterized by the spatial resolution Δx_D, which determines, to a great extent, the value of $\Delta\Omega_2$ [see equation (8.1)]. The detector's angular resolution is determined as $2\theta_D = \Delta x_D / R_D$, where R_D is the sample–detector distance, and the value of x_D changes sharply in relation to the type of detector. We shall go into more detail later when discussing the properties of present-day detectors (Section 8.2.2). Here, it is noteworthy that for the sequential mode of data collection, $\Delta(2\theta_D)$ is determined primarily by the collimation system of scattered radiation and for this reason may be greatly diminished. In the case of the position-sensitive detector (Figure 8.2b) the aforementioned value is determined by the properties of the detector itself because the detector faces no collimation system.

Any detector is featured by a dynamic range of measured intensities. This range may be characterized by two parameters, namely, the background level I_b (its own and background scattering) and the greatest possible detected intensity I_{max}. These parameters are important for small-angle scattering studies because the scattering intensity frequently changes by several orders of magnitude within a narrow range of values of s.

A separate part of any small-angle scattering experiment is the on-line connection of detectors with computers and the quantitative analysis of the measurement accuracy. The latter is absolutely necessary in order to apply successive procedures of data treatment (Chapter 9) and to carry out structure determinations (Chapter 3–7). Up-to-date detectors combined with computers provide an analysis of the accuracy of the intensity and the scattering angle measurements.

It is very often necessary to measure the scattering intensity on an absolute scale, i.e., to measure the ratio between the incident and scattered intensities. Direct measurements are very difficult because this ratio is usually too high. For this purpose one uses secondary standards (see Section 3.3.2), such as a platelet of polyethylene. The quotient of the scattered and primary beams must be determined with a high degree of accuracy. Thus, the above-mentioned general characteristics of small-angle scattering instruments permit one to compare various X-ray and neutron instruments.

8.2. Laboratory X-Ray Instruments

Most small-angle scattering structure research is done with laboratory instruments that utilize commercial sealed tubes as the X-ray source. In the last decades several patterns and designs of laboratory instruments have been developed for scientific investigations and practical purposes.

8.2.1. X-Ray Tubes

Structure analysis is conducted with the characteristic X-ray radiation of metallic atoms, where the excitation is provided by electrons of energy 10^4–10^5 eV, and this is the operating principle for many of the high-vacuum sealed tubes available commercially. These tubes are produced with various metallic anodes, of different sizes of focus and differing powers. The wavelength of the characteristic radiation applied in small-angle scattering experiments is usually variable within the range of 0.71 Å (Mo) to 2.3 Å (Cr). As a rule, wavelengths of greater value are not used owing to the high absorption factor. However, a carefully thought-out approach enables longer wavelengths to be employed. In practice one uses K_α-radiation; other characteristic lines and white radiation are cut off by a monochromator system. X-ray tubes with anodes of Mo(0.71 Å), Cu(1.54 Å), Ni(1.65 Å), Fe(1.93 Å), Cr(2.3 Å), as well as some others, are widely used in routine laboratory work.

The sizes of focal spots are varied from tens of micrometers to tens of millimeters and mainly determine a construction of collimation systems. There are two types of X-ray tube: the first has isometric, point focal spots, the second linear focal spots. In the latter case the width is a factor of 50–100 smaller than the height. The isometric focal spots are used for point collimation systems (Section 8.2.3) and the linear sources for slit collimation systems (Section 8.2.4).

As a rule, the power of an X-ray tube ranges from several watts to several kilowatts. However, the specific X-ray power is much higher for a small focal spot, because in this case conditions for anode cooling are more favorable. Hence, the most powerful laboratory X-ray sources are tubes with rotating anodes (up to 100 kW). The intensity of an X-ray beam originating from rotating anode tubes is some dozen times greater than under routine conditions. The vacuum is provided by constantly running pumps, and so the stability of this type of X-ray source is much less than in sealed-off tubes.

8.2.2. X-Ray Detectors

Nowadays there are several widely used methods and a great number of models and devices to detect soft X rays (photon energies between 5 and 20 keV), which are used for small-angle scattering studies (see Timothy and Madden, 1983). This branch of experimental technique is progressing very rapidly. For this reason we shall treat only fundamental possibilities of X-ray detection as a part of small-angle scattering devices. As we noted above (Figure 8.2), the type of detector determines one of two principally different methods of measurement: the sequential or parallel mode of data collection.

First, photographic film has been used extensively as a detector for X rays. This is a parallel mode of data collection with high spatial resolution (about 10 μm), but unfortunately with low counting efficiency (~ 1%), poor accuracy of intensity measurements, and low dynamic range. Since the 1950s radiation counters have been widely employed, such as scintillation and proportional detectors of various designs over large ranges of energy. Energy resolution for soft X rays is 15–50% for such detectors. These counters are efficient and very convenient for data collection and computer calculation and are the basis of various sequential-mode small-angle scattering diffractometers. Semiconductor detectors have been designed, and their energy resolution is so high (100–200 eV) that they can be used as energy dispersive detectors.

In the small-angle scattering instrument with a sequential mode of data collection, a detector moves step by step and one may successively measure the number of photons at every angle. Spatial resolution of a detector is determined by a collimation system of the scattering beam and if necessary, may be very high. In such a case the precision of the scattering-angle determination is also very high because it is a function of the mechanical accuracy of the detector's motion. The statistical accuracy of the intensity measurements depends on the nature of the object, and errors do not usually exceed 1%. The weak features of all instruments with a sequential mode of data collection are the extended time needed for measurements, the necessity of high stability of the radiation source, and the large dose of sample radiation exposure.

Finally, the 1970s saw the introduction of one- and two-dimensional position-sensitive detectors in X-ray and neutron structure studies (Hendrix, 1985), which are able to perform simultaneous measurements of the scattering intensity and scattering angle. Thus in position-sensitive detectors we observe a remarkable combination of the positive properties of photographic film and the one-channel counter. However, for small-angle scattering experimentation one faces specific problems, such as spatial resolution of a detector, dynamic range of the intensity, heterogeneity of the properties of the detector along its surface, stability operation, setting of the zero-angle point, and so on. It is appropriate to note the averaged characteristics of the gas-filled proportional chambers: the counting efficiency for the energy interval 1–50 keV is 100–10%, spatial resolution 0.05–1 mm, counting rate 0.1–10 MHz, and energy resolution 10–30%.

Solid-state silicon imagers were recently applied to X-ray detectors and imaging (Allinson, 1982; Borso, 1982). These solid-state devices have some advantages, such as stability of performance, small dimensions, low operating voltage and current. Detectors of such a type exhibit high spatial resolution, large dynamic range, absence of "beam" lag, and high linearity, properties that are very favorable in small-angle X-ray scattering studies. There are two basic forms of silicon imagers: the photodiode array and the change-coupled device. The former exibits higher speed, the latter essentially lower background. Area imagers have been designed but today, mainly line imagers used. Averaged spatial resolution for currently available line imagers is about 15 μm, and the speed is about 20 MHz. It should be noted that solid-state imagers are being developed very rapidly and many new devices based on new principles will be designed in the near future.

8.2.3. Point Collimation System

The natural way to obtain a narrow beam in a small-angle scattering camera is to set the circular diaphragms suitably, the point collimation system (Figure 8.1). This scheme permits perfect approximation of plane-wave conditions used in most theoretical calculations pertaining to scattering-phenomena research. Apart from the above-mentioned two diaphragms that form the narrow primary beam, a third diaphragm is arranged which shields scattering by the second one. This allows measurements very close to the projection of the primary beam. At times, a focus of the X-ray tube acts as the first diaphragm.

These cameras are easily designed and constructed. Calculation of the geometrical parameters of point-collimation cameras requires little effort and conditions governing optimum setting of the diaphragms are indicated in some papers (Gerasimov, 1970; Mildner and Carpenter, 1984). These cameras combined with fine focus X-ray tubes and photofilms, as a detector, have been commonly used for many years. One of the first versions was built by Kiessig (1942). Point collimation cameras are characterized by very weak incident beams the power of which decreases dramatically with increase in spatial resolution. Photographic materials usually used with these cameras are of low efficiency. Thus, for a long time point collimation cameras were used for studies of strong scattering objects, mainly oriented ones.

Today, point collimation cameras have been revived but in combination with powerful X-ray sources and new types of detectors. These new cameras are built for synchrotron radiation sources and powerful X-ray rotating anode tubes. As detector one can use both two-dimensional position-sensitive detectors and photographic film.

Such an instrument was designed and constructed at the National Center for Small-Angle Scattering Research (Oak Ridge, USA) and has been in routine operation for several years (Hendricks, 1978). This remarkable device comprises a high-power rotating anode X-ray tube, a pyrolytic graphite monochromator on the incident beam, a point collimation system, and a two-dimensional position-sensitive detector. The instrument permits one to utilize the two wavelengths, 1.542 Å and 0.707 Å. The sizes of the X-ray focal point, the sample, and the resolution element of the detector are each approximately 1 mm × 1 mm, which led to the formation of long distances between the tube, the sample, and the detector. The whole length of the camera is 10 m. The distances between the focal point and sample and between the sample and detector may each be varied in increments of 0.5 m up to 5 m. The two aforementioned wavelengths yield a resolution range of s equal to 3×10^{-3}–0.6 Å^{-1}. Maximum flux of the beam at a specimen is 10^6 photons/s. The camera has one remarkable feature: a large fixed scattering sample area measuring 1 mm × 1 mm. The large camera length makes it most convenient for setting various special sample chambers. The active area of the detector is 20 cm × 20 cm.

Another instrument possessing the same spatial resolution — but not so large — was designed recently by Yoda (1984) and consists of a microfocus rotating anode X-ray tube, a point collimation system, and a linear position-sensitive detector. Monochromatization and formation of a beam are achieved using a bent glass mirror and a bent quartz single crystal. This fine focusing system yields scattering angles from 0.5 to 100 mrad. The sample-to-detector distance may be varied from 0.5 up to 1.6 m. A linear detector (active area, 100 mm × 10 mm) can rotate around the beam, so one may study scattering by oriented objects.

8.2.4. *Slit Collimation System*

Isotropic small-range scattering is studied with the aid of cameras with a slit collimation system and these are very widespread. Most laboratories are equipped with such devices. Therefore we shall deal with them in detail. The application of slit systems increases the light-grasp of a device by at least 100–1000 times, contrary to point collimation. However, slit collimation leads to a dramatic distortion of the scattering intensity function. Taking these collimation distortions into account is one of the essential data treatment procedures (Section 9.4). These corrections must be combined with those arising from the other apparatus distortions, such as polychromaticity of the radiation source, instability of the primary beam, and statistical errors (Section 9.1).

We shall now examine the formation of the primary beam in a slit goniometer (Figure 8.3). Even in the simplest small-angle scattering goniometers one cannot be restricted by two slits. A third slit is necessary for decreasing scattering by the edges of the second one. Many variants of such a scheme are implemented in laboratory practice. In particular, commercial versions (e.g., Rigaku Denki, Japan) are widely known. Collimators of this type are used in AMUR cameras designed in the Institute of Crystallography of the USSR Academy of Sciences. This device permits measurements to be carried out in both the sequential (Sosfenov *et al.*, 1969) and the parallel (Mogilevsky *et al.*, 1984) modes of data collection.

Figure 8.4 shows the X-ray optical scheme of the AMUR diffractometer in a one-channel variant. The goniometer is placed on a massive optical bench on which the tube, monochromator, primary beam collimator, specimen holder, and movable bench with scattered beam collimator and scintillation detector are mounted. The angle divergence

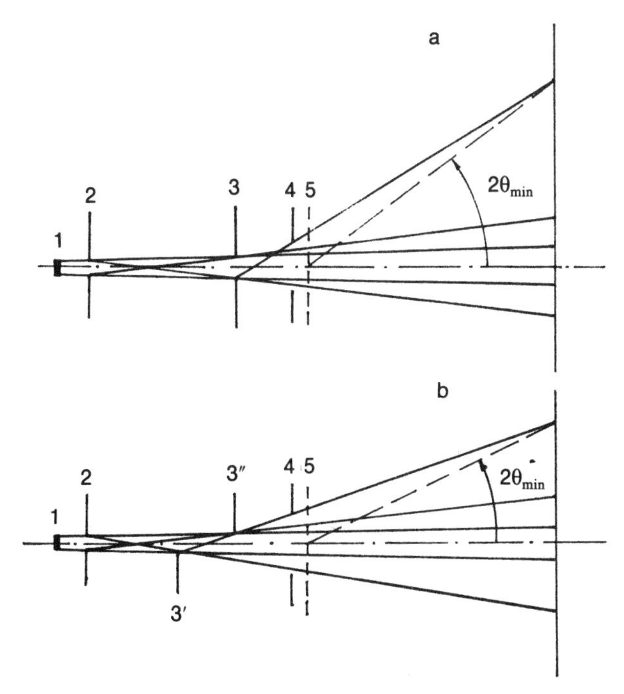

Figure 8.3. Routine scheme for a three-slit collimator of a primary beam (a) and its modification using separated edges of the second slit (b). Designations are the same as in Figure 8.1.

of the primary beam ranges from 15″ to 50′. This is attained by changing the width of the first and second slits within the limits of 0.02 to 2 mm and by changing the primary beam collimator length from 150 to 600 mm. The scattering angle variation is restricted to the area $-3°$ to $+20°$, the precision of angle setting being no worse that 5–10″. The electron circuit of the scintillation counter has a background of less than 0.1 pulses/s.

In the second variant of this instrument with parallel mode of data collection, a linear position-sensitive detector is used. The detector (Baru *et al.*, 1978; Aultchenko *et al.*, 1983) is a proportional chamber with seven anode filaments and a delay line. Spatial resolution of the detector is about 0.1 mm, which permits electronic resolution of 0.01° for one channel of an analyzer (for a sample-to-detector distance of 700 mm). The detector has an active area of 100 mm × 10 mm and a counting rate of 100 kHz. The camera is also supplied with the Bonse–Hart attachment (see below) for studying very large particles.

A somewhat different component for diminishing the background scattering is incorporated in a "block" collimation system, which is more widely known as the Kratky camera (Kratky, 1954).

In this system (Figure 8.5) a beam is formed by two blocks B_1 and B_2 and also by the entrance slit S (A is a focal plane). The shape of the primary beam projection on plane R is an asymmetric triangle. Point m_0 shows the shift of the triangle's summit in relation to the plane H. The magnitude m_0 determines the critical value $2\theta_{min}$. If one succeeds in achieving coincidence of the projection of O_1 and O_2, there will be no parasitic scattering above the plane H. This ought to be attributed to the fact that scattering by the edges k_1 and k_2 is displayed only below H and the primary beam is not in contact with edge k_2'.

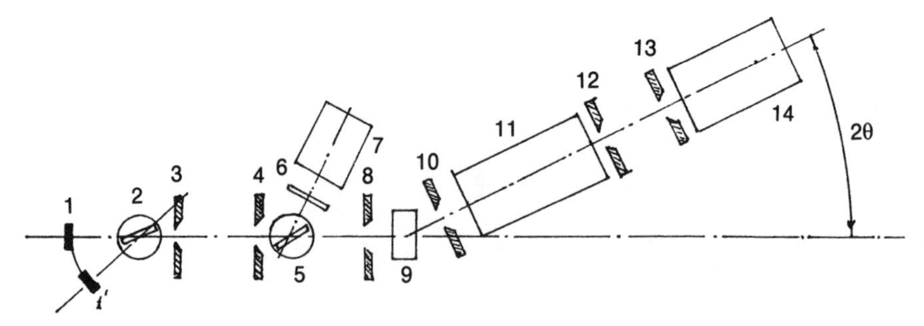

Figure 8.4. Schematic view of the AMUR-1 goniometer (after Sosfenov *et al.*, 1969): (1) focal spot; (2) crystal monochromator; (3) first slit; (4) second slit; (5) crystal for primary beam control; (6) absorber; (7) primary beam detector; (8) third slit; (9) sample; (10) first receiving slit; (11) vacuum chamber; (12), (13) receiving slits; (14) scattered radiation detector.

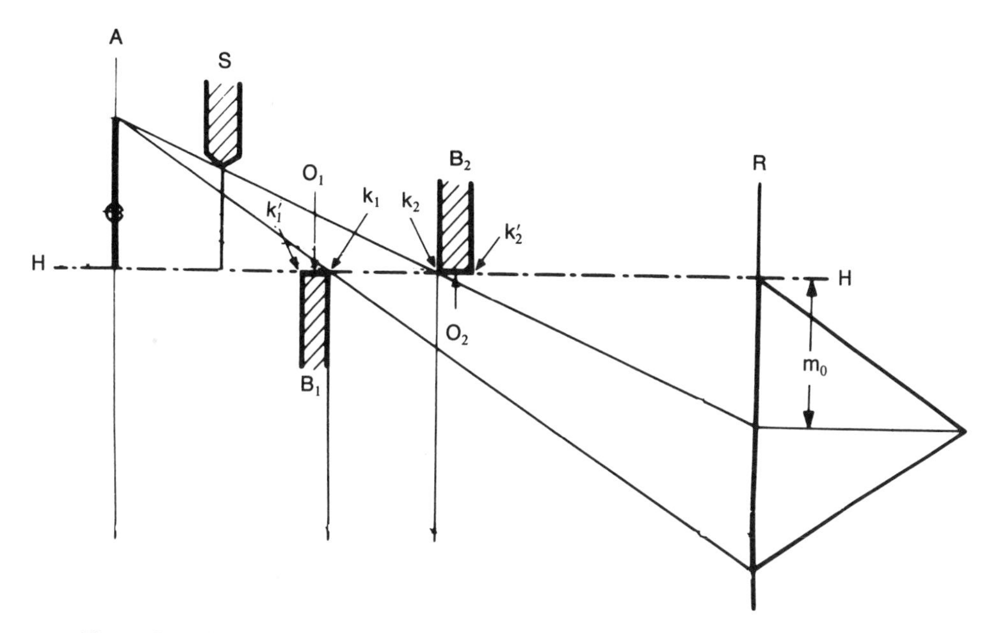

Figure 8.5. Schematic drawing of the "block" collimation system (after Kratky and Stabinger, 1984). Explanations in the text.

These collimators allow one to study space heterogeneities up to 10^4 Å. It is advantageous that the Kratky camera reveals small sizes, namely, the collimator is about 150 mm and the sample-to-detector distance is 200–400 mm. The collimator and the scattering beam path are evacuated. Many laboratory-made and commercially available devices are constructed according to this scheme, and undoubtedly the Kratky camera is the most widely used device for small-angle X-ray scattering studies throughout the world. The comprehensive description of modern devices of this type is given by Kratky and Stabinger (1984).

Another technique for obtaining narrow beams was proposed by Bonse and Hart (1965) and is used for measurements at very small angles (Figure 8.6). The X-ray beam reflects many times from two parallel faces formed by a groove in a perfect single crystal (such as germanium). This scheme allows one to obtain a very narrow monochromatic beam with a divergence of several arc seconds. Scattered radiation falls on the second crystal — the analyzer — which can be rotated about an axis perpendicular to the plane of the paper. A rotation of the analyzer through angle 2θ permits the scattering intensity to be measured only for exactly the same angle. The intensity of the incident beam in the Bonse–Hart geometry is sufficiently weak but it does not depend on the spatial resolution

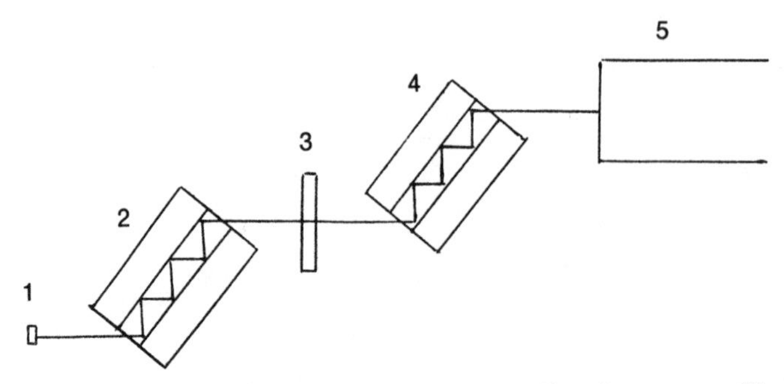

Figure 8.6. General outline of a very-high-resolution collimation system: (1) radiation source; (2), (4) crystal forming the beam and analyzing the scattering intensity; (3) sample; (5) detector.

achieved. Thus, from some value of the spatial resolution (and above) this method gives a more powerful primary beam than the slit collimation system. The actual critical spatial resolution depends on the wavelength used, but the Bonse–Hart camera is usually preferred for several thousand angstroms. It should be noted that it is not clear how position-sensitive detectors can be used in the Bonse–Hart camera. Therefore, for such high spatial resolution, the sequential mode of data collection must be employed.

8.3. Synchroton Radiation Instruments

In the case of a synchrotron source, X rays are created by an accelerated motion of electrons or positrons with relativistic velocities. This X radiation limits the energy of the accelerated electrons and is known as synchrotron radiation. Being harmful for accelerating purposes, it is nowadays a powerful source of electromagnetic waves in the far-ultraviolet and X-ray range of the spectrum. Synchrotron X radiation has considerably extended the possibilities of X-ray application in small-angle scattering studies. This has also led to a significant change in the design and construction of such devices.

8.3.1. Main Characteristics of Synchrotron Radiation

This source of X rays has created a new era in structure investigations of the condensed state of matter (Koch, 1983; Stuhrmann, 1982). A remarkable combination of several properties of the beam plays a determining role in solving structure problems:

1. Very high intensity.
2. Very broad continuous spectral range.
3. Narrow angular divergence.
4. Small focal size.
5. High degree of linear (circular) polarization.
6. Regularly pulsed time structure.
7. Ultrahigh vacuum environment.

These properties permit one to undertake basic and applied research, unthinkable without the use of synchrotron radiation. For small-angle scattering the main advantages are high intensity, narrow angular divergence, and broad spectral range. The last property allows implementation of a conceptually new technique — complex contrast variation (Section 4.3.3), which involves use of an anomalous dispersion of X rays.

8.3.2. Monochromatization and Focusing of X Rays

We have seen (Section 8.1.1) that some degree of monochromatization is necessary for all kinds of small-angle intensity measurements. When we use the characteristic radiation from metals bombarded by electrons as an X-ray source, monochromatization is achieved partly naturally within the source itself. Usually one utilizes the K lines, while the other (less intensive) characteristic lines and the relatively weak white spectrum are eliminated by various routine methods, such as selection of special operating conditions for an X-ray tube, which gives the maximum ratio of K-line intensity to white radiation; choice of selective filters; and energy discrimination of photons.

For synchrotron radiation characterized by a broad continuous spectral range of X rays, however, monochromatization is a most important procedure. Isolation of a narrow band of the spectrum is accomplished by using the reflection from a single-crystal face, when the Bragg conditions $n\lambda = 2d \sin\theta$ are fulfilled for one or another wavelength. A continuous change in λ at the monochromator output is achieved by varying the angle θ. Now Ge(1111) and Si(220) single crystals are frequently utilized and they enable one to obtain the interval of wavelengths from 0.6 Å (Si, 9°) up to 3.25 Å (Ge, 30°) over a convenient range of reflection angles. This range of λ covers most of the small-angle scattering needs, apart from cases when anomalous scattering is used. For softer radiation it is advisible to use crystals with greater interplanar spacing.

The spectral width following monochromatization by a single crystal is determined by both the crystal properties and the geometrical parameters. A working equation that gives the resolution of a monochromator in a practical situation is

$$\frac{\Delta\lambda}{\lambda} = [(d^2F_{hkl})^2 + (\text{ctg}\ \theta\Delta\theta_g)^2]^{1/2}$$

where d_{hkl} is the Bragg spacing of the planes in use, F_{hkl} is the structure amplitude for the reflection, and $\Delta\theta_g$ is the angular aperture allowed by the slit system and source dimension. When using the two-crystal monochromator, $\Delta\lambda/\lambda$ often attains the value 10^{-4}. But this high-wavelength resolution leads to a significant loss of intensity, so a very high-resolution system is not used often.

As a rule, small-angle devices operating with synchrotron radiation are supplied with focusing systems. One often uses an asymmetric cut crystal monochromator, the Fankuchen scheme (Figure 8.7). The crystal face (2) is cut at an angle φ (7–10°) to the major set of planes. In this case the monochromatic beam is foreshortened and it is important when one deals with preparations available in small amounts. This method also allows high harmonics to be eliminated. A linear beam is obtained by frequently employing two crystal monochromators: plane and cylindrically bent. The latter provides not only monochromatization, but also a focusing effect, while the beam density flow increases.

For a synchrotron beam, divergence at the horizontal plane may be sufficiently reduced by focusing, which is often accomplished by reflection from a mirror surface using the total reflection effect of X rays. The refractive index for X rays is given by

$$n = 1 - \frac{\rho e^2\lambda^2}{2\pi mc^2}$$

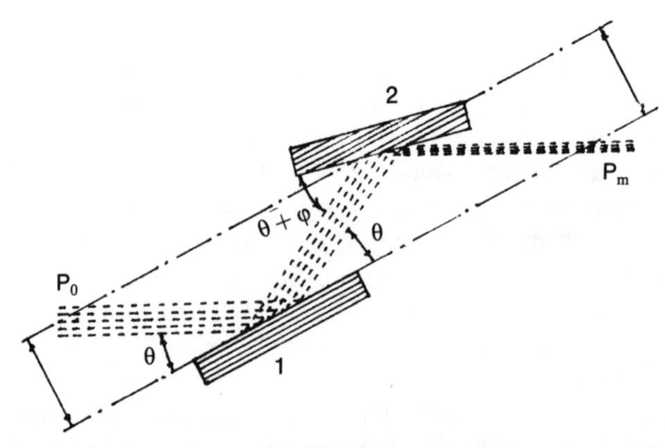

Figure 8.7. Monochromatization and contraction of the primary beam using two single crystals (1) and (2). Crystal (2) is asymmetric and its face cut at angle φ (7–10°) to a set of reflecting planes; P_0 denotes white radiation and P_m the monochromatic beam.

where e and m are the charge and mass of the electron, respectively, ρ is the electron density of a mirror, and c is the velocity of light. The second term in this expression is small in comparison with unity, thus at sufficiently small angles the refractive beam does not exist. For angles of incidence smaller than the critical value, only the total reflection beam will exist. A simple equation relates the critical reflection angle to the wavelength θ_c:

$$\theta_c = \lambda \left(\frac{\rho e^2}{\pi m c^2} \right)^{1/2}$$

Hence for a given angle of incidence, total reflection occurs only for wavelengths greater than λ_{cr}. Thus, total reflection is used to cut off the short-wavelength part of the beam. More frequently, mirrors from fuse quartz, for which θ_c is equal to 0.00263λ (Å), are utilized. For gold plating, θ_c is 2.4 times greater.

If one wishes to obtain the "point" image of a beam, both horizontal and vertical focusing must be used. For a point source the mirror surface should be ellipsoidal. But for synchrotron radiation facilities in which the distances are sufficiently long cylindrically bent mirrors can be used, which consist of a set of flat mirrors. It is recommended that these mirrors be placed behind the monochromators, if possible, because white radiation very rapidly damages the plating and the mirror's surface. It was shown experimentally that single crystals are more stable to the white X-ray spectrum than glass or quartz mirrors with or without metal plating.

8.3.3. Small-Angle Synchrotron Instruments

Devices intended for synchrotron radiation should combine a traditional collimation system and detectors with effective monochromatization and focusing. It is obvious that the whole construction should be bounded subject to specific conditions governing a given storage ring displacement (Koch and Bordas, 1983). Most instruments employ position-sensitive detectors or, in some cases, films. The only exception is the Bonse–Hart collimation system, usually assembled with a one-channel detector.

The instrument in the Laboratoire pour l'Utilisation du Rayonnement Electromagnétique (LURE, Orsay, France) was designed (Vachette, 1985) to provide maximum flexibility and ease of use (Figure 8.8). The monochromator is an 18-cm-long triangle of germanium using the (111) reflection in the Fankuchen compressing geometry with a cut angle of about 8° (or 10°). The angle of reflection can be adjusted by a step motor and is monitored by an absolute optical encoder. A small eccentric wheel

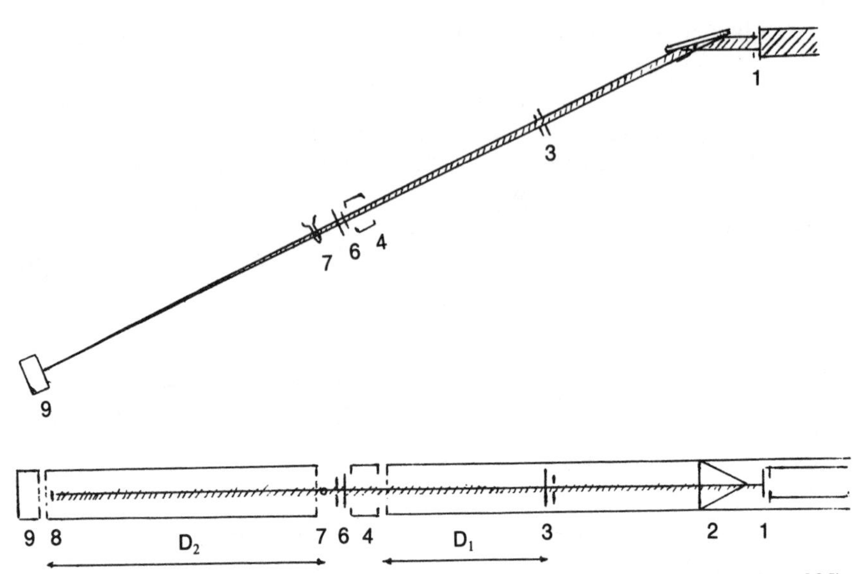

Figure 8.8. Scheme of the small-angle D24 instrument at LURE (after Vachette, 1985): (1) slits; (2) monochromator: bent germanium crystal; (3) collimating slits; (4) ionization chamber; (5) filter holder; (6) guard slits; (7) thermostatted sample holder; (8) beam stop; (9) linear position-sensitive detector.

at the top of the triangle allows bending of the crystal and subsequent focusing of the beam. The crystal intercepts more than 40 mm of the white beam in the horizontal direction; the width at half maximum focus is between 0.8 and 1.1 mm, depending on the focal distance.

All the elements downstream of the monochromator are positioned on a 3.8-m-long granite bench. About 2.5 mm of the beam in the vertical direction is intercepted by a 40-cm-long gold-coated bent mirror and refocused in 0.5 mm. The sample-to-detector distance can be varied from 20 cm up to 2 m. Position-sensitive detectors with delay-line readout are used, linear as well as annular. The annular detector is 50 mm in radius. The smallest angle is equal to $2\text{–}3 \times 10^{-3}$ rad. Usually one uses a wavelength of 1.6 Å.

A new diffractometer (Ameniya *et al.*, 1983), based on the rich experience of European laboratories, has been built at the Photon Factory (Tsukuba, Japan). Apart from focusing in a horizontal plane by an asymmetrically cut germanium single crystal, vertical focusing is accomplished with the aid of a 1.4-m-long bent quartz mirror. The mechanical and optical parts of the instrument are manufactured very carefully and a point-focused beam can be obtained with variable wavelengths and high-quality monochromatization. The main characteristics of the diffractometer are: spatial resolution up to 1000 Å,

1 Å $< \lambda <$ 2 Å, $\Delta\lambda/\lambda \sim 10^{-3}$, sample-to-detector distance 30–300 cm, beam intensity 10^{11}–10^{12} photon/s, and beam sizes in the sample plane 5–10 mm × 2–3 mm.

Specific instruments are designed for small-angle studies using anomalous scattering. For this purpose it is necessary to use a higher wavelength — up to 10 Å. The X15 diffractometer that was built several years ago (Stuhrmann and Gabriel, 1983) in the European Molecular Biology Laboratory (Hamburg, FRG) incorporates a monochromator that contains two parallel distanced flat crystals, each of which can rotate and operate independently. This system deflects the monochromatic beam 1.2 m above the level of the incident beam (it greatly diminishes background scattering). A wavelength range 0.6 Å $< \lambda <$ 3.2 Å can be achieved. The instrument is assembled with a two-dimensional position-sensitive detector 200 cm × 200 cm in area with 2 mm resolution. The sample-to-detector distance may be varied from 0.5 up to 7 m. The overall length of the instrument is 15 m.

A new instrument for anomalous scattering studies has been developed by Stuhrmann (1985). The focusing system (Figure 8.9) comprises two mirrors and one single-crystal monochromator. The first, a toroid mirror (1 m long and 8 cm wide), provides for focusing and preliminary monochromatization by cutting off the wavelengths $\lambda <$ 1 Å. The second, a face mirror, consists of two changeable plates, one of quartz and the other with a gold coating. The latter does not change the spectrum appreciably, and the former cuts off all wavelengths shorter than 2.5 Å. Beyond the second mirror the beam again becomes horizontal. In the vertical plane monochromatization is achieved by reflection from a single crystal. To obtain the desired wavelength range (0.8 Å–8 Å) it is necessary to use three crystals: Ge(111), InSb, and quartz sequentially. The minimum and maximum sample-to-detector distances are 35 cm

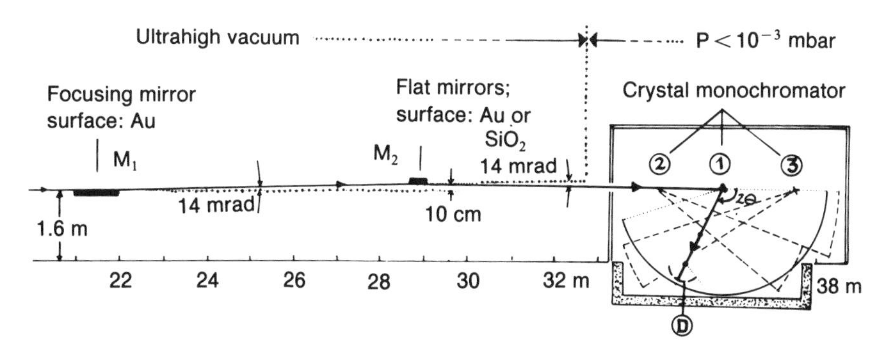

Figure 8.9. Main features of the instrument for small-angle anomalous scattering studies (after Stuhrmann, 1985): M_1 denotes the toroid mirror, M_2 the plane mirror, and D the detector.

and 3 m, respectively. Spatial resolution of the instrument is about 1000 Å.

Some other small-angle scattering cameras for synchrotron radiation have been designed in the last decade. One of the first was built at storage ring VEPP-3 in Novosibirsk (Vazina *et al.*, 1975) for studying the kinetics of muscle contraction. Now, one of the main hopes in the application of synchrotron radiation is the study of dynamic phenomena in biological and crystalline systems. Special cameras and data acquisition systems have been developed for such investigations (Bordas *et al.*, 1980; Koch and Bordas, 1983). These devices are intended for time-resolved X-ray scattering measurements and the time resolution achieved by now is of the order of a few microseconds and higher.

8.4. Small-Angle Neutron Scattering Apparatus

The optical schemes of neutron and X-ray cameras are very similar. However, there are several specific features which totally change the design and construction of neutron instruments even in comparison with X-ray facilities for synchrotron radiation. These features include attachment of a collimator to the reactor zone, monochromatization principles for neutrons, low neutron velocity, specific aspects of the interaction of neutrons with a sample, and small attenuation of neutron beams, among others. The number of small-angle neutron scattering instruments does not essentially exceed one dozen, each representing a large technical installation sometimes several dozen meters long.

All instruments for neutron scattering studies are supplied with monochromators and position-sensitive detectors, and with one or two minicomputers, frequently on-line, connected to a large computer. Instruments are equipped with special sample chambers (low and high temperatures, high pressure, electric and magnetic fields, etc.). Each instrument is available for a wide community of users, frequently from many countries. These instruments are continually being refined, in particular the monochromators and detectors, and their connections to computers. For this reason we shall review only some of these instruments from various countries and describe their main features.

8.4.1. Neutron Sources, Monochromatization, Detectors

Powerful sources of neutron beams are provided by nuclear reactors. These reactors are of two types: stationary reactors that permit one to obtain the continuous-in-time flux of thermal neutrons, and pulse reactors giving very powerful pulses of neutron beams with relatively low average power.

The problem we face is to form a spectrum of the incident beam and the beam geometry. As the reactor radiation contains γ rays and fast neutrons, and a spectrum of emitted neutrons is polychromatic, filtration and monochromatization of the beam is required. Filtration of γ rays and fast neutrons is accomplished using a neutron cut-off system. The idea behind this device is based on the phenomenon of the mirror reflection of neutrons by the walls of a neutron-guide tube. In such a system all neutrons with $\lambda < \lambda_{cr}$ and γ rays are extinguished by shield walls, and only neutrons with $\lambda > \lambda_{cr}$ reflect from the walls of a neutron-guide tube and move ahead toward the sample. A monochromatic system depends on the type of reactor. In the case of stationary reactors, single crystals, mechnical selectors, and magnetic monochromators of polarized neutrons are utilized. With pulse reactors one uses the time-of-flight method, which permits the flux of the beam to be optimized.

In neutron systems there is a possibility of considerably increasing the wavelength used. As a matter of fact, the wavelength spectrum in the reactor's outlet reveals the Maxwellian distribution of the velocities with a maximum at 1.1–1.6 Å. The use of the cold moderators of neutrons permits the maximum of the wavelength distribution to be shifted toward longer wavelengths. Now, with the cold sources, it is possible to use neutrons with wavelengths 5 Å and 10 Å, and even up to 20 Å, for structure investigations. This is one of the greatest advantages of neutron scattering, namely, it allows one to altogether exclude double Bragg reflections, providing the utilized wavelength satisfies the condition $\lambda > 2d_{max}$. This situation cannot be implemented practically in X-ray research owing to the very high absorption of radiation with long wavelengths.

On the other hand, focusing systems widely used in X-ray experimentation are almost completely unusable in the case of neutrons because of the large size of the reactor's active zone. For scattered neutrons, gas-filled (BF_3 or ^3He) Geiger counters are used as detectors. The counting efficiency is very high and reaches 90–95%. Several multidetector systems have been designed recently, in two variants: as a system of individual counters or a position-sensitive single chamber. However, in both cases the resolution is no better than several millimeters, which is the reason that small-angle neutron scattering diffractometers are so large. The multichannel detector system using ^3He-filled counters was described by Agamalyan *et al*. (1983).

8.4.2. Collimation Systems and Instruments

We consider several models of small-angle neutron instruments. The National Center for Small-Angle Scattering Research (USA), a good example of an institution active in this field, is equipped with five instruments: three for neutron studies and two for X-ray studies. The characteristics of one of them are described in Section 8.2.3; another is the Kratky camera with a linear position-sensitive detector.

The main small-angle neutron scattering instrument is a 30-m-long camera (Koehler and Hendricks, 1979). The instrument utilizes point collimation geometry; monochromatization is achieved by a flat graphite crystal at a primary beam; scattered radiation falls on a two-dimensional position-sensitive detector with active area 64 cm × 64 cm, and resolution element dimensions 5 mm × 5 mm. The wavelength has two fixed values, 4.75 and 2.38 Å; $\Delta\lambda/\lambda$ is 6%. The source-to-sample distance is 10 m, the sample-to-detector distance 1.5–18 m. A spatial resolution covers the 5×10^{-3} Å^{-1}–0.6 Å^{-1} range of s.

Another neutron facility (Christen *et al.*, 1980) that is designed for measurements at very small scattering angles is built as a two-crystal spectrometer using silicon single crystals for the monochromator and analyzer. It utilizes the slit collimation system for primary and scattered beams (4 cm × 2 cm on the sample surface) and a one-channel detector; wavelength $\lambda = 2.6$ Å and $\Delta\lambda/\lambda$ is 2%. The scattering intensity can be measured for values of s from 1×10^{-3} Å to 0.2×10^{-2} Å^{-1} with resolution 2.5×10^{-4} Å^{-1}. The features of the instrument include devices for measurements at low temperatures and in magnetic fields.

A new instrument of this type with extremely high angular resolution (several arc seconds) was described by Rauch and Schwann (1984). It permits the use of large wavelengths to attain a spatial resolution up to 10,000–100,000 Å and may be applied for studying some biological objects, critical phenomena, and phase transitions in polymers.

Several neutron facilities have been designed in the Laue–Langevin Institute (Grenoble, France). Probably the best known is the D11 instrument (the first model), built in the early 1970s (Ibel, 1976). The basic scheme of the instrument, shown in Figure 8.10, includes a rotating selector with helical grooves for monochromatization of the cold-source neutron beam, an almost point collimation system of the incident beam, and a two-dimensional detector of scattered radiation. Some features of the device are evident from the scheme. The wavelength range lies between 4.5 and 20 Å (cold source), and $\Delta\lambda/\lambda$ is equal to 0.50 and 0.09. The size of a beam on a sample is 3 cm × 5 cm. Electronic resolution of the detector is 1 cm × 1 cm. The instrument also permits measurement of the scattering intensity at middle and large angles. The overall length of this giant instrument is almost 100 m.

The "Membrana II" instrument (Agamalyan *et al.*, 1984) is in operation at Gatchina (Leningrad district, USSR). A scheme of the instrument is presented in Figure 8.11. The use of a magnetic monochromator of the polarized neutrons and a multichannel detector with extremely low background level in every counting channel (2–4 pulses

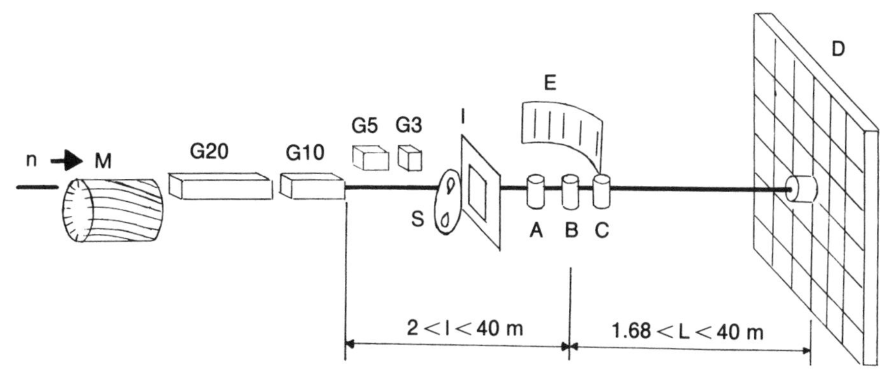

Figure 8.10. General view of the D11 small-angle camera (Laue–Langevin Institute, Grenoble, France): n denotes the beam flux from the cold source; M the drum of the mechanical slot detector; G3, G5, G10, G20 movable neutron guides allowing different collimation lengths of 3 m, 5 m, 10 m, and 20 m, respectively; S the chopper; I the diaphragm; A, B, C different sample positions; E 32 single counters for diffuse scattering, and D the multidetector.

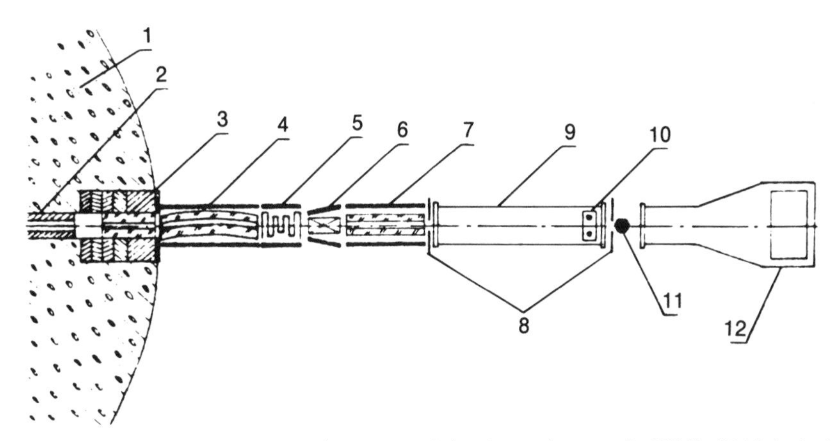

Figure 8.11. The "Membrana II" instrument (after Agamalyan *et al*., 1984): (1) biological shielding of reactor; (2) collimator; (3) striaght neutron guide; (4) curved neutron guide polarizer; (5) magnetic resonator; (6) adiabatic spin flipper; (7) straight neutron guide analyzer; (8) cadmium slits; (9) vacuum chamber; (10) monitoring unit; (11) sample holder; (12) detector system.

per hour) allow one to measure the small-angle scattering intensity even for weakly scattering objects in spite of a relatively small neutron flux. The main characteristics of the instrument are as follows: In the wavelength range 2.2 Å $< \lambda <$ 5 Å monochromatization $\Delta\lambda/\lambda$ achieves a magnitude of 0.07. Intensity on a sample is about 10^4 n/cm^2 s; the size of the beam on a sample is 0.7 cm × 6 cm; the source-to-sample and sample-to-detector distances are equal to 8 m. The number of counters is 41; the angle shift between neighboring counters is 2×10^{-3} rad. The values of s range from 10^{-3} to 0.6 $Å^{-1}$.

Pulse reactors will play a more active role as sources for small-angle studies. At the Joint Institute for Nuclear Research (Dubna, USSR) such an instrument has been built (Figure 8.12) using the IBR-2 pulse-source beam (Vagov *et al.*, 1983). A wide spectrum of neutrons (time-of-flight technique) and a ring position-sensitive detector are the principal features of the instrument and permit one to achieve a sufficiently high beam intensity and to preserve the advantages of a point collimation system (for isotropic samples only). However, spatial resolution is not so high: $8 \times 10^{-3}\,Å^{-1} < s < 2 \times 10^{-1}\,Å^{-1}$. A special procedure for the calculation of function $I(s)$ from experimentally measured data has been developed.

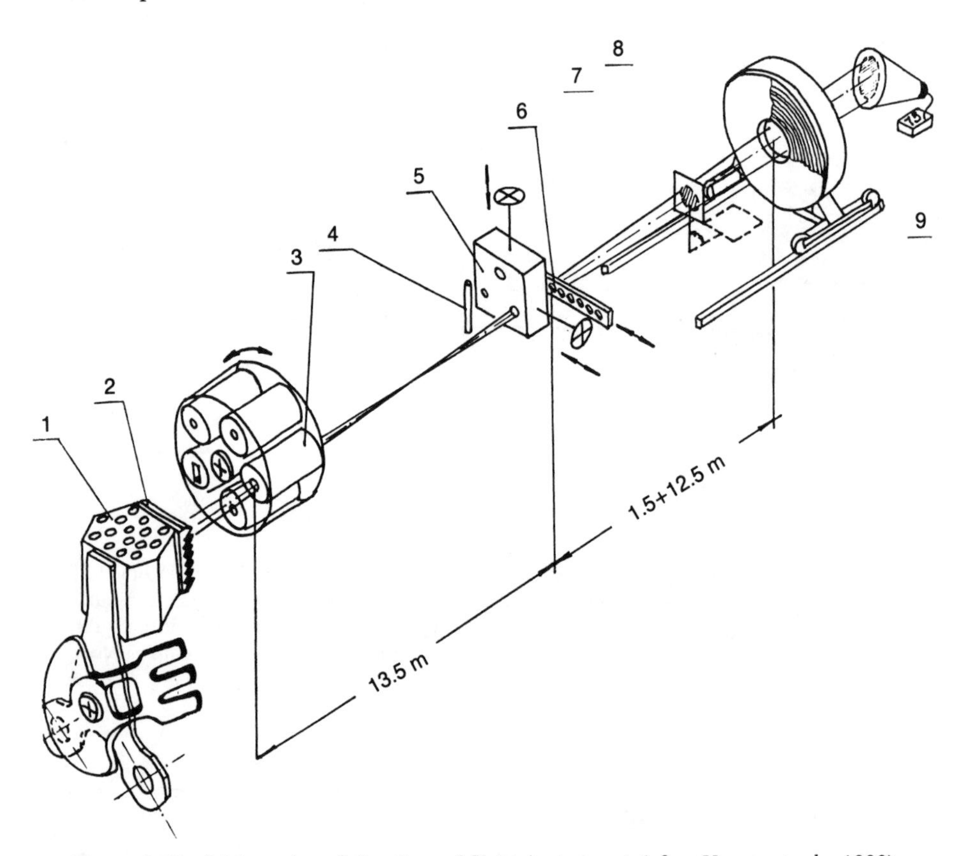

Figure 8.12. Main units of the time-of-flight instrument (after Vagov *et al.*, 1983): (1) active zone of the IBR-2 reactor; (2) moderator; (3) K1 changeable collimator; (4) monitoring, detector; (5) K2 adjustable collimator; (6) sample; (7) vanadium scatterer; (8) ring multiwire detector of scattered neutrons; (9) direct beam detector.

Aleferd and Conrad (1984) have described another instrument of this type, designed especially for research on biological molecules and cell organelles. The neutron spectrum ranges from 1 to 8 Å (maximum at 5.5 Å). The values of s are from 0.01 to 1 $Å^{-1}$ and the spatial resolution is 0.012 Å; the sample-to-detector distance varies from 0.3 to 2 m. The size of the two-dimensional position-sensitive detector is 60 cm × 60 cm, and the sample area is several square millimeters.

Some up-to-date approaches to the use of new small-angle neutron scattering facilities were presented at the Neutron Scattering in the "Nineties" Conference (1985).

8.5. Conclusion

Since the 1970s two new experimental possibilities for structure investigations have arisen, namely, the use of synchrotrons (storage rings) and nuclear reactors as powerful sources of X-ray and neutron beams, respectively. Both are used extensively in small-angle scattering research. These circumstances have limited small-angle scattering research to nuclear physics research centers in two senses: both locally and intellectually. In several well-known nuclear institutes (including Brookhaven, Oak Ridge, Stanford, Hamburg, Jülich, Grenoble, Orsay, Dursebury, Tsukuba, Novosibirsk, Gatchina and Dubna) very high-quality small-angle scattering facilities attached to synchrotron and neutron beams have been designed and constructed. However, most small-angle scattering investigations are still done using traditional sources, namely, X-ray tubes.

The instruments used with the three types of radiation sources are of various types and construction, and have been adapted to different pure and applied directions of research. Thus, there are instruments for general use and for particular purposes, e.g., for utilizing anomalous scattering, for oriented samples, and for time-resolved experiments.

Each experimentalist now has the possibility of selecting the optimum variant of an instrument for solving his particular problem. The choice of instruments is determined by two factors — technical and financial — because small-angle scattering facilities are very expensive, especially in the cases of synchrotron and neutron sources. The availability of such facilities is also an important consideration.

A comparison of various approaches to small-angle research is no easy matter. We shall stress only a few aspects of this question. The luminosity of X-ray synchrotron devices is about 100 times greater than for conventional X-ray tubes, but while one may utilize the latter on a continuous basis, synchrotron devices can usually be operated only for very short periods of time. In the case of neutron scattering we have no laboratory sources and equipment. For this reason one usually tries to carry out preliminary measurements with X rays before commencing neutron experiments. Small-angle neutron scattering installations, and thus research work, are more expensive than setups involving synchrotron radiation. It should be noted that incident beams in neutron installations are several orders of magnitude weaker than X-ray ones. However, owing to long wavelengths, large samples, optimum monochromatization, and other features, one can attain a scattering intensity in small-angle neutron experiments no worse than in X-ray studies.

On the whole, the development of small-angle scattering experiments now features an unusual dual situation. A certain proportion of the most complicated, purely scientific, and very important applied investigations have been carried out at synchrotron-radiation and nuclear-reactor centers, but most studies have been conducted with the aid of laboratory small-angle X-ray devices. This situation is reminiscent of a short period in the history of nuclear physics, in the late 1930s and early 1940s, when accelerators started to operate and most research was conducted at laboratory sources of radiation and with cosmic rays. In small-angle scattering studies this situation will last for some time yet.

9

Data Treatment

As shown in Parts II and III, the methods of structural interpretation of small-angle scattering data deal mainly with an ideal, absolutely precise intensity curve, namely, the function $I(s)$. In experiments, however, one determines not this curve but a certain set of points $J_e(s_i)$, which contain various distortions. Here it is necessary to take into account statistical noise, measurement-interval termination, scattering by various details of the experimental device and by the solvent, "smearing" of the ideal curve profile as a result of the finite dimensions of the beam and detector, polychromaticity of the radiation, and so on. Therefore, in an experimental data analysis the immediate task is, as a rule, transition from the set $J_e(s_i)$ to the function $I(s)$.

The present chapter deals with the principal methods of experimental data processing, which enables the transition to be performed. In Section 9.1 the sources of experimental data distortions are examined and mathematical expressions for writing down these distortions given. Section 9.2 is devoted to the preliminary data handling. In Section 9.3 various methods of statistical noise reduction are considered and their comparative analysis carried out. Sections 9.4 and 9.5 describe desmearing procedures, taking into account finite dimensions (Section 9.4) and polychromaticity of the beam (Section 9.5), which are involved in the solution of the corresponding integral equations. In Section 9.6, consideration is given to methods for minimizing distortions arising in both processing and interpreting small-angle data owing to termination of the experimental intensity measurement interval.

Methods involving simultaneous performance of several basic data evaluation procedures are widely applied. In Section 9.7 we examine techniques for uniting individual procedures (smoothing and collimation corrections, corrections for polychromatism) and methods that make it possible to eliminate all distortions simultaneously and allow immediate restoration of function $I(s)$ [or even $D(r)$].

9.1. General Scheme of Small-Angle Data Processing

Some idea about the general scheme of small-angle data evaluation can be obtained by first considering the main sources of distortions in the ideal scattering curve encountered in research. When possible, we shall give the equations corresponding to the influence of a given source of errors while assuming that all the others are absent.

9.1.1. Instability of Experimental Conditions

Errors that are very difficult to formulate mathematically include various changes in the recording conditions during experimentation. Among such errors only the instability of the radiation source itself can be easily taken into account. Corrections for fluctuations in the primary beam intensity I_0 can be introduced by measuring the scattering intensity of the standard sample before each experiment or at each experimental point, and then conducting appropriate experimental data renormalization. However, these are still several reasons for various distortions during an experiment, some related to the experimental device (spontaneous mechanical changes in instrument geometry, instability of beam pathway evacuation, electronics instability) and changes in the sample itself (temperature, destruction under radiation). It is rather difficult to discover such errors or to take them into account by means of mathematical operations. To make sure that no serious changes of such a nature have occurred during recording one normally carries out several successive experiments ("runs") with the same sample. Agreement of the results of these experiments may serve as a criterion for their reliability. When averaging such series of experiments, "bad" runs (or parts thereof) may be eliminated. The criteria of a "bad" run may be both quite subjective and statistically motivated (see Section 9.2).

9.1.2. Additive Scattering Components

When investigating a two-phase system (e.g., particle solutions), we are often interested in the scattering $I(s)$ by one phase (particles). The experimental curve also contains scattering by another phase (main matrix, solvent) $I_s(s)$ which, in most cases, is assumed to be additive (there is no interference between particles and solvent). In addition, the primary beam is scattered by camera units (collimating slits, sample holder, etc.), also contributing somewhat to the total scattering, which will be denoted by $I_a(s)$. To avoid these undesirable contributions, one should measure the scattering by a pure solvent $I_2(s)$ and subtract it from the scattering by the sample $I_1(s)$. These two additive contributions, $I_s(s)$ and $I_a(s)$, are eliminated by this procedure. If one takes into account the corrections for various absorption coefficients of the sample and solvent, μ_1 and μ_2, as well as for the primary beam intensity I_{01} and I_{02}, the sought scattering intensity can be written as

$$I(s) = \frac{\mu_1}{I_{01}} I_1(s) - \frac{\mu_2}{I_{02}} I_2(s)$$

In this case, as noted earlier (see Section 2.3.2), $I(s)$ corresponds to the excess scattering density.

9.1.3. Influence of Beam and Detector Dimensions

In most current small-angle instruments, slit collimation is used to increase the intensity of primary and scattered radiation. In addition, detectors of diffracted radiation also have finite dimensions. This leads to curve smearing over a certain interval of angles defined by the geometrical parameters of the collimation system, and the sizes and sensitivity of the detector. As a rule, such smearing results in serious distortions of the true curve profile and, consequently, leads to the necessity of restoring the scattering curve that would correspond to the ideal ("point") conditions of an experiment. The methods for restoring the "point" curve, i.e., of introducing the collimation corrections, are very important, since devices with point collimation for the time being have rather limited application (see Section 8.3.3).

We now examine the causes of collimation distortion and derive the general relationship between the ideal and distorted scattering curves. Figure 9.1 shows a scheme for a diffraction experiment in which the

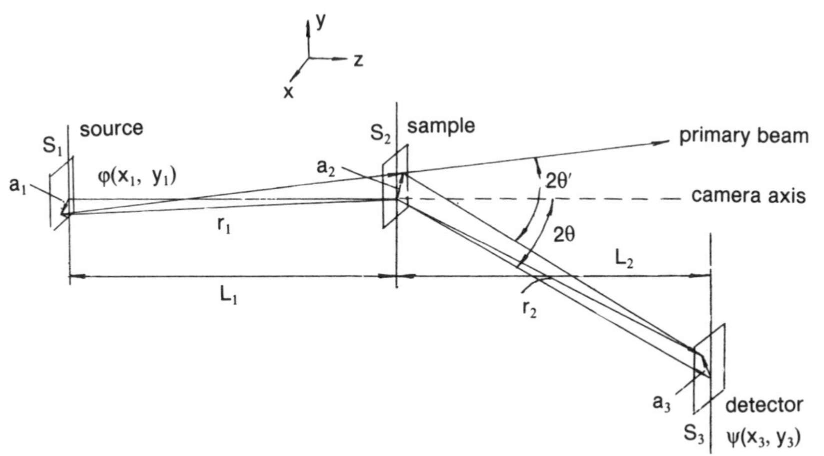

Figure 9.1. Smearing effects caused by slit collimation (scheme). Coordinates of the vectors: $\mathbf{a}_1(x_1, y_1, -L_1)$, $\mathbf{a}_2(x_2, y_3, 0)$, $\mathbf{a}_3[x_3 + L_2\,\mathrm{tg}(2\theta), y_3, L_2\cos(2\theta)]$.

radiation source, sample, and detector are considered as objects of finite dimensions. The recording system is in a position for which the actual scattering angle for point collimation is equal to 2θ. We consider an arbitrary detector-recorded ray, which may not pass through the center of the source, sample, and detector. If its path is described by vectors $\mathbf{r}_1$, $\mathbf{r}_2$, and $\mathbf{a}_2$, then the actual scattering angle is given by

$$2\theta' = \text{arc sin}\left\{\frac{|[(\mathbf{r}_1 - \mathbf{a}_2)(\mathbf{r}_2 - \mathbf{a}_2)]|}{|\mathbf{r}_1 - \mathbf{a}_2|\ |\mathbf{r}_2 - \mathbf{a}_2|}\right\}$$

The coordinates of vectors $\mathbf{r}_1$, $\mathbf{r}_2$, and $\mathbf{a}_2$ are shown in Figure 9.1 (the origin of coordinates is at the center of the sample). For small-angle scattering, the distances between the source, sample, and detector greatly exceed their dimensions, and the scattering angle is small enough to assume that $\text{tg}(2\theta) \approx 2\theta$ and $\sin(2\theta) \approx 2\theta$. In such a case

$$2\theta' \approx [y'^2 + (2\theta - x')^2]^{1/2}$$

where

$$y' = y_3/L_2 + y_1/L_1 - y_2(L_1^{-1} + L_2^{-1})$$

and

$$x' = x_3/L_2 + x_1/L_1 - x_2(L_1^{-1} + L_2^{-1})$$

At a predetermined angle 2θ, the detector will also record rays for which vectors $\mathbf{a}_1$, $\mathbf{a}_2$, and $\mathbf{a}_3$ lie respectively within the bounds of the source, sample, and detector. The full scattering intensity will be given by integration over all possible directions of these rays. It should be taken into account that at various points on the source and detector the intensity of radiation and the effectiveness of its recording may differ. In terms of the intensity distribution in the primary beam $\varphi(x_1, y_1)$ and the detector's effectiveness $\psi(x_3, y_3)$, the general expression for the recorded intensity is expressed in the form

$$\tilde{I}(2\theta) = A \int_{S_1} \int_{S_2} \int_{S_3} \varphi(x_1, y_1)\psi(x_3, y_3) \times I([y'^2 + (2\theta - x')^2]^{1/2})dx_1 dx_2 dx_3 dy_1 dy_2 dy_3 \quad (9.1)$$

where S_1, S_2, and S_3 are the corresponding areas of the source, sample, and detector, and A is a normalizing constant.

Equation (9.1) may be reduced to a more convenient form. If we assume that $\varphi(x_1, y_1) = \varphi_x(x_1)\varphi_y(y_1)$ and $\psi(x_3, y_3) = \psi_x(x_3)\psi_y(y_3)$ (a generally accepted assumption which, in practice, is fulfilled with a sufficient degree of accuracy for all experimental installations) and introduce the variables s, $t = 4\pi y'/\lambda$, and $u = 4\pi x'/\lambda$, then

$$\tilde{I}(s) = \int_{-\infty}^{\infty}\int_{-\infty}^{\infty} W_w(u)W_l(t)I([t^2 + (s-u)^2]^{1/2})dtdu \qquad (9.2)$$

Here, the weighting functions $W_w(u)$ and $W_l(t)$ are determined by the instrument geometry as well as functions φ and ψ, and obey the normalization conditions

$$\int_{-\infty}^{\infty} W_w(u)du = \int_{-\infty}^{\infty} W_l(t)dt = 1 \qquad (9.3)$$

Integration over t is carried out in the y direction ("slit-height smearing") and over u in the x direction ("slit-width smearing"). The possibilities of calculating the weighting functions for actual devices and solving equation (9.2) for function $I(s)$ are examined below.

9.1.4. Beam Polychromaticity

We have so far assumed that the incident radiation is strictly monochromatic. It may well be, however, that in addition to the main radiation of wavelength λ_0 the incident beam contains another radiation of wavelength λ_1. Then, at some given recording angle 2θ, scattering with both $s = 4\pi\theta/\lambda_0$ and $s_1 = 4\pi\theta/\lambda_1$ will be recorded, in which case

$$\tilde{I}(s) = I(s) + \gamma I(s/\lambda) \qquad (9.4)$$

where $\lambda = \lambda_1/\lambda_0$, and γ is the relative intensity of radiation with wavelength λ_1.

However, if the incident beam contains a set of wavelengths characterized by a spectral function $W_\lambda(\lambda)$, then, assuming that the sample has no anomalously scattering atoms (Section 9.4.2), for the measured radiation we have

$$\tilde{I}(s) = \int_0^{\infty} W_\lambda(\lambda)I(s/\lambda)d\lambda \qquad (9.5)$$

where the weighting function is normalized by the condition

$$\int_0^{\infty} W_\lambda(\lambda)d\lambda = 1 \qquad (9.6)$$

In this way the polychromaticity of the radiation, similar to finite-size collimation systems, leads to scattering curve profile "smearing."

Elimination of the polychromatic effect involves as well the solution of the corresponding integral equation (9.5).

9.1.5. Statistical Errors

The small-angle scattering intensity is measured for some discrete values of the scattering angle. Therefore, each experiment results not in a continuous function $I(s)$, but in a set of discrete values $I(s_i)$ corresponding to definite values s_i (the errors in the values of s_i are usually negligibly small). The vaues of $\tilde{I}_i$ are, generally speaking, random numbers, i.e.,

$$\tilde{I}_i = \tilde{I}(s_i) = I(s_i) + \varepsilon_i$$

where ε_i is the statistical error at point s_i. These errors are independent random numbers, their magnitudes depending on the accuracy of the experiment. To check agreement between successive experiments and to restore the true scattering curve profile, the methods of statistical data analysis, considered below, must be employed.

9.1.6. General Expression for Experimental Intensity

The sources of distortion under consideration can be combined and generalized in the form

$$J(s) = \int_{-\infty}^{\infty}\int_{-\infty}^{\infty}\int_{0}^{\infty} W_w(u)W_l(t)W_\lambda(\lambda)I\{[t^2 + (s-u)^2]^{1/2}\lambda^{-1}\}d\lambda dt du \tag{9.7}$$

and

$$J_e(s_i) = \mu_1[J(s_i) + \varepsilon_i] + \mu_2[J_a(s_i) + \varepsilon_{ai}] \tag{9.8}$$

where $J_e(s_i)$ is the experimental scattering intensity by a sample, $J_a(s_i)$ the additive component (buffer) distorted by the same integral effects, μ_1 and μ_2 the normalization constants for absorption, primary beam intensity, and so on, while ε_i and ε_{ai} are the measurement errors.

In accounting for this relationship, one can formulate the main stages of small-angle scattering data processing:

1. Preliminary processing (normalization, analysis of the agreement of runs and their averaging for a sample and buffer).
2. Smoothing of averaged sets and subtraction of an additive component (buffer).

Generally speaking, this succession is preferable since the buffer subtraction before smoothing may lead to a substantial error increase, especially at large scattering angles, where $J_e(s)$ only slightly exceeds $J_a(s)$.

However, some approaches considered below demand a preliminary buffer subtraction.

3. Desmearing for the dimensions and polychromaticity of the beam.

Apart from the procedures for solving the general integral equation (9.7), one may (see, for instance, Kratky *et al.*, 1960) interpret various distortions quite independently. Then, equation (9.7) breaks down into three individual equations and the corrections should be introduced in the following successive order (Glatter and Zipper, 1975):

3a. Collimation correction for width, i.e., solution of the convolution equation with respect to $F(s)$,

$$J(s) = \int_{-\infty}^{\infty} W_w(u)F(s-u)du \tag{9.9}$$

3b. Collimation correction height, involving solution of the equation with respect to $I_n(s)$,

$$F(s) = \int_{-\infty}^{\infty} W_l(t)I_n(\sqrt{s^2+t^2})dt \tag{9.10}$$

3c. Correction for beam polychromaticity, i.e., solution of the equation with respect to $I(s)$,

$$I_n(s) = \int_0^{\infty} W_\lambda(\lambda)I(s/\lambda)d\lambda \tag{9.11}$$

A successive approach to the solution of equation (9.7) is applied rather often, since it is found to be possible, under the given experimental conditions, to neglect some effects, usually the beam width and/or polychromaticity, and to take more accurate account of the remaining effects.

Therefore, the examination of distortion sources led to a general scheme for experimental data processing — a succession of problems to be solved in order to restore the scattering curve $I(s)$ from a set of experimental data $J_e(s_i)$. In subsequent sections consideration will be given to the main methods of solving the formulated problems.

9.2. Preliminary Data Processing

Preliminary processing is reduced to normalization (to account for differences in source intensity and for absorption of various samples) and averaging of results obtained in repeated measurements of the same sample. The normalization is a standard procedure and causes no difficulties. Problems arising in the course of data averaging are related to the fact that individual experimental points or entire runs (called "rejections" and "bad runs") can, owing to apparatus or sample distortion, fall out of line. On averaging, such errors can substantially distort the result.

We shall assume that K individual measurements on the same sample have been carried out, each at N points s_i. The experimental intensities are then given by $J_{ik} = N_{ik}/T_{ik}$, where N_{ik} is the number of pulses recorded at the ith point of the kth run ($i = 1, \ldots, N$; $k = 1, \ldots, K$) while T_{ik} is the measurement time at this point. The values of J_{ik} contain statistical noise and can be expressed in the form $J_{ik} = J_i + \varepsilon_{ik}$, where J_i is the exact scattering-intensity value at the given point, and ε_{ik} is a random value distributed, for sufficiently large N_{ik} (> 30), according to the normal law with zero mathematical expectation and dispersion $\sigma_{ik} = J_i/\sqrt{N_{ik}} \approx J_{ik}/\sqrt{N_{ik}}$ (Brandt, 1970). The probability that $|\varepsilon_{ik}|$ may exceed $\omega \cdot \sigma_{ik}$ is given by the error integral

$$P(\omega) = 1 - \phi(\omega) = 1 - \sqrt{2/\pi} \int_0^{\omega} \exp(-t^2/2)dt$$

and equals 32% for $\omega = 1$, 4.5% for $\omega = 2$, and 0.3% for $\omega = 3$. It is evident that deviations in the random value from its mathematical expectation with $\omega \geqslant 3$ are hardly probable, otherwise it may be assumed that something went wrong or a systematic error has been introduced.

Since the exact value of J_i is unknown, one should use its estimate $\bar{J}_i$ when analyzing rejections. In this case, the dispersion of the random value, $\bar{\varepsilon}_{ik} = J_{ik} - \bar{J}_i$, will be equal to $\sigma_{ik} = (\sigma_{ik}^2 + \bar{\sigma}_i^2)^{1/2}$, where $\bar{\sigma}_i$ is the estimated dispersion of $\bar{J}_i$. Therefore, analysis of the difference $J_{ik} - \bar{J}_i$ allows one, with some degree of confidence determined by the value of ω, to judge whether the given measurement is a rejection. As an evaluation of $\bar{J}_i$ one can choose a value averaged over all the runs; in this case

$$\bar{\sigma}_i = \bar{J}_i \Big/ \left(\sum_{j=1}^{K} N_{ij} \right)^{1/2}$$

Such an algorithm was considered by Zipper (1972) and, for $K > 3$, gives sufficiently reliable results. The use of *a priori* data on the scattering-curve smoothness makes it possible to estimate $\bar{J}_i$ by extrapolating (e.g., polynomially) over the neighboring experimental points (see Section 9.3). This allows one to limit oneself when analyzing the rejections in a run to the data on this run.

Agreement between the runs can be most easily examined using the χ^2-criterion. Since the difference between the kth and mth measurements at the ith point, $J_{ik} - J_{im}$, is a random value with center 0 and dispersion $(\sigma_{ik}^2 + \sigma_{im}^2)^{1/2}$, the expression

$$\chi_{km}^2 = \sum_{i=1}^{N} \frac{(J_{ik} - J_{im})^2}{\sigma_{ik}^2 + \sigma_{im}^2} \tag{9.12}$$

is the sum of squares of normalized random values with center 0 and dispersion 1. It obeys a χ^2-distribution with N degrees of freedom which, for $N > 30$, passes asymptotically to a normal distribution (Brandt, 1970). Therefore, for instance, the quantity $R_{km} = (\chi_{km}^2/N)^{1/2}$ is distributed normally with center 1 and dispersion $(2/N)^{1/2}$, and if, say, the expression $|R_{km} - 1|$ does not fall in the confidence interval $\omega \cdot (2/N)^{1/2}$, then one of the runs (the kth or mth) is most probably a "bad" one. Since the "bad" run will yield substantial disagreement with respect to all the "good" ones, such a mutual analysis carried out for all the runs will easily reveal it.

The methods discussed above enable automatic sorting out of rejections and bad runs with the aid of computers. It should be noted, however, that the results were obtained using statistical laws; they are of a probabilistic nature and depend on the chosen value of the confidence interval. In practice, use is made of values $\omega = 2.5$–3, which agree with the 98.8–99.7% level of confidence.

9.3. Experimental-Data Smoothing

A discrete set of normalized averaged values of intensities $\bar{J}_i$ $(i = 1, 2, \ldots, N)$ contains certain errors $\bar{\sigma}_i$ that remain after averaging. The presence of statistical errors hampers the introduction of instrumental corrections. (The procedures connected with the solution of integral equations are incorrectly stated problems and may result in sharply increased errors.) Therefore, there often arises a problem of minimization of statistical noise, i.e., of experimental data smoothing.

The possibility of smoothing the $\bar{J}_i$ set is based on some preliminary knowledge of the scattering-curve behavior. We know that the curve represents a smooth function, several experimental points falling on each of its maxima. Such a belief is based on both purely physical considerations (small-angle curves are smeared owing to statistical isotropy of a sample as well as collimation and polychromatic effects) and mathematical confirmation. Actually, scattering by a solvent is, as a rule, close to constant, while we have already seen that scattering curves from monodisperse systems obey the sampling theorem (3.22). The experiment is planned in such a way that the angular step Δs is substantially smaller than the sampling value π/D. Such an approach enables the application of smoothing procedures. In the case of polydisperse systems the curves become even smoother, as a result of averaging over the shape and dimensions of particles.

The problem of smoothing lies in the restoration of a smooth curve $J(s)$ from a set of points. The smoothing of small-angle data can be carried out by several methods, which differ from one another in the mode of definition of the required smoothness of the curve.

9.3.1. Algebraic Polynomials

We assume that the smoothness of the curve $J(s)$ is no lower than m, that is, over the whole area of its definition all its derivatives should exist up to and including order $m+1$. Then, in the neighborhood of any point s_0 in (s_{min}, s_{max}) of $J(s)$ one can introduce the Taylor expansion

$$J(s) = J(s_0) + \sum_{k=1}^{m} \frac{1}{k!}(s - s_0)^k J^{(k)}(s_0) + \varepsilon(s, s_0)$$

where the approximation accuracy is

$$|\varepsilon(s, s_0)| \leqslant |(s - s_0) J^{(m+1)}(\xi)/(m+1)!|, \qquad \xi \in [s_0, s]$$

Therefore, function $J_0(s)$ may be approximated at intervals over the entire area of definition, with a certain accuracy, by polynomials of mth order. This enables one to obtain a smoothed value of $\tilde{J}$ at each point, fitting the experimental data around this point by an algebraic polynomial of this order.

The least-squares method (LSM) is used to achieve the approximation (see, for example, Lawson and Hanson, 1974). As a rule, smoothing is carried out over the midpoint, i.e., one selects the $2p+1$ values of the argument $s_{i-p}, \ldots, s_{i+p}$ and the polynomial of the best approximation

$$P_m(s) = \sum_{k=0}^{m} a_k (s - s_i)^k$$

is sought from the values at these points. The value of $P_m(s_i)$ is taken as the smoothed value $\tilde{J}^{(p)}(s_i)$.

According to the LSM, the best approximating polynomial is that which ensures the minimum of the functional

$$\phi = \sum_{j=-p}^{p} \frac{1}{\sigma_{i+j}^2} [\bar{J}(s_{i+j}) - P_m(s_{i+j})]^2 \tag{9.13}$$

The coefficients a_k of such a polynomial are obtained by solving the system of $m + 1$ normal linear equations

$$\frac{\partial \phi}{\partial a_k} = 0, \qquad k = 0, 1, \ldots, m \tag{9.14}$$

In a particular case of measurements with identical accuracy and constant increment ($\sigma_i = \sigma$, $\Delta s = \text{const}$) it is possible to write a convenient explicit expression for the smoothed values. Thus, when smoothing over polynomials with $m = 3$ we have, for each inner point $p + 1 \leqslant i \leqslant N - p$,

$$\tilde{J}^{(p)}(s_i) = \sum_{j=-p}^{p} \frac{9p^2 + 9p - 3 - 15j^2}{(2p+1)(4p^2 + 4p - 3)} \bar{J}(s_{i+j}) \tag{9.15}$$

For $i \leqslant p$ and $i > N - p$, extrapolation over cubic parabolas obtained at the $(p + 1)$th and $(N - p)$th midpoints is carried out. In the general case, system (9.14) should be solved at all the inner points. It is clear that in any case the results of smoothing will depend substantially on the relationship between the values of m and p; therefore when using such a method it is necessary to have criteria for selecting these parameters. The accuracy of the determination of an approximating polynomial, and consequently the values of $\tilde{J}^{(p)}(s_i)$, increases as p becomes larger at predetermined m. However, in this case, the aproximation accuracy $\varepsilon(s, s_i)$ becomes worse. Therefore, the optimum value of p for smoothing each actual point should depend on the behavior of function $J(s)$ in the vicinity of this point as well as on the angular step and the measurement accuracy. General recommendations like "$2p + 1$ should be 2–3 times more than m" are not always applicable. A simple method has been proposed by Rolbin *et al.* (1980) for determining the optimum number of neighbors for a given degree of polynomial. First, it should be noted that, proceeding from practical considerations, in small-angle scattering it is convenient to limit oneself to the value $m = 3$, which suffices to display the characteristic features of the curve profile of $J(s)$. (In fact, the cubic parabola allows localization of both maxima and minima, and points of inflection, when employing approximation of intervals.) Rolbin *et al.* (1980) considered smoothing over an angular interval in which the behavior of function $J(s)$ (such as the width of its maxima) changes only slightly. The case is characteristic for small-angle scattering experiments, while the breakdown into such angular sections is usually performed by the experimentalist himself. It may be assumed, therefore, that an optimum (e.g., in the sense of the mean-square deviation of smoothed values from the true ones) number p_{opt} does exist for the process of smoothing the entire curve with a constant number of neighbors. This value is sought by considering the expression

$$R_\sigma(\bar{J}, \tilde{J}^{(p)}) = \left\{ \frac{1}{N} \sum_{i=1}^{N} \frac{[\bar{J}(s_i) - \tilde{J}^{(p)}(s_i)]}{\sigma_i^2 + \tilde{\sigma}_i^2} \right\}^{1/2} \tag{9.16}$$

It should be taken into account that, as in equation (9.12), the random value $R_\sigma(\bar{J}, J)$ is, asymptotically, normally distributed with center 1 and dispersion $(2/N)^{1/2}$. It is clear that for small p the value of $R_\sigma[\bar{J}, \tilde{J}^{(p)}]$ should also be small (the degree of smoothing is small and the smoothed values differ only slightly from the initial values). Then it increases while tending to unity, and if the mean accuracy of aproximation becomes worse so that it is comparable with the experimental accuracy, then R_σ exceeds 1 and increases further as ε becomes worse. Figure 9.2 shows three possible types of behavior of function R_σ. Curve 1 corresponds to too large an angular step [the approximation accuracy may be comparable with the experimental accuracy even for the minimum value of p ($p_{\min} = 2$ for $m = 3$)]. Curve 2 is obtained when the profile of function $J(s)$ is either devoid of characteristic features (e.g., scattering by a solvent) or they are completely obliterated by statistical noise. Such curves provide evidence that the experiment is wrong; then, in order to reveal the profile of function $J(s)$ reliably, it is necessary either to

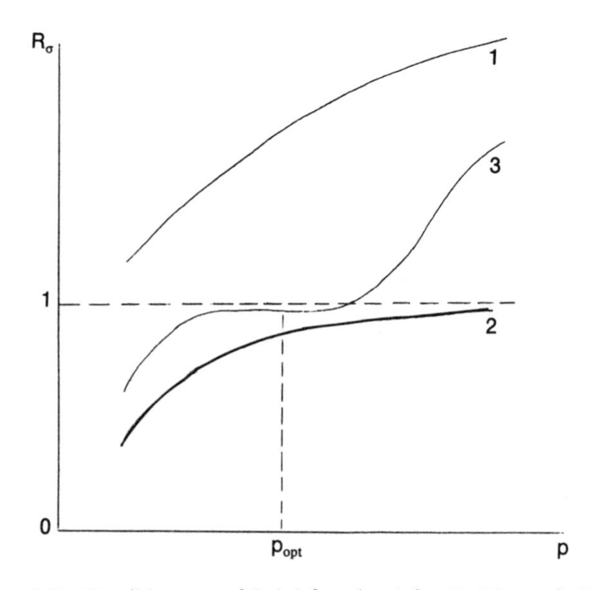

Figure 9.2. Possible types of $R_\sigma(p)$ function (after Rolbin *et al.* 1980).

decrease the angular step (case 1) or increase the accuracy (case 2). In a correct experiment a type 3 curve should be obtained, and the optimum value can be estimated by determining the position of the point of inflection on this curve. In such a way, analysis of function $R_\sigma(p)$ permits one not only to find the optimum number of neighbors, but also to estimate the experimental quality. Residual errors in $\tilde{J}^{(p)}(s_i)$ can be calculated by the law of error propagation for system (9.14). The level of statistical-noise reduction can be estimated using linear filter theory. Any smoothing according to the LSM corresponds to a nonrecursive linear filter (Savitzky and Golay, 1964)

$$\tilde{J}_i = \sum_{j=-k_1}^{k_2} \beta_j \bar{J}_{i+j}, \qquad \sum_{j=-k_1}^{k_2} \beta_j = 1 \tag{9.17}$$

[in equation (9.15), $k_1 = k_2 = p$ and $\beta_j = b_j$]. The level of errors is decreased by the ratio (Monroe, 1962)

$$B_p \approx R_\sigma^2[\bar{J}, J]/R_\sigma^2[\tilde{J}^{(p)}, J] \approx \sum_{j=-p}^{p} b_j^2$$

The values of B_p for different p at $m = 3$ are given in Table 9.1. If the decrease of errors is insufficient, i.e., if the set $\tilde{J}^{(p)}(s_i)$ does not also ensure the required smoothness, the smoothing can be carried out again. The optimum number of neighbors for such smoothing $p_{\text{opt}}^{(2)}$ can be found by a similar procedure, substituting $\bar{J}_i$ by $\tilde{J}_i^{(p)}$. In practice, repeated smoothing with the optimum number of neighbors allows one to remove the statistical noises almost completely while still preserving details of the true profile of the curve. The effectiveness of such a technique is shown by Robin *et al.* (1980) in model examples, one of which is shown in Figure 9.3.

It should also be noted that the small-angle scattering curves often represent sharply decreasing functions. Absolute errors in large $\bar{J}_i$ may prevail when determining the polynomial parameters during the minimization of functional (9.13). In such cases it is more convenient to smooth the small-angle curves on the scale which equalizes the values of the absolute intensity (e.g., on a logarithmic scale).

Table 9.1. Reduction Factors of the Variance for Polynomial Regression with $p = 3$

p	2	3	4	5	6	7	8	9	10
B_p	0.486	0.333	0.255	0.208	0.175	0.151	0.133	0.119	0.108

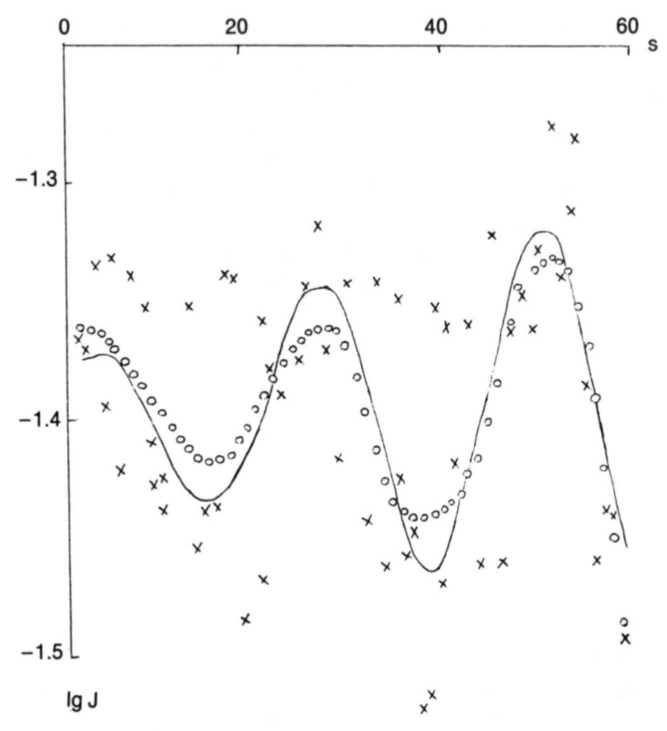

Figure 9.3. Optimum polynomial smoothing of a model curve (after Rolbin *et al.*, 1980). The solid line refers to function $J(s)$, crosses to values of $J(s_i)$ with statistical noise, $\sigma = 10\%$, and circles to the smoothed curve ($m = 3$, $p_{\text{opt}} = 9$, $p_{\text{opt}}^{(2)} = 4$).

9.3.2. Spline Functions

Another method for which the *a priori* smoothness of function $J(s)$ is employed for minimization of statistical errors is the construction of an aproximating spline. Spline functions have recently gained wide usage in the solving of some applied problems, particularly in small-angle scattering. Therefore, we will now give a brief account of spline theory.

For applied purposes, one usually uses polynomial splines of the third degree — cubic splines, which are determined as follows. We assume that the net of values of the argument $s_1 < s_2 < \cdots < s_N$ is predetermined. Then, the cubic spline is the function $S_3(s)$ satisfying the following conditions: (1) $S_3(s)$ is a polynomial of third degree in each interval $[s_i, s_{i+1}]$, $i = 1, \ldots, N$ and constant when $s < s_1$ and $s > s_N$; (2) $S_3(s)$ and its derivatives $S_3'(s)$ and $S_3''(s)$ are continuous, while $S_3^{(3)}(s)$ may exhibit discontinuities at the net knots.

The most important property of cubic splines is the following. We assumed that values of f_i are defined at the net knots s_i, that function $f(s)$ must be plotted from these values, and that this situation yields the functional

$$\phi = \int_{s_1}^{s_N} [f''(s)]^2 ds \tag{9.18}$$

minimum at the predetermined boundary conditions (the values of the first or second derivative of $f(s)$ at $s = s_1$ and $s = s_N$). Spline theory proves that the spline $S_3(s)$ (the interpolation cubic spine; see Greville, 1969) supplies a solution to such a problem.

Thus spline interpolation of a discrete function enables one to construct a curve possessing minimum integral curvature (9.18). Similarly, the cubic spline will yield the functional

$$\phi_\lambda = \int_{s_1}^{s_N} [f''(s)]^2 ds + \lambda \sum_{i=1}^{N} p_i [f(s_i) - f_i]^2 \tag{9.19}$$

minimum, where $\lambda \geqslant 0$ and $p_i > 0$. Such a spline is called approximating (or smoothing). One can easily see that it will depend on the value of the Lagrange factor λ; at $\lambda = 0$ we have the interpolation spline, for a polynomial approximation according to the LSM.

For spline parametrization it is convenient to use B-splines. Cubic B-splines are those defined over five net knots according to the relationship

$$B(s) = 4 \sum_{i=k}^{k+4} \frac{(s_i - s)_+^3}{\omega_k'(s_i)}$$

where

$$s_+ = \max(0, s) \quad \text{and} \quad \omega_k(s) = \prod_{i=k}^{k+4} (s - s_i)$$

It is evident that $B(s)$ differs from zero only in the interval $s_k \leqslant s \leqslant s_{k+4}$. The B-splines are linearly independent and may serve as a basis in spline space (Greville, 1969). For some modification of marginal B-splines any spline at $s_1 < s < s_N$ can be represented as a linear combination of N B-splines:

$$S_3(s) = \sum_{i=1}^{N} c_i B_i(s) \tag{9.20}$$

A set of B-splines for a uniform network with $N = 6$ is given in Figure 9.4.

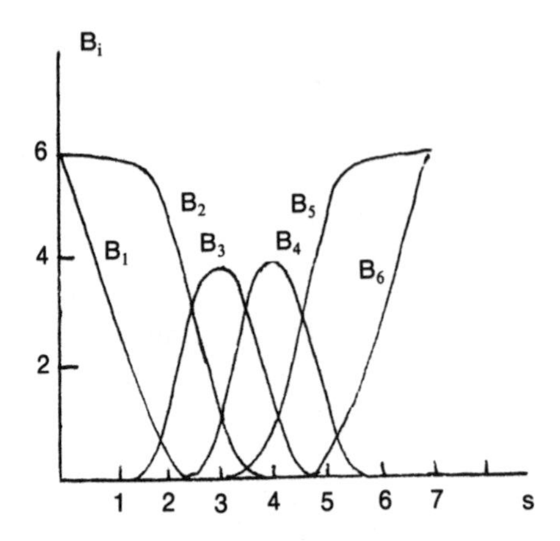

Figure 9.4. A complete set of *B*-splines for a uniform net with six knots.

In this case, coefficients c_i minimizing functional (9.19) are obtained by solving the system of normal linear equations

$$\sum_{i=1}^{N} (\mu_{ki} + \lambda\beta_{ki})c_i = \lambda\gamma_k, \qquad 1 \leqslant k \leqslant N \tag{9.21}$$

where

$$\mu_{ki} = \int_{s_1}^{s_N} B''_k(s)B''_i(s)ds, \quad \beta_{ki} = \frac{1}{N}\sum_{j=1}^{N} p_j B_k(x_j)B_i(x_j)$$

and

$$\gamma_k = \frac{1}{N}\sum_{j=1}^{N} p_j f_j B_k(s_j)$$

If f_i implies the value $\bar{J}_i$ and p_i the value $1/\sigma_i$, we have a smoothing procedure which differs from that considered in Section 9.3.1, namely, superposition of an additional condition — the requirement of minimum integral curvature. The spline constructed in such a way is the smoothed function $\bar{J}(s)$.

Here, the degree of smoothness will depend on parameter λ. Its optimum value may also be sought using the χ^2-criterion: smoothing with different values of λ until the deviation

$$R(\bar{J}, c) = \left\{\frac{1}{N}\sum_{j=1}^{N}\frac{1}{\gamma_j^2}\left[\sum_{i=1}^{N} c_i B_i(s_j) - \bar{J}(s_j)\right]^2\right\}^{1/2}$$

is found to lie within the confidence interval $[1-\varepsilon, 1+\varepsilon]$, where

$\varepsilon = \omega(2/N)^{1/2}$. If $R(\bar{J}, c) > 1 + \varepsilon$, then λ should be increased; if $R(\bar{J}, c) < 1 - \varepsilon$, then λ should be decreased.

Hence the spline-function apparatus permits smoothing of experimental data by varying the relationship between the "smoothness" requirements of the curves and their deviation from some experimental set of data. It is necessary, however, to choose the parameter λ carefully so as to control stability when solving system (9.21). Otherwise, serious errors may occur in the spline parameter and consequently give rise to strong oscillations along the smoothed curve.

9.3.3. Frequency Filtering Method

The Fourier cosine image of the curve for scattering from a monodisperse system, regardless of collimation conditions, will differ from zero only in the finite frequency range defined by the maximum size D of a particle (see Section 3.3.4). Termination of the spectrum function imposes a rather rigid condition, which demands a certain "smoothness" of this function; therefore it can be used for smoothing of the experimental data.

We shall assume that the experimental values $\bar{J}_{di} = J_d((i-1)\Delta s)$ of the difference curve are obtained by direct subtraction of the averaged scattering intensity of the solvent from the averaged values $\bar{J}_i$ of the sample scattering intensity. The discrete function thus defined can be expressed as the Fourier series

$$\bar{J}_{di} = \frac{A_0}{2} + \frac{A_{N-1}}{2}\cos[\pi/(N-1)] + \sum_{p=1}^{N-2} A_p \cos\left(\frac{\pi p^2}{N-1}\right) \tag{9.22}$$

where the coefficients A_p of the Fourier cosine image are calculated from the experimental data. Then, the smoothed values $\tilde{J}_{di}$ can be obtained from the relationship (Damaschun *et al*., 1971, 1974)

$$\tilde{J}_{di} = \frac{A_0}{2} + \sum_{p=1}^{\bar{M}} A_p \cos\left(\frac{\pi p^2}{N-1}\right) + \sum_{p=\bar{M}+1}^{M} A_p \exp[-a(p-M)^2]\cos\left(\frac{\pi p^2}{N-1}\right) \tag{9.23}$$

where $\bar{M} = Ds_{max}/\pi$ corresponds to the boundary frequency, while the exponential damping factor is 10^{-2}–10^{-3} for $p = M$ ($\bar{M} < M < n-1$). Such a procedure removes higher harmonics, which fit the statistical noise in the experimental curve. The parameters a and M are so selected as to enable function $\tilde{J}_{di}$ to be devoid of oscillations arising from sharp termination of series (9.22) at $p = \bar{M}$. At the same time, harmonics that may occur as a result of scattering by a particle (with $p < \bar{M}$) remain unchanged. If the curve was measured with a variable angular increment, this involves no serious changes with the exception of making equations (9.22) and (9.23) more complicated.

Such a method, however, implies extrapolation of experimental data to $s = 0$ and $s \to \infty$ when calculating coefficients A_p. Walter *et al.* (1975) have shown that this approach can be applied mainly to scattering curves with a decrease in intensity that is not less than three orders of magnitude. The modified frequency filtering (Müller and Damaschun, 1979) makes use of the same principle of smoothing but in a different way. The frequency function

$$C(x) = \frac{1}{\pi} \int_0^\infty \bar{J}_d(s) \cos(sx)\, ds$$

is multiplied by function $F(x)$ — the convolution of a step function $\Pi(x - D)$ and some function $W(x)$, which smooths the discontinuity at the point $x = D$. This corresponds, in reciprocal space, to the convolution of $\bar{J}_{di}$ with the product of the Fourier image of functions $\Pi(x - D)$ and $W(x)$, and can be written as a linear filter. For the constant increment $\Delta s = \beta(\pi/D)$, $\beta < 1$ and the Hamming window

$$H_j = \mathscr{F}_c[W(x)] = \begin{cases} \cos(j\pi/2J), & |j| \leqslant J \\ 0, & |j| > J \end{cases}$$

we obtain linear filter (9.17) for $k_1 = k_2 = J$,

$$\mu_j = \frac{\sin\{j\pi\beta/[1 - 2/(\beta J)]\}}{j\pi \cos^2(j\pi/2J)} \qquad (9.24)$$

while $F(D) = 0.999$. The parameter J determines the degree of smoothing. Müller and Damaschun (1979) employed test examples to show that it is convenient to choose J such that the value J_β should be 10–25. Such a filter, as well as filter (9.23), minimizes the influence of harmonics with frequency greater than D; however, it can still be applied to curves with sufficiently small intensity decrease.

9.3.4. Problem of Optimum Smoothing

As far as possible, optimum smoothing of experimental data should minimize the level of random errors for minimum distortions of the true curve $J(s)$. All the methods considered, regardless of the mode of definition of the required curve smoothness, involve some parameters that vary the degree of smoothing. The preference for applying one or another method of smoothing depends on the possibility of seeking the optimum values of these parameters.

A number of papers (see, e.g., Müller and Damaschun, 1979; Walter *et al.*, 1977) are devoted to a comparative analysis of smoothing techniques; however, only fixed values of the smoothing parameters have been used in these studies. Here, we summarize the advantages and disadvantages of the various methods as well as possibilities of carrying out optimum smoothing with each of them.

9.3.4.1. Range of Application

Smoothing over cubic polynomials and splines can be applied to any function with the smoothness not lower than $m = 3$. Frequency filtering can be employed if the Fourier image of function $J(s)$ is finite. It is clear that the region of applicability of the first two methods is wider. In particular, they enable one to smooth the scattering data by nonparticulate systems as well as to carry out separate smoothing of scattering data from a sample and buffer (in frequency filtering, a preliminary buffer subtraction is necessary).

9.3.4.2. Determination of Optimum Parameters

Methods of using the χ^2-criterion, considered in Sections 9.3.1 and 9.3.2, make it possible to determine automatically the smoothing parameters p_{opt} and λ_{opt} for polynomials and splines. It seems that p_{opt} can be determined more easily, since the range of p is fixed (for $m = 3$, $p_{min} = 2$ and $p_{max} < (N - 1)/2$, but in practice one assumes, as a rule, that $p_{max} =$ 15–20 for a sample and 20–25 for a buffer). When smoothing with the aid of frequency filtering the main parameter (D) is assumed to be known. However, errors in its determination as well as the necessity of selecting values of a [equation (9.23)] or J [equation (9.24)] also require smoothing optimization (in principle, this can be achieved by applying the χ^2-criterion).

9.3.4.3. Account of Data Accuracy

For polynomials and splines the experimental errors σ_i are taken into account explicitly in the course of smoothing. The residual errors of smoothed values may be obtained by means of the error propagation law when solving systems (9.14) and (9.21). An account of errors σ_i in frequency filtering is a more complicated task. A general estimation of

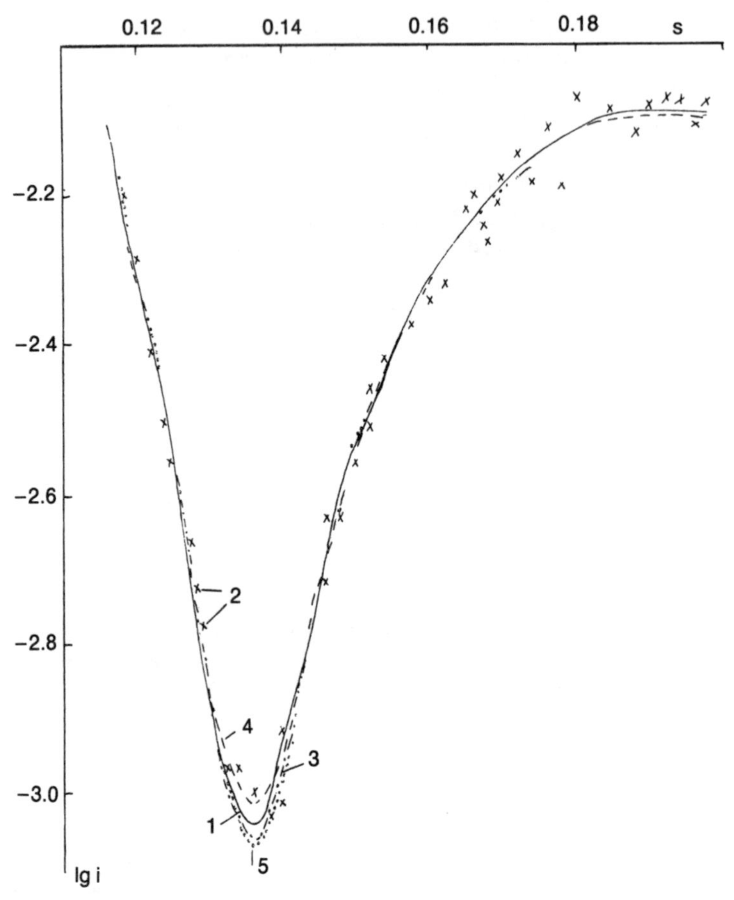

Figure 9.5. Comparison of different smoothing procedures: (1) true curve $J(s)$; (2) values $\tilde{J}(s_i)$, $\sigma = 5.4\%$. Smoothing by: (3) frequency filtering, (4) polynomials, (5) splines.

noise-level minimization for this method can be obtained with the aid of coefficient R_p according to the theory of linear filters (see Section 9.3.1).

It should be noted that, on the whole, the method of polynomial smoothing according to the LSM with optimum choice of parameters has been shown by Rolbin *et al*. (1980) to be the simplest; it allows simultaneous checking of experimental quality. The methods of spline functions and frequency filtering also permit one to obtain reliable results, but they appear more difficult in their realization and search for the optimum degree of smoothing. Figure 9.5 shows a comparison of various methods when smoothing a typical small-angle scattering curve.

9.4. Collimation Corrections

An account of collimation distortions is one of the most complicated tasks in small-angle experimental data evaluation. The introduction of collimation corrections is connected with the solution of the corresponding integral equations examined in Section 9.1. These are Fredholm equations of the first kind and various mathematical methods have been developed for their solution (see, for instance, Tikhonov and Arsenin, 1979). In this section we shall consider the main approaches in the successive introduction of corrections for the slit width according to the equation

$$J(s) = \int_{-\infty}^{\infty} W_w(u)F(s-u)du \tag{9.25}$$

and for the slit height

$$F(s) = \int_{-\infty}^{\infty} W_l(t)I[(s^2+t^2)^{1/2}]dt \tag{9.26}$$

Methods for solving the general integral equation (9.7) are analyzed in Section 9.7.

9.4.1. Weighting Functions

In Section 9.1.3, weighting functions $W_w(u)$ and $W_l(t)$ were defined by the collimation system geometry, intensity distribution in the primary beam, and detector sensitivity. In principle, they can be determined experimentally after measuring the scattering intensity in the absence of a sample. In fact, in this case $J(s) \approx \delta(s)$ and, as seen from equation (9.2), $J(u, t) \approx W_w(u)W_l(t)$. However, the results of such measurements will be distorted by parasitic scattering from camera units. Moreover, measurement of the slit-height weighting function appears to be hampered by the fact that the counter cannot always move in the perpendicular direction. Therefore, it is important to calculate the weighting functions analytically.

Hendricks and Schmidt (1967, 1973) worked out general methods of calculating weighting functions for various geometries of collimation systems. They provided explicit expressions for the widely used cameras with slit collimation according to Kratky and Beeman (see Section 8.2.3). The calculation and the relationships themselves, although rather complicated, are quite general. For cameras with rectangular slits it is sometimes convenient to apply the simplified equation

$$W(x)=\begin{cases}1/a, & |x|\leqslant(a-l)/2\\ [(a+l)/2-|x|]/al, & (a-l)/2\leqslant|x|\leqslant(a+l)/2\\ 0, & |x|>(a+l)/2\end{cases} \tag{9.27}$$

for both $W_w(u)$ and $W_l(t)$. Here, the point source and uniform detector sensitivity are assumed to exist. Quantity a is the projection of the beam width (or height) in the detector plane, and l is the width (or height) of the detector slit. Calculation of weighting functions for more complicated cases can be carried out in line with the program described by Buchanan and Hendricks (1971).

9.4.2. Slit-Width Correction

The slit width is considerably less than the height, and is so small that this effect can often be neglected, assuming $W_w(u)=\delta(u)$ and, correspondingly, $J(s)=F(s)$. However, in some cases width corrections must be introduced. This is due to the fact that further introduction of a height correction is a process associated with sharpening of all the characteristic features of a function, and it is rather sensitive to the slightest deviation of $F(s)$ from $J(s)$. In addition, the smearing effect for the height becomes weaker as the scattering angle increases, while the width influence does not depend on s (see Section 9.6). Therefore, at large values of s, the width and height effects may be of the same order.

The Fourier transform is the simplest technique for solving equation (9.25). Since the integral on the right side is the convolution of $W_w(s)$ and $F(s)$ (see Section 1.2), then, taking Fourier transforms of both sides of equation (9.25), we obtain

$$\mathscr{F}[J(s)]=\mathscr{F}[W(s)]\mathscr{F}[F_w(s)]$$

or

$$F(s)=\mathscr{F}^{-1}\{\mathscr{F}[J(s)]/[\mathscr{F}[W_w(s)]\} \tag{9.28}$$

This "exact" solution is, however, rather sensitive to errors in $J(s)$. The control developed by Tikhonov and Arsenin (1979) enables one to obtain a stable solution (see below, Section 9.5).

Another simple method was proposed by Kratky *et al*. (1960) and Sauder (1966). The Taylor expansion of $F(s)$ in the vicinity of s gives, to within third-order terms,

$$J(s)\approx\int_{-\infty}^{\infty}W_w(u)\left[F(s)-F'(s)\cdot u+F''(s)\cdot\frac{u^2}{2}-F^{(3)}(s)\frac{u^3}{6}\right]du$$

For frequently occurring even functions $W_w(u)$, the kth-order moments

$$M_k = \int_{-\infty}^{\infty} u^k W_w(u) du$$

are equal to zero for odd k. Taking into account normalization condition (9.3) and assuming that $J''(s) \approx F''(s)$ holds for sufficiently narrow slits, we have

$$F(s) \approx J(s) - \frac{M_2}{2} J''(s) \tag{9.29}$$

This relationship is well known. Taylor and Schmidt (1967) used a polynomial approximation to calculate $J''(s)$, while Schelten and Hossfeld (1971) applied spline differentiation for the same purpose. Such techniques increase solution stability with respect to errors in $J(s)$, but do not remove the main restrictions: requirement of parity of $W_w(u)$, Taylor-expansion termination, and, most important, the condition of equality of the second derivatives $J''(s)$ and $F''(s)$ that can be fulfilled satisfactorily only for very narrow slits. Therefore, such an approach can be applied when width effects are weak.

The method proposed by Rolbin *et al.* (1977) is more general in character and makes use of a standard technique for seeking the solution in the form of a power series. If $F(s)$ is expressed as the nth-degree polynomial

$$F(s) = P_n(s) = \sum_{i=0}^{n} a_i s^i$$

then, as seen from equation (9.25), $J(s)$ is also an nth-degree polynomial,

$$J(s) \approx D_n(s) = \sum_{i=0}^{n} b_i s^i$$

Here, a_i and b_i are related by the system of linear equations of rank $n + 1$ with triangular matrix

$$b_{n-k} = M_0 a_{n-k} + \sum_{p=1}^{k} (-1)^p C_{n-k+p}^{p} M_p a_{n-k+p} \tag{9.30}$$

where M_p are the weighting-function moments, $k = 0, 1, 2, \ldots, n$, while C_n^k are binomial coefficients (Korn and Korn, 1961). The polynomial coefficients are determined by the $J(s)$ approximation using a least-squares method over the interval $(s + a_w, s + b_w)$, where $a_w \leqslant u \leqslant b_w$ is

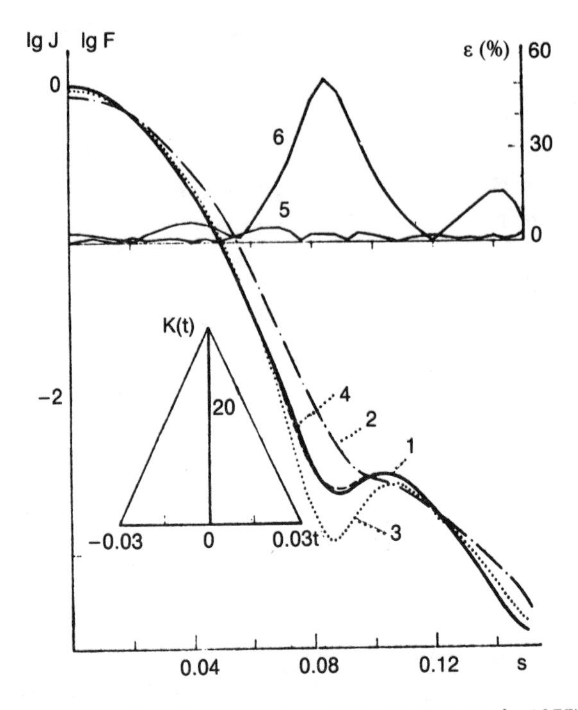

***Figure* 9.6.** Slit-width collimation corrections (after Rolbin *et al*., 1977): (1) function $F(s)$; (2) function $J(s)$; (3), (4) restoration according to equations (9.29) and (9.30), respectively; (5), (6) relative deviations of curves (3) and (4) from (1); $W_w(u)$ denotes the weighting function applied.

the interval within which $W_w(u)$ differs from zero. In this case the degree n is increased until the predetermined approximation accuracy reaches

$$\varepsilon = \left\{\frac{1}{m-1}\sum_{i=1}^{m}\frac{1}{\sigma_i^2}[J(s_i) - D_{n+1}(s_i)]^2\right\}^{1/2}$$

where m is the number of points $J(s_i)$ in the interval $(s + a_w, s + b_w)$. This technique is of general type and allows the solution accuracy to be checked. Rolbin *et al*. (1977) have shown that it yields reliable results in the case of slit widths for which the use of equation (9.29) leads to considerable errors (Figure 9.6).

9.4.3. Slit-Height Correction

Slit-height effects substantially exceed distortions caused by the slit width. Correspondingly, a considerable number of algorithms have been designed for solving equation (9.26). We shall now examine the most frequently used methods, which take into account the special features of the kernel of this integral equation.

It should first be noted that equation (9.26) will not change if any weighting function $W_l'(t)$ is symmetrized according to the equation $W_l(t) = [W_l'(-t) + W_l'(t)]/2$. Hence, without loss of generality, $W_l(t)$ can be regarded as an even function and equation (9.26) can be written in the form

$$F(s) = 2\int_0^\infty W_l(t)I[(s^2 + t^2)^{1/2}]dt \tag{9.31}$$

The explicit solution of equation (9.31) was obtained by Kratky *et al.* (1951) and can be written as

$$I(s) = -\frac{1}{\pi}\int_0^\infty \frac{F'[(s^2 + x^2)^{1/2}]}{(s^2 + x^2)^{1/2}} g(x)dx \tag{9.32}$$

where $g(x)$ satisfies the equation

$$\int_0^{\pi/2} W_l(r\sin\theta)g(r\cos\theta)d\theta = \pi/2 \tag{9.33}$$

for any r. On the face of it, such a solution seems devoid of any sense — we have simply replaced one integral equation by another. The principal difference between them lies in the fact that equation (9.31) must be solved for each experimental curve, while the solution of equation (9.33) depends only on the weighting function $W_l(t)$ and can be used for all curves $J(s)$ measured with a given setup geometry.

In some cases, the explicit solution assumes a simplified form. Thus, if the weighting function is Gaussian, i.e., $W_l(t) = p\pi^{-1/2}\exp(-p^2t^2)$, then (Kratky *et al.*, 1951)

$$I(s) = \frac{e^{-p^2s^2}}{p\sqrt{\pi}}\int_0^\infty \frac{N'[(s^2 + x^2)^{1/2}]}{(s^2 + x^2)^{1/2}} dx$$

where

$$N(x) = e^{-p^2s^2}F(x)$$

As p tends to zero, the weighting function tends to the constant $W_l(t) = W_l(0)$ and

$$I(s) = \frac{1}{\pi W_l(0)}\int_0^\infty \frac{F'[(s^2 + x^2)^{1/2}]}{(s^2 + x^2)^{1/2}} dx$$

This is the case of an "infinitely high" slit, which is realized in practice when the interval in which $W_l(t) = \text{const}$ reaches values of t such that, at any s, $F[(s^2+t^2)^{1/2}]$ appears to be negligibly small (Guinier and Fournet, 1947; Du Mond, 1947). A number of papers (see, for instance, Heine and Roppert, 1962; Schmidt, 1965; Schelten and Hossfeld, 1971) deal with the numerical realization of such cases.

A general approach to the solution of equation (9.33) is proposed by Fedorov (1968) and then developed by Deutch and Luban (1978). If this equation involves the variables $z=r^2$ and $y=z\cos^2\theta$, then it assumes the form

$$\int_0^z H(y)G(z-y)dy=\pi, \qquad z\neq 0$$

where $H(t^2)=W_l(t)/t$ and $G(t^2)=g(t)/t$. Application of the convolution theorem for Laplace transforms (Korn and Korn, 1961) now yields, by analogy with equation (9.28),

$$g(x)=\pi x\left[\frac{1}{2\pi i}\int_{c-i\infty}^{c+i\infty}\frac{\exp(ux^2)}{u\tilde{H}(u)}du\right] \qquad (9.34)$$

where

$$\tilde{H}(u)=2\int_0^\infty W_l(t)\exp(-ut^2)dt$$

These relations enable function $g(x)$ to be computed. Besides, it was shown by Deutch and Luban (1978) that integration by parts of equation (9.32) gives

$$I(s)=F(s)+\frac{1}{\pi}\int_0^\infty [F[(s^2+x^2)^{1/2}]-F(x)]\frac{\partial}{\partial x}\left[\frac{g(x)}{x}\right]dx \qquad (9.35)$$

which is more convenient for practical calculations, since no differentiation is required.

Fedorov *et al.* (1968) and Deutch and Luban (1978b) developed numerical approaches for solving equation (9.35); Schmidt and Fedorov (1978) and Luban and Deutch (1980) derived explicit expressions for frequently occurring types of functions $W_l(t)$. Thus, for the already considered trapezoidal weighting function (9.27),

$$g(x)=\begin{cases}\mu\leqslant 1/\sqrt{2}, & \begin{cases}1+\mu, & 0\leqslant u\leqslant\mu\\ (1+\mu)[1+(u-\mu)/(1-\mu)], & \mu<u\leqslant\mu\sqrt{2}\end{cases}\\[2ex] \mu>1/\sqrt{2}, & \begin{cases}1+\mu, & 0\leqslant u\leqslant\mu\\ (1+\mu)[1+(u-\mu)/(1-\mu)], & \mu<u\leqslant 1\\ 2(1+\mu), & 1<u\leqslant\mu\sqrt{2}\end{cases}\end{cases}$$

where $\mu = (a - l)/(a + l)$ and $u = x/(a + l)$. The asymptotic behavior of $g(u)$ at $u > 1$ is given by the straight line $g(u) = \pi x$. Luban and Deutch (1980) tabulated functions $g(x)$ for various $W_l(t)$; this makes possible practical application of the general approach under consideration.

Another approach is the numerical solution of equation (9.31). We shall now consider one of the most widely employed numerical methods for solving such equations — the matrix-inversion technique (some other numerical methods are examined in Section 9.7).

The matrix-inversion technique involves numerical integration of equation (9.31), e.g., by the trapezoidal rule. In this case

$$F(s_i) \approx 2 \sum_j W_l[(s_j^2 - s_i^2)^{1/2}]\Delta t_j I(s_j)$$

where $s_j^2 = s_i^2 + t_j^2$ and $\Delta t_j = (s_j^2 - s_i^2)^{1/2} - (s_{j-1}^2 - s_i^2)^{1/2}$; the values of s_i and s_j may be chosen at experimental points. Since the integration interval in equation (9.31) is finite owing to interval termination where $W_l(t) \neq 0$ (the case of an infinite slit has already been considered), each experimental value is given as

$$F(s_i) = \sum_{j=1}^{N} A_{ij} I(s_i), \qquad i, j = 1, \ldots, N \tag{9.36}$$

Hence we obtain a system of N linear equations with N unknowns; its solution

$$I(s_j) = \sum_{i=1}^{N} B_{ij} F(s_i) \tag{9.37}$$

($\mathbf{B} = \mathbf{A}^{-1}$) makes it possible to restore the sought function $I(s)$.

Such a method may be regarded as the simplest and most convenient way of solving equation (9.31) for an arbitrary weighting function. However, its application involves a number of problems. The major difficulty is that the solution of equation (9.36) exhibits numerical instability with respect to errors in $F(s)$ (see, e.g., Strobl, 1970). In practice, therefore, algorithms for the realization of this method should somehow or other ensure solution stability.

This problem was solved most easily by Schedrin and Feigin (1966). Integration by parts of the right-hand side of equation (9.31) yields

$$F(s) = 2 \int_0^{\infty} K(s, t) I'[(s^2 + t^2)^{1/2}] dt \tag{9.38}$$

where the new kernel of the equation is given by

$$K(s, t) = t \int_0^t W_l(x) dx / (s^2 + t^2)^{1/2}$$

When equation (9.38) is solved by the method considered here, we obtain a relationship similar to equation (9.37), not for the quantities $I(s_i)$ but for their derivatives. Further transition to $I(s)$ involves integration over the quantities $I(s_i)$, substantially improving solution stability. Another method of stabilization was proposed by Vonk (1971). In this approach one artificially decreases the number of unknowns in system (9.36) by n (in practice, $n = 3$–5). Initially, quantities $I(s_i)$ $(i = 1,\ n + 1,\ 2n + 1, \ldots)$ are assumed to be unknown. The values of $I(s_i)$ for the other subscripts i are expressed as linear combinations of unknown values using parabolic interpolation (that is, *a priori* information on the smoothness of curve $I(s)$ is taken into account). Such a system of N equations with N/n unknowns is solved by means of a least-squares method, which makes the solution more stable. Then, the process is repeated for variables $I(s_i)$ with $i = 2, n + 2, 2n + 2, \ldots,$ and so on. In this case, therefore, the solution smoothness is required for implicit stabilization.

The matrix-inversion technique was considered in general by Mazur (1971). For a solution, the Taylor expansion of $I(s)$ in the neighborhood of s_i is used and a general method has been developed for seeking shift operators that make it possible to obtain a stable solution. The algorithms implemented for this method have also been worked out.

Matrix-inversion methods have extensive application, due mainly to their reliability and simplicity in realization. It should, however, be remembered that their use, as well as the use of an analytical approach, requires preliminary smoothing of experimental data. The above procedures of solution stabilization are capable, at best, of not allowing any increase in statistical noise, but they do not permit its decrease. For instance, an increase of n in the method of Vonk (1971) may lead only to systematic errors in $I(s)$. Many studies deal with comparison of various methods of slit-height desmearing. Thus, Walter *et al.* (1977) compared the methods of Heine and Roppert (1962), Schmidt (1965), and Vonk (1971), and the iteration method (Lake, 1967; see Section 9.7.1), as applied to a model curve typical for small-angle scattering. Figure 9.7 presents an example of the application of these methods for restoration of the curve for scattering from a sphere with $R = 100$ and two concentric density layers: ρ_1 at $r \leqslant 60$ and $\rho_2 = -\rho_1/4$ at $60 \leqslant r \leqslant 100$. The exact curve was distorted following equation (9.31) with the "infinite-height" condition; then, statistical noise with $\sigma = 3\%$ was introduced in the smeared curve. Afterward, smoothing (according to the frequency filtering method), extrapolation of curve $F(s)$ to large angles (see Section 9.6), and then its desmearing were carried out. It is seen that the algorithms under consideration give approximately the same and quite satisfactory data. On the other hand, the results of Walter *et al.* (1977) show that all these methods are unstable with respect to statistical errors and experimental-data smoothing is a necessary procedure in this case.

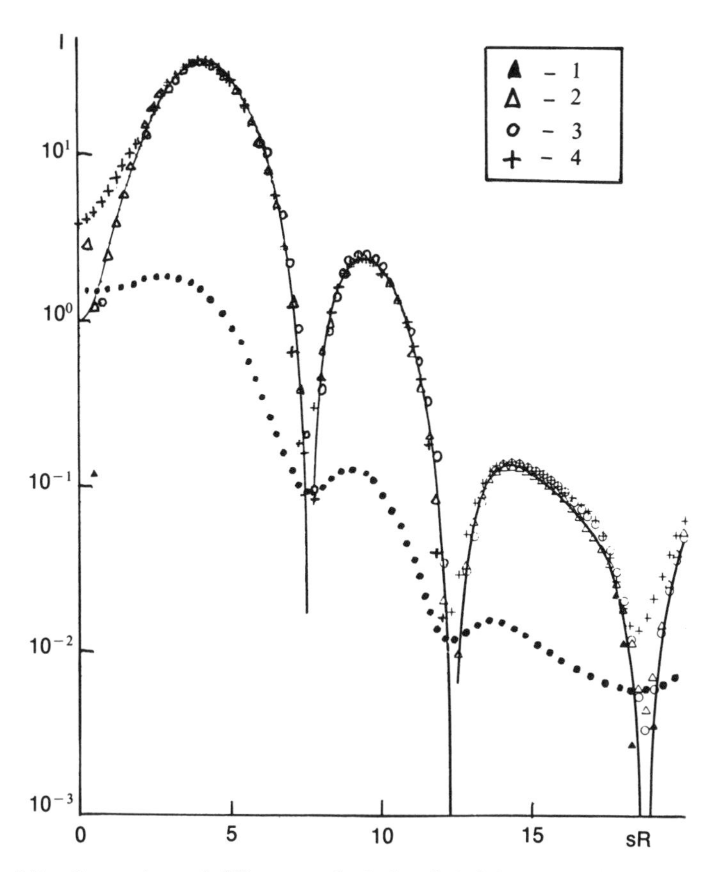

Figure 9.7. Comparison of different methods for slit-height desmearing (after Walter *et al.*, 1977). The solid curve refers to $I(s)$ and the dotted curve to $F(s)$. Restored curves follow: (1) Heine and Roppert (1962); (2) Schmidt (1965); (3) Vonk (1971); (4) Lake (1967).

9.5. Corrections for Polychromaticity

Special methods of introducing corrections for polychromatic radiation are known to exist in considerably smaller numbers than algorithms that account for collimation distortions. This is explained by the fact that, until recently, the main radiation sources for small-angle investigations have been X-ray tubes, for which the high-intensity monochromatic beam was due to the narrow characteristic lines of X-ray spectra. A simplified algorithm for X-ray studies was suggested by Zipper (1969), who, by assuming that the spectrum contains radiations with two wavelengths ($\lambda_1 = \lambda_\alpha$, $\lambda_2 = \lambda_\beta$) [see equation (9.4)], developed a method that takes into account the characteristic K_β-line.

At present, research carried out with synchrotron radiation and neutron reactors (see Chapter 8) has attained more and more importance in small-angle scattering. Here, the radiation spectra are continuous and very high monochromatization decreases substantially the brightness of the primary beam. A method has been proposed by Svergun and Semenyuk (1985, 1986) to introduce corrections for a polychromaticity beam at an arbitrary radiation spectrum $W_\lambda(\lambda)$. On taking the Mellin transform of both sides of equation (9.11) and using the convolution theorem for this transform (Korn and Korn, 1961), we obtain

$$\tilde{I}_n(\xi) = \tilde{I}(\xi)/\tilde{W}(\xi)$$

where

$$\tilde{I}(\xi) = \int_0^\infty I(s)s^{\xi-1}ds$$

is the Mellin image of function $I(s)$, while $\tilde{I}_n(\xi)$ and $\tilde{W}(\xi)$ are the Mellin transforms of functions $I_n(s)$ and $\lambda W_\lambda(\lambda)$, respectively. Hence,

$$I(s) = \frac{1}{2\pi i}\int_{c-i\infty}^{c+i\infty} \frac{\tilde{I}_n(\xi)}{\tilde{W}_\lambda(\xi)} s^{-\xi} d\xi$$

This solution is, however, unstable. According to Tikhonov and Arsenin (1977) the stable, regularized solution can be written as

$$I_\alpha(s) = \frac{1}{2\pi i}\int_{c-i\infty}^{c+i\infty} \frac{\tilde{W}_\lambda^*(\xi)\tilde{I}_n(\xi)s^{-\xi}d\xi}{\tilde{W}_\lambda^*(\xi)\tilde{W}_\lambda(\xi) + \alpha M(\xi)} \tag{9.39}$$

where

$$\tilde{W}_\lambda^*(\xi) = \int_0^\infty \lambda W_\lambda(\lambda)\lambda^{1-\xi}d\lambda$$

$\alpha > 0$ is the regularization parameter, and $M(\xi)$ is the stabilizer. Svergun and Semenyuk (1986) used a stabilizer of type $M(\xi) = |\xi|^{2p}$, $p > 0$, and chose an optimum value of α by the discrepancy method (similar to the χ^2-criterion for smoothing). It was shown in model examples that the regularized solution allows reliable restoration of function $I(s)$ at sufficiently large distortions arising from polychromatic effects and in the presence of errors in both functions $I_n(s_i)$ and $W_\lambda(\lambda)$. The reduction of termination effects of $I_n(s)$ will be dealt with in Section 9.6.

It should be noted that a similar method may be used to obtain the regularized solution of equation (9.28) for slit-width desmearing. The difference consists only in replacing Mellin images by Fourier images; function $\tilde{W}_w^*(\xi)$ will be the complex-conjugate function of $\mathscr{F}[W_w(s)]$.

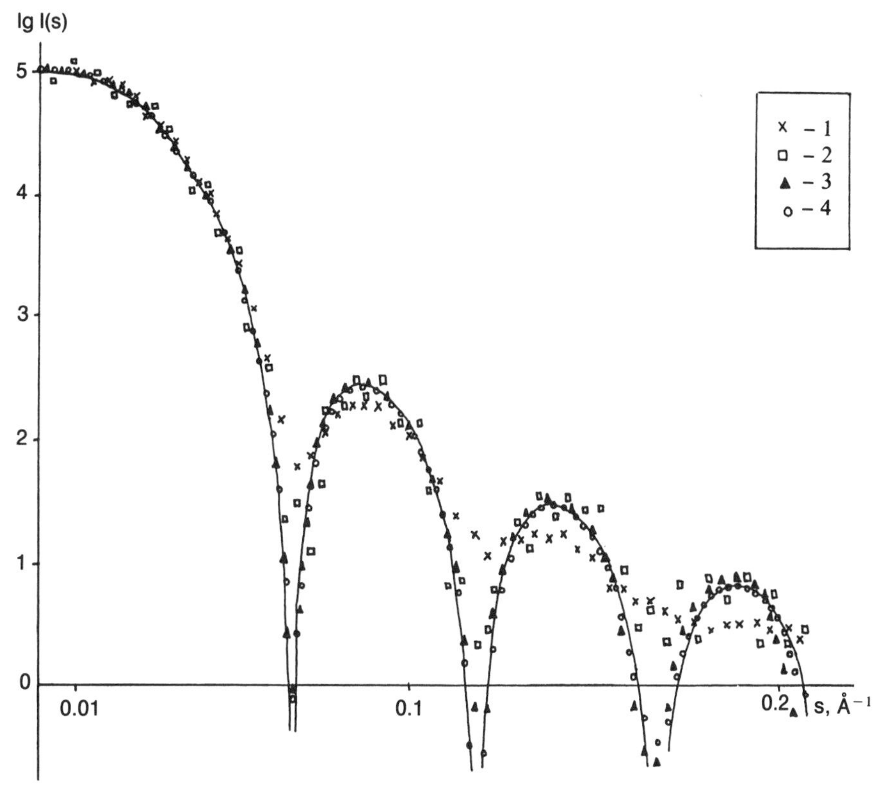

Figure 9.8. Corrections for beam polychromaticity. The solid curve denotes $I(s)$: (1) $I_n(s)$ with statistical noise, $\sigma = 3\%$; restored curves according to: (2) iteration procedure of Lake (1967); (3) Glatter (1980b); (4) Svergun and Semenyuk (1986).

Corrections for polychromatic effects may also be achieved in terms of general algorithms for the simultaneous elimination of instrumental distortions, which are considered further in Section 9.7. Figure 9.8 presents the method proposed by Svergun and Semenyuk (1986) with an iteration procedure (Section 9.7.1) and Glatter's technique of indirect Fourier transform (Section 9.7.2).

9.6. Termination Effects

All the equations defining smearing of the curve $I(s)$ due to instrumental distortions are expressed in general form; integration is carried out from zero to infinity. In reality, this is not achieved for two reasons: (1) all the weighting functions normally differ from zero only over finite ranges of the argument, $a_w \leqslant u \leqslant b_w$, $0 \leqslant t \leqslant b_l$, $a_\lambda \leqslant \lambda \leqslant \lambda \leqslant b_\lambda$, and (2) experimental measurements are performed within the finite range of scattering angles. We now examine how these factors can be taken into account.

The definition as to the ranges of the weighting functions allows one to estimate the various effects as a function of the scattering angle s. Actually, by examining the ranges within which the variables u, t, and λ vary, we obtain

$$\Delta s_w = b_w - a_w$$

$$\Delta s_l = (s^2 + b_l^2)^{1/2} - s$$

$$\Delta s_\lambda = s(b_\lambda - a_\lambda)/b_\lambda a_\lambda$$

for the intervals over which integration is carried out in equations (9.25), (9.31), and (9.11). Therefore, the slit-width effects do not depend on the angle and slit-height effects decrease with increasing angle, while polychromatic effects increase linearly. Figure 9.9 shows Δs as functions of s.

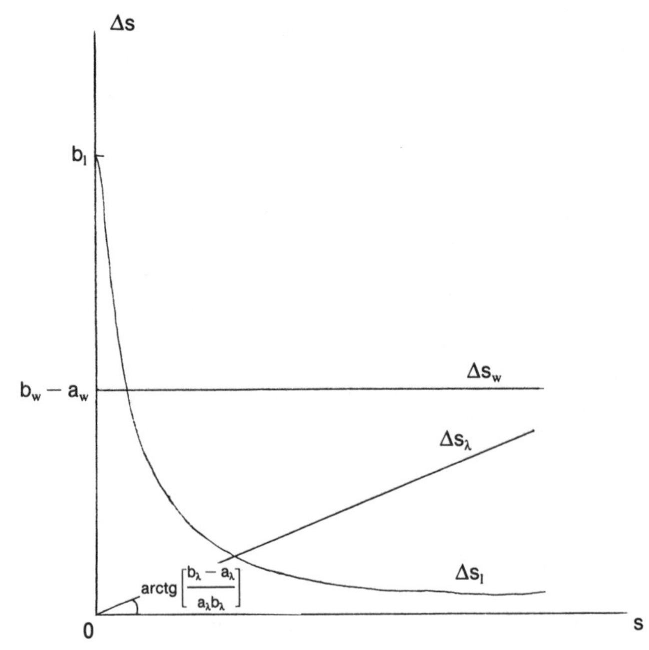

***Figure* 9.9.** Integration intervals Δs as functions of the scattering angle for different smearing effects.

Experimental-data measurement in the finite interval $s_{min} \leqslant s \leqslant s_{max}$ gives rise to some difficulties when desmearing. If we wish to restore curve $I(s)$ in the same interval, then near its edges the integration ranges will be determined not by the weighting-function definition, but by the value of s_{min} or s_{max} [in the case of correction for height, this holds only for the right edge of the curve $I(s)$]. Hence termination effects arise that may often distort substantially the edges of the restored curve. Therefore, using various methods of $I(s)$ restoration one extrapolates the experimental curve for $s < s_{min}$ and $s > s_{max}$. The most widespread methods are extrapolation according to Guinier (Section 3.3.1) in the region of small angles and proportional to s^{-3} (the Porod law for smeared curves) in the region of large s. If these methods are inapplicable, one may use, for instance, polynomial extrapolation.

The problem of termination effects when introducing slit-height corrections was considered by Rolbin *et al.* (1981) using the matrix-inversion technique. It has been shown on model examples that termination effects are negligibly small if $s_{max} \geqslant 2b_l$; on the other hand, if the experimental value satisfies $s_{max} < 2b_l$, then to reduce termination effects it suffices to define correctly the general course of the curve $F(s)$ in the interval $(s_{max}, 2b_l)$, e.g., by measuring several experimental values in this interval.

The termination effects for large s exert a very strong influence when correcting for polychromatic radiation. In the method of Svergun and Semenyuk (1986) in Section 9.5, the analytical reduction of these effects is achieved: assuming $I(s) \approx c_4/s^4$ when $s > s_{max}$, function $J(\xi) + c_4(s_{max})^{\xi-4}/(4-\xi)$ will be involved in equation (9.39) instead of $J(\xi)$.

Problems related to termination are also encountered when calculating integrals of the type

$$f(r) = \int_0^\infty s^2\varphi(s)\frac{\sin(sr)}{sr}\,ds \tag{9.40}$$

where $\varphi(s)$ is a function determined from experimentation. Transformations of such a kind are performed when evaluating the characteristic function, the value of D, the radial density distribution (see Sections 2.4.2 and 5.1). Extrapolation of $\varphi(s)$ to $s = 0$ is carried out, as a rule, according to the Guinier approximation, and the problem of termination is left only for large angles. Taking into account the fact that $\varphi(s)$ decreases rapidly enough when s tends to infinity, one can write

$$\tilde{f}(r) = f(r) + \Delta f(r, s_{max})$$

where $\tilde{f}(r)$ is the function defined by equation (9.40) over a finite interval

$(0, s_{\max})$, while

$$\Delta f(r, s_{\max}) = \frac{\sin(s_{\max} r)}{s_{\max} r} \int_{s_{\max}}^{\infty} \xi\varphi(\xi)\delta\xi + \cos(s_{\max} r) \int_{s_{\max}}^{\infty} \xi^2\varphi(\xi)\delta\xi$$

Hence $\tilde{f}(r)$ is obtained by superimposing "termination waves" on $f(r)$ — an effect analogous to the Gibbs phenomenon for Fourier transforms (Hamming, 1962). The "temperature" factor (Waser and Schomaker, 1953) is often used to eliminate this effect, i.e., instead of $\tilde{f}(r)$ the function

$$f_\alpha(r) = \int_0^{s_{\max}} s^2\varphi(s)\exp(-\alpha s^2)\frac{\sin(sr)}{sr}\,ds$$

is calculated. Here, α is a parameter and its selection enables a termination-wave amplitude to be decreased while preserving the true profile of curve $f(r)$. A method was developed by Rolbin *et al.* (1980a) that allows termination-wave reduction without selecting parameters. It has been shown that the operator

$$\Omega[\tilde{f}(r)] = \frac{1}{2tr} \int_{r-t}^{r+t} \xi \tilde{f}(\xi)d\xi \qquad (9.41)$$

where $t = \pi/s_{\max}$, is equivalent to the Lanzos σ-factor (Hamming, 1962). It has been shown by Schedrin and Simonov (1969) that this factor permits one to obtain better results than those obtained with the temperature factor for some optimum value of α.

It is easily seen that the filter (9.41) fits averaging over the termination-wave period $2t = 2\pi/s_{\max}$ and will not distort considerably the details of the profile of $f(s)$ if it is a slowly varying function in comparison with $\exp(isr)$. In practice, therefore, this situation enables termination effects in integral (9.40) to be eliminated when the scattering curve is measured with sufficiently high resolution.

The filter

$$\Omega_l[\tilde{f}_l(r)] = \frac{l+2}{(r+t)^{l+2} - (r-t)^{l+2}} \int_{r-t}^{r+t} \xi^{l+1}\tilde{f}_l(\xi)d\xi \qquad (9.42)$$

is a generalization of equation (9.41) and may be used to eliminate the termination effects in Hankel transformations

$$f_l(r) = \int_0^{\infty} s^2\varphi(s)j_l(sr)ds$$

which are used in the multipole theory of small-angle diffraction (see Sections 5.3 and 5.4.3).

9.7. Simultaneous Elimination of Various Distortions

We have considered methods for the successive elimination of instrumental distortions, when each procedure accounts for one or other type of distortion (smoothing, collimation corrections, and polychromaticity). Many procedures involve parameters defining the results of their application; some of the examined methods possess numerical instability and require regularization. Therefore, the question arises as to whether it is possible to somehow combine various procedures (or all of them) which have to be carried out in data evaluation. Algorithms realizing these possibilities will be examined in the present section.

9.7.1. Iteration Methods

Iteration algorithms are frequently used to solve integral equations (Tikhonov and Arsenin, 1977). In the main, they can be formulated as follows: if an equation of the type

$$\mathbf{AI} = \mathbf{J} \tag{9.43}$$

is defined, where $\mathbf{A}$ is a linear operator [such as the integral transformation operator (9.7)], then it can be solved using the iteration process

$$\mathbf{I}^{(k+1)} = \mathbf{I}^{(k)} + \boldsymbol{\tau}_k(\mathbf{J} - \mathbf{AI}^{(k)}) \tag{9.44}$$

where $\boldsymbol{\tau}_k$ is some function (operator, parameter) that forces the method to converge.

Lake (1967) considered the possibilities of such an approach for solving the general integral equation (9.7) (only collimation distortions were taken into account). It turned out that for the best convergence of the method in the regions of extrema of function $I(s)$ it is convenient to choose $\tau_k = I(s)/J^{(k)}(s)$. This gives the iterative equation

$$I^{(k+1)}(s) = I^{(k)}(s)\,\frac{J(s)}{J^{(k)}(s)} \tag{9.45}$$

The function $I^{(0)}(s) = J(s)$ is used as an initial approximation. Owing to its generality and simplicity, this algorithm has found wide application.

A discrepancy factor monitoring the results of two successive iterations can be used as a criterion for interrupting the process (normally, the number of iterations does not exceed five or six).

It should be noted, however, that such an iteration process is unstable with respect to random errors, and may lead to systematic errors in the minima of the curve $I(s)$ [see Walter *et al*. (1977), Deutch and Luban (1978), and Figures 9.7 and 9.8]. A modification of method (9.45), improving its stability and convergence, was proposed by Glatter (1974); however, this approach requires that a number of parameters be selected. Schelten and Hossefeld (1971) noted that iteration methods yield reliable results only for curves measured with a small angular increment and a rather high degree of accuracy.

9.7.2. Orthogonal Expansions

This is another general method for solving integral equations of type (9.43). We assume $I(s)$ is expressed as a linear combination of some orthogonal functions $\varphi_\nu(s)$:

$$I(s) \approx I_\varphi(s) = \sum_{\nu=1}^{n} c_\nu \varphi_\nu(s) \tag{9.46}$$

Substitution of equation (9.46) into equation (9.43) then yields that $J(s)$ is a superposition of functions $\psi_\nu(s)$

$$J_\nu(s) \approx \sum_{\nu=1}^{n} c_\nu \psi_\nu(s)$$

where $\psi_\nu = \mathbf{A}[\varphi_\nu]$. Coefficients c_ν may be found with the aid of the least-squares method, by minimizing the deviation of function $J_\nu(s)$ from the experimental set $J(s_i)$.

Such an approach was used by Schelten and Hossefeld (1971) to solve problems associated with collimation corrections. The curve $I(s)$ was represented by a linear superposition of cubic B-splines. The variational problem of searching for the best approximation coefficients was here reduced to the solution of system (9.21) replacing $B_k(s_j)$ by the functions

$$\beta_k(s_j) = \int_{-\infty}^{\infty} \int_{-\infty}^{\infty} B_k\{[(s_j - u)^2 + t^2]^{1/2}\} W_l(t) W_w(u) dt\, du$$

Such a procedure allows smoothing and collimation corrections to be dealt with simultaneously (a correction for polychromaticity can also be achieved). The algorithm proves to be not much more complicated than in the case of smoothing with B-splines. However, there are definite shortcomings inherent to the method. The main problem is that, in contrast to smoothing, the number n of splines should be smaller than the number N of experimental points (for solution stability). Therefore, application of the method may lead to systematic errors in these sections where $I(s)$ exhibits considerable curvature.

Another example of using orthogonal expansion is the method of "indirect Fourier transform" (Glatter, 1977a,b). In this case, with the aid of series (9.46), not function $I(s)$ but the distance distribution function $p(r)$ is expressed in the form

$$p_\nu(r) = r^2\gamma_\nu(r) = \sum_{\nu=1}^{n} c_\nu\varphi_\nu(r) \tag{9.47}$$

This expansion is carried out over a certain finite interval $0 \leqslant r \leqslant D$ (if the system is monodisperse, D fits the maximum particle size).

With allowance for equation (9.7), the general integral relation between $p(r)$ and $J(s)$ can be written as

$$J(s_i) = T_4T_3T_2T_1[p_\nu(r)] + \varepsilon(s_i) \tag{9.48}$$

with a sequence of integral transformations: T_1 is transition to reciprocal space according to equation (2.20), T_2 is distortion for polychromatic effect (9.11), T_3 and T_4 are collimation distortions (9.26) and (9.25). The expansion coefficients c_ν may be defined anew by minimization of the root-mean-square deviation

$$\phi = \sum_{i=1}^{N}\left[J(s_i) - \sum_{\nu=1}^{n} c_\nu\varphi_\nu(s_i)\right]^2 \tag{9.49}$$

where

$$\psi_\nu(s) = T_4T_3T_2T_1[\varphi_\nu(r)]$$

However, the solution of the system of linear equations following from the minimum of functional (9.49) may be unstable. To stabilize it, Glatter (1977a,b) proposed minimizing the functional

$$\phi_\alpha = \phi + \alpha N_c \tag{9.50}$$

where

$$N_c = \sum_{\nu=1}^{n-1} (c_{\nu+1} - c_\nu)^2$$

[the "stabilizer," according to Tikhonov and Arsenin, (1977)] introduces the requirement for a smooth solution, while parameter $\alpha > 0$ defines the degree of smoothness. The solution of this problem is reduced to the system of linear normal equations

$$(B_{\mu\nu} + \alpha K_{\mu\nu})c_\nu = d_\mu \tag{9.51}$$

where

$$B_{\mu\nu} = \sum_{i=1}^{N} \psi_\mu(s_i)\psi_\nu(s_i)/\sigma^2(s_i)$$

and

$$d_\mu = \sum_{i=1}^{N} J(s_i)\psi_\mu(s_i)/\sigma^2(s_i)$$

while matrix $K_{\mu\nu}$ is determined as

$$K = \begin{bmatrix} 1 & -1 & & & 0 & & \\ -1 & 2 & -1 & & & & \\ & -1 & 2 & -1 & & & \\ \cdots & \cdots & & \cdots & \cdots & & \\ \cdots & & & \cdots & \cdots & \cdots & \\ & & & -1 & 2 & -1 & \\ 0 & & & & & & \\ & & & & -1 & 1 & \end{bmatrix}$$

Glatter (1977b) also used cubic B-splines as a system of orthogonal functions, as was done by Schelten and Hossfeld (1971). However, Glatter's approach possesses a number of advantages. First, the profile of function $p(r)$ is not as complicated as that of $I(s)$. Thus approximation of $p(r)$ by a superposition of a sufficiently small number n of orthogonal functions is more justified. Furthermore, functions $\psi_\nu(s)$ are determined on the entire numerical axis and this automatically eliminates termination effects [that is, calculation of coefficients c_ν using the experimental interval $(s_{\min}, s_{\max})$ leads to extrapolation for $s < s_{\min}$ and $s > s_{\max}$]. In particular, this permits one to determine the values of R_g and $I(0)$, making use of function $p(r)$ without resorting to extrapolation according to Guinier. This approach often yields more accurate results (see Section 3.3.1).

In such an approach the result will depend on three parameters: the value of D, the number of splines n, and the multiplier α. It is therefore necessary to determine or, at least, estimate them. The value of D for monodisperse systems may be estimated experimentally with some error (see Section 3.3.4) and can be varied when using the method. It is evident that decrease in D will lead to errors, while an increase yields a decrease in solution stability. For polydisperse systems, parametrization of the size distribution function $D_N(R)$ in line with equation (9.47) is also possible; in this case D is the size of the largest particles in the system (see Section 7.2.2).

Problems of the optimum choice of n and α are closely interrelated. The fact is that resolution in real space is defined by the quantity $\Delta r = \pi / s_{max}$; consequently, the number of splines would not exceed $n_{max} = D/\Delta r = s_{max} D/\pi$ (the same result follows from the sampling theorem; see Section 3.2.2). In practice, however, such restriction results in extremely small values of n and, consequently, makes it impossible to represent $p(r)$ adequately. Regularization defined by parameter α enables one to use values $n > n_{max}$ (the number of splines may, in practice, reach $n = 30$). A point-or-inflection method was proposed to choose the optimum value of α (Figure 9.10); we have already encountered such a procedure when choosing the value of p_{opt}; see Figure 9.2. In this case the impossibility of determining α_{opt}, i.e., the absence of the characteristic shape of functions $N_c(\alpha)$ and ϕ_α, supplies evidence for the incorrect choice of parameters D and/or n.

A similar approach was proposed for eliminating termination effects when calculating the spherically symmetric scattering density distribution [the radial distribution function $\rho(r)$ was represented in line with

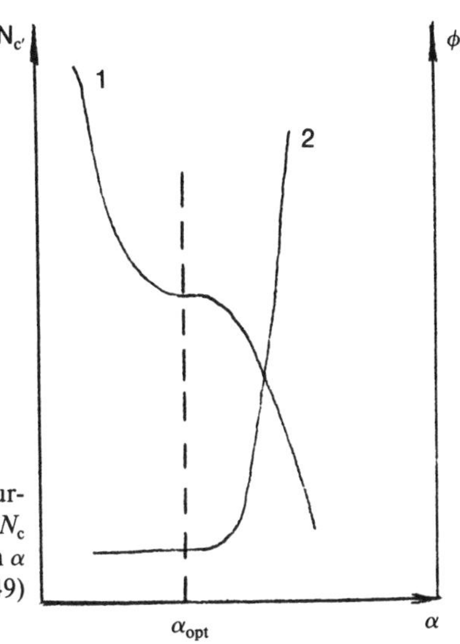

Figure 9.10. Choice of α_{opt} for an indirect Fourier transform (after Glatter, 1977a). Values of N_c (curve 1) and ϕ_α (curve 2) are obtained for given α by substituting coefficients c_v into equations (9.49) and (9.50).

equation (9.47); see Glatter (1977a)], for special cases of scattering by lamellar and cylindrical particles [expansion of functions $p_t(r)$ and $p_c(t)$, respectively (Glatter, 1980c)], and also for analyzing scattering by polydisperse systems [expansion of size distribution function $D_N(R)$; Glatter (1980a), see Section 7.2.2]. These techniques are utilized in a program complex, ITP-81, which, at present, is employed extensively since it enables simultaneous account to be taken of all instrumental distortions at high stability with respect to experimental errors and termination effects. A model example of using the method is presented below (see Section 9.7.4).

The indirect Fourier transform method possesses great applied importance and may provide new possibilities. For instance, it was shown by Glatter and Laggner (1983) with the aid of model examples that application of this method permits one to estimate the radii of gyration and particle shape when scattering curves are measured with "white" poychromatic synchrotron radiation (see Section 8.3.2). This allows synchrotron sources to be used to their maximum extent when carrying out kinetic experiments.

However, some shortcomings and limitations of the method should be noted. As already seen, the number of splines n is an arbitrarily chosen value. At the same time, the solution may depend strongly on this value, especially with the accuracy not being very high and the small angular interval of the measured experimental data; see, for example, Laggner and Müller (1978). Therefore, to obtain reliable results it is sometimes necessary to apply the procedure for various values of n. Moreover, the choice of B-splines as a system of orthonormal functions is also somewhat arbitrary. For example, it is easily seen from equation (9.47) that, since the splines are defined on the grid with increment equal to $\Delta r_n = D/(n-1)$, the scattering curve corresponding to such an expansion will be quasi-periodic with "wavelength" $2\pi/\Delta r_n = 2\pi(n-1)/D$, i.e., we have serious distortions in the asymptotic behavior of the approximating scattering intensity.

9.7.3. Use of the Sampling Theorem

We have seen in the previous section that the choice of the number n of independent parameters enabling the description of a scattering curve [or function $p(r)$] is a serious problem. The attempt to obtain the most accurate description of the sought function compels us to increase n; this, however, worsens the stability of the results and makes it necessary to resort to additional regularizing procedures which, in turn, require selection of stabilizers and regularization parameters. Therefore, a problem arises concerning the search for a number of independent variables, necessary and sufficient for the parametrization.

It was shown in Section 3.2.2 that, according to the sampling theorem, an ideal scattering intensity curve $I(s)$ for a monodisperse system is fully determined by its values at $s_k = k\pi/D = k\Delta s$. We assume the curve is known in the interval $s_{\min} \leqslant s \leqslant s_{\max}$. Then, by assuming that the asymptotic behavior of $I(s)$ as s tends to infinity is described by equation (3.28), we obtain, according to equation (3.20) for any s in this interval (Taupin and Luzzati, 1982),

$$I(s) = \frac{1}{s}\left\{\sum_{k=1}^{k_{\max}} W_k \phi(s, k, D) + \sum_{k=k_{\max}+1}^{\infty} \left[\frac{c_0}{(k\Delta s)^3} + c_4 k\Delta s\right] \phi(s, k, D)\right\} \quad (9.52)$$

where

$$W_k = k\Delta s I(k\Delta s)$$

(on the assumption that the condition $s_{\min} < \pi/D$ is fulfilled). One can easily see that function $J(s)$ containing instrumental distortions can be expressed in the same way, but function

$$\psi(s, k, D) = T_4 T_3 T_2[\phi(s, k, D)/s]$$

should replace $\phi(s, k, D)/s$ in equation (9.52). Such a description contains $n = (s_{\max} D/\pi) + 2$ independent parameters (degrees of freedom). Its advantage consists of the fact that we can be sure description (9.52) with such a number of parameters [which is deliberately less than, say, the number of splines in the methods of Schelten and Hossfeld (1971) and Glatter (1977b)] reproduces adequately the scattering curve [if the assumption about the asymptotic behavior of $I(s)$ is correct]. Quantities W_k, c_0, and c_4 are unknowns that can be determined by the standard least-squares procedure leading to a well-known system of normal equations. The accuracy of their determination is estimated with the aid of the error propagation law.

Taupin and Luzzati (1982) further generalized such an approach to measurements of scattering curves at different solvent densities. If each basic curve (4.18) is expressed in line with equation (9.52), we obtain a system of equations in $3n$ unknowns. The solution of this system allows determination of the shape scattering curve, the inhomogeneity scattering curve, and the interference term [it should be noted, however, that the asymptotic trend (3.28) is hardly fulfilled for functions $I_{cs}(s)$ and $I_s(s)$]. As an example, some calculations were performed with the distorted scattering curve of low-density lipoprotein (Luzzati *et al*., 1979); data evaluation and determination of structural parameters of the particle were carried out.

A similar approach (with the same number of parameters) was developed by Moore (1980). Function $r\gamma(r)$ was expanded as a trigonometric series (3.22), where the expansion coefficients (as already noted in Section 3.2) coincide with the quantities W_k. These coefficients are determined in the same way as in Glatter's technique, but there is no regularization of the solution, D being the only parameter. When using this algorithm, a number of calculations are conducted with different values of D and D_{opt} is to be chosen. Figure 9.11 illustrates the results obtained by Moore (1980) for an experimental set of neutron data (Langer *et al*. 1978). These results assert that the value $D = 100$ Å is, in this case, a reliable estimate. It should be noted that when choosing 100 Å $< D <$ 150 Å the values of the structural parameters calculated in the interval $D <$ 100 Å vary only slightly. A comparison with the indirect Fourier transform method shows that Moore's approach normally gives results similar to those of the former approach with maximum α in the stable range. From a theoretical point of view, the method of Moore (1980) is more convenient and reliable than that of Glatter (1977b), but if termination effects are large (the number of degrees of freedom too small) the use of an indirect Fourier transform with optimum regularization enables better results to be obtained.

General probems arising in connection with the application of the sampling theorem for small-angle data processing are considered by Gerber and Schmidt (1983), who show that, for a fixed duration of an experiment, it does not matter whether the measurements were carried out with angular increments $\Delta s = \pi/D$ or $\Delta s \ll \pi/D$. Using the sampling

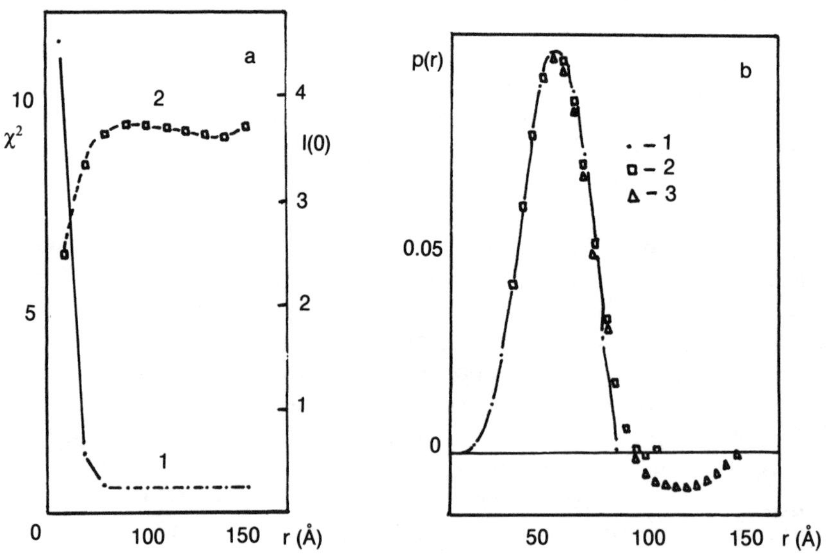

Figure 9.11. Application of Moore's (1980) approach: (a) χ^2-criterion (1) and quantity $I(0)$ (2) as functions of D; (b) restored function $p(r)$ for $L = 80$ Å (1), 100 Å (2), and 140 Å (3).

theorem as a tool for data treatment, one obtains similar final results for both sets of data. This means that in an experiment the value of Δs should be chosen as small as possible; this enables the inequality $\Delta s \leqslant \pi/d$ to be fulfilled.

Methods making use of the sampling theorem are a reliable instrument for data evaluation. The advantageous features of these methods are: a justified parametrization and stability of the results, a small number of parameters, the possibility of calculating structural parameters and estimation of their errors (see Section 3.2.2). One should, however, bear in mind that the condition for function $\gamma(r)$ to have a finite support is necessary in order to employ these methods. Therefore, the main field of their application is the evaluation of scattering data for monodisperse systems (or such systems where an analogue of maximum distance can be introduced).

9.7.4. General Regularization Procedure

The idea of an indirect transform for the treatment of data can be realized very simply without any parametrization of the solution to be found. In fact, by changing the order of integration in equation (9.48) and integrating over the variables corresponding to the instrumental distortions, one obtains (Svergun and Semenyuk, 1987)

$$J(s) = \int_0^D K_1(s, r)p(r)dr + \varepsilon(s_i) \tag{9.53}$$

where

$$K_1(s, r) = 4\pi r^2 \int_{-\infty}^{\infty} \int_{-\infty}^{\infty} \int_0^{\infty} W_w(x)W_l(t)W_\lambda(\lambda) \times \frac{\sin([(s-x)^2 + t^2]^{1/2} r/\lambda)}{[(s-x)^2 + t^2]^{1/2} r/\lambda} d\lambda dt dx \tag{9.54}$$

is the kernel, which is fixed under the given experimental conditions and depends only on weighting functions. In this way the treatment of data for monodisperse systems is reduced to the solution of equation (9.53), which is a Fredholm equation of the first kind. [The same integral equation can easily be written for polydisperse systems with respect to function $D(R)$.] Svergun and Semenyuk (1987) used Tikhonov's regularization method (Tikhonov and Arsenin, 1977) to solve the equation. Here, the Tikhonov functional to be minimized is

$$\phi_\alpha = \sum_{i=1}^{N} \frac{1}{\sigma_i^2} [J(s_i) - \tilde{J}(s_i)]^2 + \alpha \| p \|^2$$

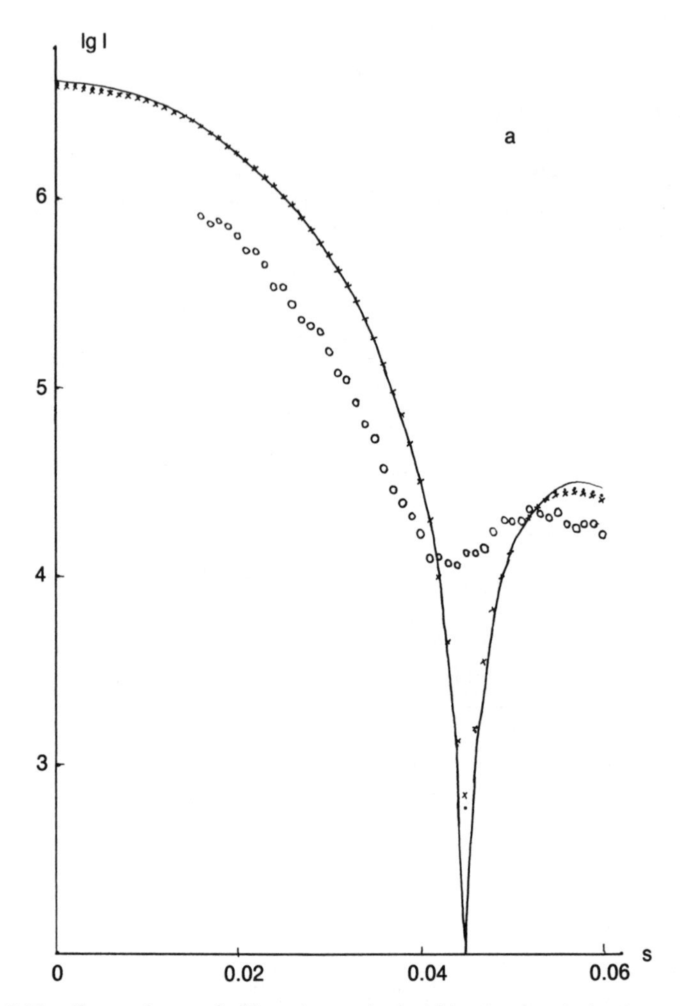

Figure 9.12. Comparison of Glatter's method with the regularization technique: (a) scattering curves: solid line refers to exact scattering curve from a sphere of radius $R = 100$; open circles to smeared curve with noise 5%, $s_{\min} = 0.016$, $s_{\max} = 0.06$; crosses and dots refer to restoration by two methods, $D = 240$; (b) distance distribution functions with notation as in (a).

where σ_i is the mean-square deviation at the point s_i, and $\tilde{J}(s)$ is the smeared curve corresponding to the calculated function $p(r)$. The stabilizer is taken in the form

$$\| p \|^2 = \int_0^D [p(r)]^2 dr + k \int_0^D [p'(r)]^2 dr$$

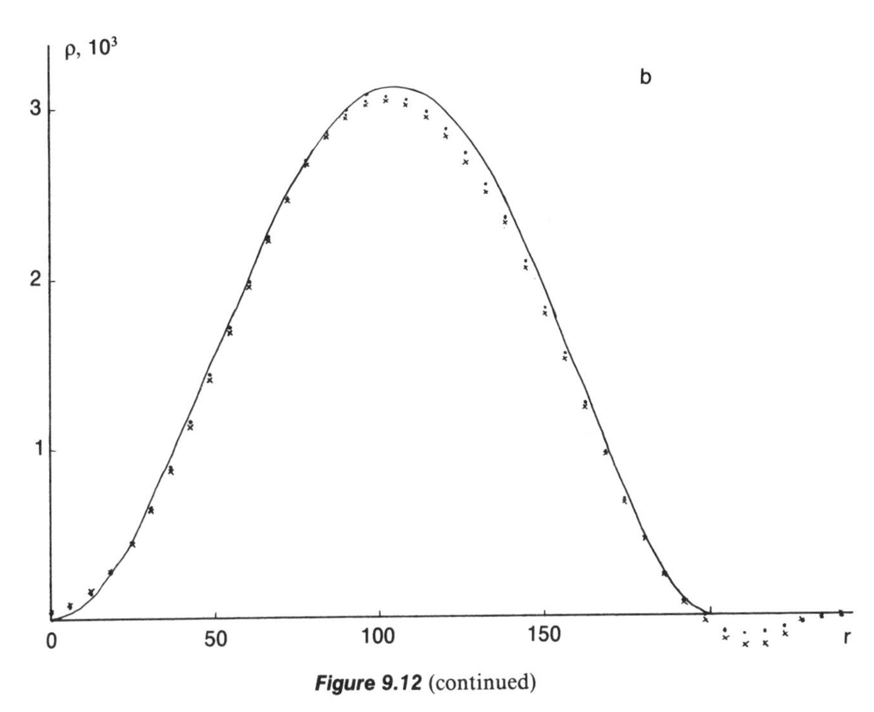

Figure 9.12 (continued)

which implies that the solution and its first derivative are finite; the constant k can be selected from metric considerations. Svergun and Semenyuk (1987) tested several methods for choosing the regularization parameter α: the discrepancy method (here it is, in fact, the χ^2-test), the quasi-optimum method ($\| \alpha \partial p_\alpha / \partial \alpha \| = \min, \alpha > 0$), and the point-of-inflection method ($-\partial \| p_\alpha \| / \partial \alpha = \min, \alpha > 0$). It has been proved that a rough estimation of α by the discrepancy method followed by its refinement with the aid of the other two methods enables reliable determination of the value of α.

One can easily see that the described approach is similar to the indirect transform technique. Figure 9.12 presents a model example in which the two methods are compared. They give almost identical results and are very stable with respect to statistical errors and termination effects. At the same time, the regularization procedure can be more convenient to apply. Actually, the fact that no parametrization is used, although it increases the number of unknowns, makes the method more general. Moreover, the regularization parameter can easily be chosen by a straightforward routine, while corresponding parameters in ITP should to some extent be selected "by hand." Therefore, the regularization procedure combines the advantages of indirect transform methods with the simplicity of standard techniques.

9.8. Conclusion

The mathematical apparatus for experimental small-angle scattering data evaluation has recently undergone serious changes. Methods of mathematical programming, orthonormal expansions (B-splines, polynomials, trigonometrical functions), and methods for the regularization of ill-posed problems are now widely applied. Modern powerful computers make it possible to employ various numerical methods instead of searching for an exact analytical solution to a problem. Therefore, numerical methods are now the main instruments in the treatment of data.

The different approaches to the solution of similar problems have resulted in the appearance of investigations dealing with their comparison. There is now a striking tendency to develop general algorithms for the optimization of a solution. Some of the reviewed methods (smoothing over polynomials and splines, solution of integral equations, regularization techniques) are standard routines of applied mathematics. On the other hand, contemporary research involves algorithms which take into account the features of small-angle scattering data. In particular, methods making use of the sampling theorem give a fair representation of a scattering curve as well as strict estimates of errors in the results.

The authors have not tried to provide a unique answer as to which methods should be applied for data evaluation. We have simply strived to elucidate the possibilities and limitations of the principal methods now used in the evaluation of small-angle scattering data. Not all of the methods are universal, and their choice may be dictated by the specific features of the actual experiment and the potentialities of the computer employed.

References

Abramowitz, M. and Stegun, I. A., eds. (1964). *Handbook of Mathematical Functions*, Government Printing Office, Washington.

Acuña, R. J. and Craievich, A. F. (1979). *J. Non-Cryst. Solids*, **34**, 13–27.

Agamalyan, M. M., Drabkin, G. M., Deriglazov, V. V., and Krivshich, T. I. (1984). Preprint LIYaF, No. 921, Leningrad (in Russian).

Agamalyan, M. M., Drabkin, G. M., Dovzhikov, A. A., Krivshich, T. I., Lvov, Yu. M., Feigin, L. A., Gashpar, Sh., and Ronto, D. (1982). *Kristalografia*, **27**, 92–96. [Engl. transl. (1982). *Sov. Phys. Crystallogr.*, **27**, 53–55.]

Agamalyan, M. M., Panev, G. A., and Zavediya, A. A. (1983). *Nucl. Instrum. Methods*, **213**, 488–493.

Agarwal, S. C. and Herman, H. (1974). *Siemens Rev.*, **41**, 34–40.

Albere, J., Fisher, J., Rabeka, V., Rogers, L. C., and Schoenborn, B. (1975). *Nucl. Instrum. Methods*, **127**, 505–523.

Aleferd, B. and Conrad, A. (1984). *Berichte Kernforschungsanlage Jülich*, No. 1954, 100–106.

Allinson, N. M. (1982). *Nucl. Instrum. Methods*, **201**, 53–64.

Altgelt, K. and Schulz, G. V. (1960). *Makromol. Chem.*, **36**, 209–219.

Ameniya, Y., Wakabayashi, K., Hamanaka, T., Wakabayashi, T., Matsushita, T., and Hashisumo, H. (1983). *Nucl. Instrum. Methods*, **208**, 471–477.

Anderegg, J. W. (1965). In: *Small-Angle X-Ray Scattering* (Brumberger, H., ed.), pp. 243–265, Gordon and Breach, New York.

Aultchenko, V. M., Baru, S. E., Sidorov, V. A., Savinov, G. A., Feldman, I. G., Khabakhpashev, A. G., and Yasenev, M. V. (1983). *Nucl. Instrum. Methods*, **208**, 443–444.

Bacon, G. E. (1975). *Neutron Diffraction*, Clarendon Press, Oxford.

Baru, S. E., Proviz, G. I., Savinov, G. A., Sidorov, V. A., Feldman, I. G., and Khabakhpashev, A. G. (1978). *Nucl. Instrum. Methods*, **152**, 195–197.

Bassani, F. and Altarelli, M. (1983). In: *Handbook on Synchrotron Radiation*, Vol. 1A (Koch, E.-E., ed.), pp. 463–605, North-Holland, Amsterdam.

Benoit, H. and Doty, P. (1953). *J. Phys. Chem.*, **57**, 958–963.

Benoit, H., Decker, D., Higgins, J. S., Picot, C., Cotton, J. P., Farnoux, B., Jannik, G., and Ober, R. (1973). *Nature (London)*, **245**, 13–15.

Bhatia, A. B. and Thornton, D. E. (1970). *Phys. Rev. B*, **2**, 3004–3012.

Blaurock, A. E. and Wilkins, M. H. F. (1969). *Nature (London)*, **223**, 906–909.

Blaurock, A. E. and Wilkins, M. H. F. (1972). *Nature (London)*, **236**, 313–314.

Blokhin, M. A. (1957). *Physics of X Rays*, Gostehizdat, Moscow (in Russian).

Blundel, T. L. and Johnson, L. N. (1976). *Protein Crystallography*, Academic Press, New York.
Bochner, S. (1959). *Lectures on Fourier Integrals*, Princeton University Press, Princeton.
Bonse, U. and Hart, M. (1965). *Appl. Phys. Lett.*, 7, 238–240.
Bordas, J., Koch, M. H. J., Clout, P. H., Dorrington, E., Boulin, C. and Gabriel, A. (1980). *J. Phys. E*, **13**, 938–994.
Borso, C. S. (1982). *Nucl. Instrum. Methods*, **201**, 65–71.
Bowen, T. J. (1971). *An Introduction to Ultracentrifugation*, Wiley, New York.
Boyarintseva, A. K., Dembo, A. T., Rolbin, Yu. A., and Feigin, L. A. (1975). *Kristallografia*, **20**, 149–151. [Engl. transl. (1975). *Sov. Phys. Crystallogr.*, **20**, 85–86.]
Boyarintseva, A. K., Rolbin, Yu. A., and Feigin, L. A. (1977). *Doklady AN SSSR*, **237**, 709–712 (in Russian).
Bradaczek, H. and Luger, P. (1978). *Acta Crystallogr., Sect. A*, **34**, 681–683.
Bragg, W. L. and Perutz, M. F. (1952). *Acta Crystallogr.*, **5**, 277–283.
Brämer, R. (1972). *Koloid-Z. Z. Polym.*, **250**, 1034–1038.
Brandt, S. (1970). *Statistical and Computational Methods in Data Analysis*, North-Holland, Amsterdam.
Brill, O. L., Weil, C. G., and Schmidt, P. W. (1968). *J. Colloid Interface Sci.*, **27**, 479–492.
Brink, D. M. and Satchler, G. R. (1968). *Angular Momentum*, Oxford Library of Physical Sciences, 2nd ed., Clarendon Press, Oxford.
Buchanan, D. R. (1971). *J. Polym. Sci.*, **9**, 645–658.
Buchanan, M. G. and Hendricks, R. W. (1971). *J. Appl. Crystallogr.*, **4**, 176–177.
Cahn, J. W. (1962). *Acta Metall.*, **10**, 907–913.
Chabre, M., Saibil, H., and Worcester, D. L. (1975). *Proc. Brookhaven Symp.*, **27**, 77–85.
Christen, D. K., Kerehner, H. R., Sekula, S. T., and Thorel, P. (1980). *Phys. Rev. B*, **21**, 102–117.
Cook, H. E. (1970). *Acta Metall.*, **18**, 297–306.
Cotton, J. P. and Benoit, H. (1975). *J. Phys. (Paris)*, **36**, 905–910.
Cowley, J. M. (1975). *Diffraction Physics*, North-Holland, Amsterdam.
Crist, B. (1973). *J. Polym. Sci., Polym. Phys. Ed.*, **11**, 635–661.
Crowther, R. A. (1967). *Acta Crystallogr.*, **22**, 758–764.
Crowther, R. A. (1969). *Acta Crystallogr., Sect. B*, **25**, 2571–2580.
Damaschun, G. and Pürschel, H.-V. (1969). *Monatsh. Chem.*, **100**, 247–279.
Damaschun, G. and Pürschel, H.-V. (1971). *Acta Crystallogr., Sect. A*, **27**, 193–197.
Damaschun, G., Müller, J. J., and Pürschel, H.-V. (1968). *Monatsh. Chem.*, **99**, 2343–2348.
Damaschun, G., Müller, J. J., and Pürschel, H.-V. (1971). *Acta Crystallogr., Sect. A*, **27**, 11–18.
Damaschun, G., Müller, J. J., Damaschun, H., Pürschel, H.-V., Walter, G., and Kranold, R. (1974). *Stud. Biophys.*, **47**, 47–89.
Dantzig, G. B. (1963). *Linear Programming and Extensions*, Princeton University Press, New Jersey.
Deas, H. D. (1952). *Acta Crystallogr.*, **5**, 542–546.
Debye, P. (1915). *Ann. Physik*, **28**, 809–823.
Debye, P. (1927). *Physik. Z.*, **28**, 135–141.
Debye, P. (1947). *J. Phys. Colloid Chem.*, **51**, 18–32.
Debye, P. and Bueche, A. M. J. (1949). *J. Appl. Phys.*, **20**, 518–525.
Deutch, M. and Luban, M. (1978a). *J. Appl. Crystallogr.*, **11**, 87–97.
Deutch, M. and Luban, M. (1978b). *J. Appl. Crystallogr.*, **11**, 98–101.
Du Mond, J. W. M. (1947). *Phys. Rev.*, **72**, 83–84.
Earnshaw, W. C. and Harrison, S. C. (1977). *Nature (London)*, **268**, 598–602.
Edmonds, A. R. (1957). *Angular Momentum in Quantum Mechanics*, Princeton University Press, Princeton.

Engelman, D. M. and Moore, P. B. (1972). *Proc. Natl. Acad. Sci. USA*, **69**, 1997–1999.
Engelman, D. M., Moore, P. B., and Schoenborn, B. R. (1975). *Proc. Natl. Acad. Sci. USA*, **72**, 3888–3892.
Epperson, J. E., Hendricks,R. W., and Farrel, K. (1974). *Philos. Mag.*, **30**, 803–817.
Erdélyi, A. (1954). *Tables of Integral Transforms*, Vol. 1, Bateman Mathematical Project, McGraw-Hill, New York.
Eyring, H. (1932). *Phys. Rev.*, **39**, 746–748.
Farugi, A. R. and Huxley, H. E. (1978). *J. Appl. Crystallogr.*, **11**, 449–454.
Fedorov, B. A. (1968). *Kristallografia*, **13**, 763–769. [Engl. transl. (1969). *Sov. Phys. Crystallogr.*, **13**, 663–667.]
Fedorov, B. A. (1971). *Acta Crystallogr., Sect. A*, **27**, 35–42.
Fedorov, B. A. and Aleshin, V. G. (1966). *Vysokomol. Soed.*, **8**, 1506–1513 (in Russian).
Fedorov, B. A. and Denesyuk, A. I. (1978). *J. Appl. Crystallogr.*, **11**, 473–477.
Fedorov, B. A., Ptitsyn, O. B., and Voronin, L. A. (1972). *FEBS Lett.*, **28**, 188–190.
Fedorov, B. A., Ptitsyn, O. B., and Voronin, L. A. (1974). *J. Appl. Crystallogr.*, **7**, 181–186.
Fedorov, B. A., Andreeva, N. A., Volkova, L. A., and Voronin, L. A. (1968). *Kristallografia*, **13**, 770–775. [Engl. transl. (1969). *Sov. Phys. Crystallogr.*, **13**, 668–672.]
Fedorov, B. A., Kröber, R., Damaschun, G., and Ruckpaul, K. (1976). *FEBS Lett.*, **65**, 92–95.
Fedorov, B. A., Shpungin, I. L., Gelfand, V. I., Rosenblat, V. A., Damaschun, G., Damaschun, H., and Papst, M. (1977). *FEBS Lett.*, **84**, 153–155.
Fedorova, I. S. and Schmidt, P. W. (1978). *J. Appl. Crystallogr.*, **11**, 405–411.
Feigin, L. A. and Soler, I. (1975). *Kristallografia*, **20**, 494–500. [Engl. transl. (1975). *Sov. Phys. Crystallogr.*, **20**, 302–307.]
Feigin, L. A., Schmidt, P. W., and Schedrin, B. M. (1981). *Kristallografia*, **26**, 920–924. [Engl. transl. (1982). *Sov. Phys. Crystallogr.*, **26**, 523–525.]
Feigin, L. A., Gonchar, N. A., Likhtenshtein, G. I., Lvov, Yu. M., and Marakushev, S. A. (1978). *Kristallografia*, **23**, 749–755. [Engl. transl. (1978). *Sov. Phys. Crystallogr.*, **24**, 420–424.]
Fiddy, M. A. and Ross, G. (1979). *Opt. Acta*, **26**, 1139–1146.
Filipovich, V. N. (1956). *Zh. Tekh. Fiz.*, **26**, 398–416 (in Russian).
Finch, J. T. and Holmes, K. C. (1967). In: *Methods in Virology* (Moramorosh, K. and Koprowski, H., eds.), Vol. 3, pp. 351–475, Academic Press, New York.
Fischbach, F. A. and Anderegg, J. W. (1965). *J. Mol. Biol.*, **14**, 458–473.
Fisher, E. M. (1966). *J. Chem. Phys.*, **44**, 612–622.
Flory, P. J. (1969). *Statistical Mechanics of Chain Molecules*, Wiley, New York.
Franks, N. P. and Lieb, W. R. (1979). *J. Mol. Biol.*, **133**, 469–500.
Gadjiev, A. M. and Vazina, A. A. (1984). *Mol. Biol.*, **18**, 792–797 (in Russian).
Genin, Ya. V., Gerasimov, V. I., and Tsvankin, D. Ya. (1973). *Vysokomol. Soed., A*, **15**, 1798–1801 (in Russian).
Gerasimov, V. I. (1970). *Kristallografia.*, **15**, 156–159. [Engl. transl. (1970). *Sov. Phys. Crystallogr.*, **15**, 122–124.]
Gerasimov, V. I. and Tsvankin, D. Ya. (1969). *Vysokomol. Soed., Ser. A*, **11**, 2652–2665 (in Russian).
Gerasimov, V. I., Genin, Ya., and Tsvankin, D. Ya. (1974). *J. Polym. Sci., Polym. Phys. Ed.*, **12**, 2035–2046.
Gerber, Th. and Schmidt, P. W. (1983). *J. Appl. Crystallogr.*, **16**, 581–589.
Gerold V. (1967). In: *Small-Angle X-Ray Scattering* (Brumberger, H., ed.), pp. 277–317, Gordon and Breach, New York.
Gerold, V. (1977). *J. Appl. Crystallogr.*, **10**, 25–27.
Gerold, V. and Kostorz, G. (1978). *J. Appl. Crystallogr.*, **11**, 376–404.
Glatter, O. (1974). *J. Appl. Crystallogr.*, **7**, 147–153.

Glatter, O. (1977a). *Acta Phys. Austriaca*, **47**, 83–102.
Glatter, O. (1977b). *J. Appl. Crystallogr.*, **10**, 415–421.
Glatter, O. (1979). *J. Appl. Crystallogr.*, **12**, 166–175.
Glatter, O. (1980a). *J. Appl. Crystallogr.*, **13**, 7–11.
Glatter, O. (1980b). *Acta Phys. Austriaca*, **52**, 243–256.
Glatter, O. (1980c). *J. Appl. Crystallogr.*, **13**, 577–584.
Glatter, O. (1981). *J. Appl. Crystallogr.*, **14**, 101–108.
Glatter, O. and Hainisch, B. (1984). *J. Appl. Crystallogr.*, **17**, 435–441.
Glatter, O. and Kratky, O., eds. (1982). *Small-Angle X-Ray Scattering*, Academic Press, London.
Glatter, O. and Laggner, P. (1983). *J. Appl. Crystallogr.*, **16**, 42–46.
Glatter, O. and Zipper, P. (1975). *Acta Phys. Austriaca*, **43**, 307–310.
Glynn, L. and Steward, M., eds. (1981). *Structure and Functions of Antibodies*, Wiley, New York.
Golubkov, V. V. and Porai-Koshits, E. A. (1981). *Fiz. Khim. Stekla*, 7, 278–282 (in Russian).
Greschner, G. S. (1973). *Makromol. Chem.*, **170**, 203–229.
Greville, T. N. E. (1969). *Theory and Applications of Spline Functions*, Academic Press, New York.
Guinier, A. (1938). *Nature (London)*, **142**, 569–570.
Guinier, A. (1939). *Ann. Phys.*, **12**, 161–237.
Guinier, A. and Fournet, G. (1947). *J. Phys. Radium*, **8**, 345–351.
Guinier, A. and Fournet, G. (1955). *Small-Angle Scattering of X-Rays*, Wiley, New York.
Gulik-Krzywicki, T., Yates, M., and Aggerbeck, L. (1979). *J. Mol. Biol.*, **131**, 475–484.
Hamada, F., Hayashi, H. and Nakajima, A. (1978). *J. Appl. Crystallogr.*, **11**, 514–519.
Hamming, R. W. (1962). *Numerical Methods for Scientists and Engineers*, McGraw-Hill, New York.
Handbuch der Physik. X-Rays. (1957). Vol. 30, Springer-Verlag, Berlin.
Harrison, S. C. (1969). *J. Mol. Biol.*, **42**, 457–483.
Haubold, H.-G. and Martinsen, D. (1978). *J. Appl. Crystallogr.*, **11**, 592–596.
Hayashi, H., Hamada, F., and Nakajima, A. (1976). *Macromolecules*, **9**, 543–547.
Hayashi, H., Hamada, F., and Nakajima, A. (1977). *Macromol. Chem.*, **178**, 827–842.
Heine, S. and Roppert, J. (1962). *Acta Phys. Austriaca*, **15**, 148–166.
Hendricks, R. W. (1978). *J. Appl. Crystallogr.*, **11**, 15–30.
Hendricks, R. W. and Schmidt, P. W. (1967). *Acta Phys. Austriaca*, **26**, 97–122.
Hendricks, R. W. and Schmidt, P. W. (1973). *Acta Phys. Austriaca*, **37**, 20–30.
Hendricks, R. W., Schelten, J., and Lippman, G. (1977). *Philos. Mag.*, **36**, 907–921.
Hendricks, R. W., Schelten, J., and Schmatz, W. (1974). *Philos. Mag.*, **30**, 819–837.
Hendrix, J. (1985). In: *Advances in Polymer Science* (Kausch, H. H. and Zachmann, H. G., eds.), Vol. 67, pp. 59–96, Springer-Verlag, Berlin.
Hermes, C., Parak, F., and Stuhrmann, H. B. (1980). Research project at EMBL, Hamburg.
Higgins, J. S. and Stein, R. S. (1978). *J. Appl. Crystallogr.*, **11**, 346–375.
Hiragi, Y. (1979). *J. Appl. Crystallogr.*, **12**, 628.
Hoppe, W. (1972). *Isr. J. Chem.*, **10**, 321–333.
Hoppe, W. (1973). *J. Mol. Biol.*, **78**, 581–585.
Hosemann, R. and Bagchi, S. W. (1962). *Direct Analysis of Diffraction by Matter*, North-Holland, Amsterdam.
Huber, R. and Bennett, W. S. (1983). *Biopolymers*, **22**, 261–279.
Huxley, H. E., Farugi, A. R., Kress, M., Bordas, J., and Koch, M. H. J. (1982). *J. Mol. Biol.*, **169**, 469–506.
Ibel, K. (1976). *J. Appl. Crystallogr.*, **9**, 296–309.
Ibel, K. and Stuhrmann, H. B. (1975). *J. Mol. Biol.*, **93**, 255–265.
Jack, A. and Harrison, S. C. (1975). *J. Mol. Biol.*, **99**, 15–25.

Jacrot, B. (1976). *Rep. Progr. Phys.*, **39**, 911–953.
Jacrot, B. and Zaccai, G. (1981). *Biopolymers*, **20**, 2413–2426.
Kahan, L., Winkelman, D. A., and Lake J. A. (1981). *J. Mol. Biol.*, **145**, 193–214.
Kayushina, R. L., Mogilevsky, L. Yu., Isotova, T. D., Shmakova, E. V., and Khurgin, Yu. I. (1982). *Stud. Biophys.*, **87**, 281–282.
Kayushina, R. L., Rolbin, Yu. A., and Feigin, L. A. (1974a). *Kristallografia*, **19**, 724–729. [Engl. transl. (1975). *Sov. Phys. Crystallogr.*, **19**, 420–424.]
Kayushina, R. L., Rolbin, Yu. A., and Feigin, L. A. (1974b). *Kristallografia*, **19**, 1161–1165. [Engl. transl. (1975). *Sov. Phys. Crystallogr.*, **19**, 722–724.]
Kayashina, R. L., Svergun, D. I., Izotova, T. D., Mogilevsky, L. Yu., Shmakova, F. V., and Khurgin, Yu. I. (1986). *Studia Biophys.*, **112**, 189–196.
Kiessig, H. (1942). *Kolloid Z.*, **98**, 213–221.
Kirste, R. G. (1967a). *Makromol. Chem.*, **101**, 91–103.
Kirste, R. G. (1967b). *J. Polym. Sci.*, **16**, 2039–2048.
Kirste, R. G. and Oberthür, R. C. (1982). In: *Small-Angle X-Ray Scattering* (Glatter, O. and Kratky, O., eds.), pp. 387–431, Academic Press, London.
Kirste, R. G. and Wunderlich, W. (1968). *Z. Phys. Chem. (Frankfurt am Main)*, **58**, 133–147.
Kirste, R. G., Kruse Q. A., and Ibel, K. (1975). *Polymer*, **16**, 120–124.
Kirste, R. G., Kruse, W. A., and Schelten, J. (1972). *Makromol. Chem.*, **162**, 299–303.
Koch, E.-E., ed. (1983). *Handbook on Synchrotron Radiation*, Vol. 1, North-Holland, Amsterdam.
Koch, M. H. J. and Bordas, J. (1983). *Nucl. Instrum. Methods*, **208**, 461–469.
Koch, M. H. J., Parfait, R., Hass, J., Crichton, R. R., and Stuhrmann, H. B. (1978). *Biophys. Struct. Mech.*, **4**, 251–262.
Koehler, W. C. and Hendricks, R. W. (1979). *J. Appl. Phys.*, **50**, 1951.
Korn, G. A. and Korn, T. M. (1961). *Mathematical Handbook for Scientists and Engineers*, McGraw-Hill, New York.
Kortleve, G. and Vonk, C. G. (1968). *Kolloid-Z. Z. Polym.*, **225**, 124–131.
Kostorz, G. (1982). In: *Small-Angle X-Ray Scattering* (Glatter, O. and Kratky, O., eds.), pp. 467–498, Academic Press, London.
Kosturko, L. D., Hogan, M., and Dattagupta, M. (1979). *Cell*, **16**, 515–522.
Kotelnikov, V. A. and Nikolaev, A. M. (1950). *Fundamentals of Radio Engineering*, Sviazizdat, Moscow (in Russian).
Kranold, R., Heyer, W., and Walter, G. (1983). *Stud. Biophys.*, **98**, 53–60.
Kratky, O. (1954). *Z. Elektrochem.*, **58**, 49–53.
Kratky, O. (1960). *Makromol. Chem.*, **35A**, 12–48.
Kratky, O. (1966). *Pure Appl. Chem.*, **12**, 483–523.
Kratky, O. (1982). In: *Small-Angle X-Ray Scattering* (Glatter, O. and Kratky, O., eds.), pp. 361–386, Academic Press, London.
Kratky, O. and Pilz, I. (1972). *Q. Rev. Biophys.*, **5**, 481–537.
Kratky, O. and Porod, G. (1949). *Rec. Trav. Chim. Pays-Bas*, **68**, 1106–1122.
Kratky, O. and Stabinger, H. (1984). *Coll. Polym. Sci.*, **262**, 345–360.
Kratky, O. and Wawra, H. (1963). *Monatsh. Chem.*, **94**, 981–987.
Kratky, O. and Worthman, W. (1947). *Monatsh. Chem.*, **76**, 263–281.
Kratky, O., Pilz, I., and Schmitz, P. J. (1966). *J. Colloid Interface Sci.*, **21**, 24–34.
Kratky, O., Porod, G., and Kahovec, L. (1951). *Z. Electrochem.*, **B55**, 53–59.
Kratky, O., Porod, G., and Skala, Z. (1960). *Acta Phys. Austriaca*, **13**, 76–128.
Krivandin, A. V., Lvov, Yu. M., Ostrovsky, M. A., Fedorovich, I. B., and Feigin, L. A. (1981). *Doklady AN SSSR*, **260**, 485–488 (in Russian).
Krivoglaz, M. A. (1969). *The Theory of Thermal Neutron Scattering by Real Crystals*, Plenum Press, New York.
Kuhn, W. (1934). *Kolloid Z.*, **68**, 2-15.

Laggner, P. and Müller, K. (1978). *Q. Rev. Biophys.*, **11**, 371–425.
Laggner, P., Gotto, A. M., and Morrisett, J. D. (1979). *Biochemistry*, **18**, 164–171.
Lake, J. A. (1967). *Acta Crystallogr.*, **23**, 191–194.
Landau, L. D. and Lifshitz, E. M. (1985a). *Classical Theory of Fields*, Pergamon Press, Oxford.
Landau, L. D. and Lifshitz, E. M. (1985b). *Quantum Mechanics*, Pergamon Press, Oxford.
Langer, J. A., Engelman, D. M., and Moore, P. B. (1978). *J. Mol. Biol.*, **119**, 463–485.
Langridge, R., Marvin, D. A., Seeds, W. E., Wilson, H. R., Hooper, C. W., Wilkins, M. H. F., and Hamilton L. D. (1960). *J. Mol. Biol.*, **2**, 38–62.
Lawson, C. L. and Hanson, R. J. (1974). *Solving Least Squares Problems*, Princeton University Press, New Jersey.
Leontovich M. A. (1983). *Introduction to Thermodynomics*, Nauka, Moscow (in Russian).
Lesselauer, W., Cain, J. E., and Blasie, J. K. (1971). *Proc. Natl. Acad. Sci. USA*, **69**, 1499–1503.
Letcher, J. R. and Schmidt, P. W. (1966). *J. Appl. Phys.*, **37**, 649–655.
Lindberg, V. W., McGervey, J. D., Hendricks, R. W., and Triftshäuser, W. (1977). *Philos. Mag.*, **36**, 117–128.
Linder, P. and Oberthür, R. C. (1984). *Rev. Phys. Appl.*, **19**, 759–763.
Litman, G. and Good, R., eds. (1978). *Immunoglobulins*, Plenum Medical, New York.
Luban, M. and Deutch, M. (1980). *J. Appl. Crystallogr.*, **13**, 233–243.
Luzzati, V. (1960). *Acta Crystallogr.*, **13**, 939–945.
Luzzati, V., Tardieu, A., and Aggerbeck, L. (1979). *J. Mol. Biol.*, **131**, 435–473.
Luzzati, V., Witz, J., and Nicolaieff, A. (1961). *J. Mol. Biol.*, **3**, 367–378.
Luzzati, V., Tardieu, A., Mateu, L., and Stuhrmann, H. B. (1976). *J. Mol. Biol.*, **101**, 115–127.
McGillavary, C. H. and Rieck, G. D., eds. (1983). *International Tables of X-Ray Crystallography*, Vol. 3, Kynoch Press, Birmingham.
Makowski, L. (1981). *J. Appl. Crystallogr.*, **14**, 160–168.
Marguerie, G. and Stuhrmann, H. B. (1976). *J. Mol. Biol.*, **102**, 143–156.
Mateu, L., Tardieu, A., Luzzati, V., Aggerbeck, L., and Scanu, A. M. (1972). *J. Mol. Biol.*, **70**, 105–116.
Mazyr, J. (1971). *J. Res. Natl. Bur. Stand., Sect. B*, **75**, 173–187.
Mildner, F. and Carpenter, J. (1984). *J. Appl. Crystallogr.*, **17**, 249–256.
Mittelbach, P. (1964). *Acta Phys. Austriaca*, **19**, 53–102.
Mittelbach, P. and Porod, G. (1961a). *Acta Phys. Austriaca*, **14**, 185–211.
Mittelbach, P. and Porod, G. (1961b). *Acta Phys. Austriaca*, **14**, 405–439.
Mittelbach, P. and Porod, G. (1962). *Acta Phys. Austriaca*, **15**, 122–147.
Mittelbach, P. and Porod, G. (1965). *Kolloid-Z. Z. Polym.*, **202**, 40–49.
Mogilevsky, L. Yu., Dembo, A. T., Svergun, D. I., and Feigin, L. A. (1984). *Kristallografia*, **29**, 587–591 (in Russian).
Monroe, A. J. (1962). *Digital Processes for Sampled Data Systems*, Wiley, New York.
Moore, P. B. (1980). *J. Appl. Crystallogr.*, **13**, 168–175.
Moore, P. B. and Weinstein, E. (1979). *J. Appl. Crystallogr.*, **12**, 321–326.
Moore, P. B., Engelman, D. M., and Schoenborn, B. P. (1974). *Proc. Natl. Acad. Sci. USA*, **71**, 171–176.
Moore, P. B., Langer, J. A., and Schoenborn, B. P. (1977). *J. Mol. Biol.*, **112**, 119–134.
Moore, P. B., Capel, M., Kjeldgaard, M. and Engelman, D. M. (1986). In: *Structure, Function and Genetics of Ribosomes* (Hardesty, B. and Krauer, G., eds.), pp. 87–100, Spinger-Verlag, New York.
Morawetz, H. (1963). *Macromolecules in Solution*, Wiley, New York.
Müller, J. J. (1983). *J. Appl. Crystallogr.*, **16**, 74–82.

Müller, J. J. and Damaschun, G. (1979). *J. Appl. Crystallogr.*, **12**, 267–274.
Müller, J. J., Damaschun, G., and Hübner, G. (1979). *Acta Biol. Med. Ger.*, **38**, 1–10.
Müller, J. J., Schmidt, P. W., Damaschun, G., and Walter, G. (1980). *J. Appl. Crystallogr.*, **13**, 280–283.
Müller, K., Laggner, P., Kratky, O., Kostner, G., Holasek, A., and Glatter, O. (1974). *FEBS Lett.*, **40**, 213–218.
Neutron Scattering in the "Nineties" (1985). Proc. Conf. Jülich, IAEA, Vienna.
Nikiforov, A. F. and Uvarov, V. B. (1974). *Basic Theory of Special Functions*, Nauka, Moscow (in Russian).
Ninio, J., Luzzati, V., and Yaniv, M. (1972). *J. Mol. Biol.*, **71**, 217–229.
Nozik, Yu. Z., Ozerov, R. P., and Hennig, K. (1979). *Structural Neutron-Diffraction Study*, Atomizdat, Moscow (in Russian). [Engl. transl. (1986). *Neutrons and Solid State Physics* (Ozerov, R. P., ed.), Plenum Press, New York.]
Oberthür, R. C. (1978). *Makromol. Chem.*, **179**, 2693–2706.
Ostanevich, Yu. M. and Serdyuk, I. N. (1982). *Usp. Fiz. Nauk*, **137**, 85–116 (in Russian).
Pape, E. H. (1974). *Biophys. J.*, **14**, 284–294.
Pape, E. H. and Kreutz, W. (1978). *J. Appl. Crystallogr.*, **11**, 421–425.
Patel, I. S. and Schmidt, P. W. (1971). *J. Appl. Crystallogr.*, **4**, 50–55.
Peterlin, A. (1960). *J. Polym. Sci.*, **47**, 403–415.
Picot, C., Duplessix, R., Decker, D., Beniot, H., Cotton, J. P., Daout, M., Farnoux, B., Jannik, G., Nierlich, G., de Vries, A. J., and Picus, P. (1977). *Macromolecules*, **10**, 436–442.
Pilz, I. (1969). *J. Colloid Interface Sci.*, **30**, 140–144.
Pilz, I., Glatter, O., and Kratky, O. (1972). *Z. Naturforsch.*, **27b**, 518–524.
Pilz, I., Kratky, O., and Moring-Claesson, I. (1970). *Z. Naturforsch.*, **25b**, 600–606.
Pilz, I., Goral, K., Hoyaerts, M., Lontie, R., and Witters, R. (1980). *Eur. Biochem.*, **105**, 539–543.
Plavnik, G. M. (1979). *Kristallografia*, **24**, 737–742. [Engl. transl. (1979). *Sov. Phys. Crystallogr.*, **24**, 422–425.]
Plavnik, G. M. (1984). *Kristallografia*, **29**, 210–214 (in Russian).
Plavnik, G. M., Kozevnikov, A. I., and Shishkin, A. B. (1976). *Doklady AN SSR*, **226**, 630–633 (in Russian).
Plestil, J. and Baldrian, J. (1976). *Czech. J. Phys.*, **B26**, 514–527.
Pope, D. P. and Keller, A. (1975). *J. Polym. Sci., Polym. Phys. Ed.*, **13**, 533–566.
Porai-Koshits, E. A., Golubkov, V. V., Titov, A. P., and Vasilevskaya, T. N. (1982). *J. Non-Cryst. Solids*, **49**, 143–156.
Porod, G. (1948). *Acta Phys. Austriaca*, **2**, 255–292.
Porod, G. (1949). *Monatsh. Chem.*, **80**, 251–255.
Porod, G. (1951). *Kolloid. Z.*, **124**, 83–114.
Porod, G. (1952). *Kolloid. Z.*, **125**, 51–57, 109–122.
Porod, G. (1967). In: *Small-Angle X-Ray Scattering* (Brumberger, H., ed.), pp. 1–15, Gordon and Breach, New York.
Porod, G. (1972). *Monatsh. Chem.*, **103**, 395–405.
Prudnikov, A. P., Brychkov, Yu. A., and Marichev, O. I. (1981). *Integrals and Series*, Nauka, Moscow (in Russian).
Ramakrishnan, V. R. and Moore, P. B. (1981). *J. Mol. Biol.*, **153**, 719–738.
Ramakrishnan, V. R., Capel, M., Kjeldgaard, M., Engelman, D. M. and Moore, P. B. (1984). *J. Mol. Biol.* **174**, 265–284.
Ramakrishnan, V. R., Yabuki, S., Sillers, I.-Y., Schnidler, P. G., Engelman, D. M., and Moore, P. B. (1981). *J. Mol. Biol.*, **153**, 739–760.
Rauch, H. and Schwann, D. (1984). *Berichte Kernforschungsanlage Jülich.*, No. 1954, 395–402.

Renninger, A. L., Wicks, G. G., and Uhlman, D. R. (1975). *J. Polym. Sci., Polym. Phys. Ed.*, **13**, 1247-1261.
Richards, K., Williams, R., and Calendar, R. (1973). *J. Mol. Biol.*, **78**, 255–259.
Rolbin, Yu. A., Feigin, L. A., and Schedrin, B. M. (1971). *Appar. Metody Rentgenovskogo Anal.*, **9**, 46–50 (in Russian).
Rolbin, Yu. A., Feigin, L. A., and Schedrin, B. M. (1977). *Kristallografia*, **22**, 1166–1175. [Engl. transl. (1977). *Sov. Phys. Crystallogr.*, **22**, 663–667.]
Rolbin, Yu. A., Svergun, D. I., and Schedrin, B. M. (1980). *Kristallografia*, **25**, 231–239. [Engl. transl. (1980). *Sov. Phys. Crystallogr.*, **25**, 133–137.]
Rolbin, Yu. A., Kayushina, R. L., Feigin, L. A., and Schedrin, B. M. (1973). *Kristallografia*, **18**, 701–709. [Engl. transl. (1974). *Sov. Phys. Crystallogr.*, **18**, 442–444.]
Rolbin, Yu. A., Svergun, D. I., Feigin, L. A., and Schedrin, B. M. (1980a). *Kristallografia*, **25**, 1125–1128. [Engl. transl. (1981). *Sov. Phys. Crystallogr.*, **25**, 645–647.]
Rolbin, Yu. A., Svergun, D. I., Feigin, L. A., and Schedrin, B. M. (1981). *Kristallografia*, **26**, 592–595. [Engl. transl. (1981). *Sov. Phys. Crystallogr.*, **26**, 334–336.]
Rolbin, Yu. A., Svergun,. D. I., Feigin, L. A., Gashpar, Sh., and Ronto, Gy. (1980b). *Doklady AN SSSR*, **255**, 1497–1500 (in Russian).
Ronto, G., Agamalyan, M. M., Drabkin, G. M., Feigin, L. A. and Lvov, Yu. M. (1983). *Biophys. J.*, **43**, 309–314.
Ruland, W. (1971). *J. Appl. Crystallogr.*, **4**, 70–73.
Ruland, W. (1977). *Colloid Polym. Sci.*, **255**, 417–427.
Sadler, D. M. and Keller, A. (1977). *Macromolecules*, **10**, 1128–1140.
Saibil, H., Charbe, M., and Worcester, D. (1976). *Nature*, **262**, 266–270.
Sauder, W. C. (1966). *J. Appl. Phys.*, **37**, 1495–1507.
Savitzky, A. and Golay, M. (1964). *Ann. Chem.*, **36**, 1627-1639.
Schedrin, B. M. and Feigin, L. A. (1966). *Kristallografia*, **11**, 159–163. [Engl. transl. (1966). *Sov. Phys. Crystallogr.*, **11**, 166–168.]
Schedrin, B. M. and Simonov, V. I. (1969). *Kristallografia*, **14**, 502–504. [Engl. transl. (1969). *Sov. Phys. Crystallogr.*, **14**, 411–413.]
Schelten, J. (1981). In: *Scattering Techniques Applied to Supramolecular and Nonequilibrium Systems* (Chen, S.-H., Chu, B., and Nossal, R., eds.), pp. 35–48, Plenum Press, New York.
Schelten, J. and Hendricks, R. W. (1978). *J. Appl. Crystallogr.*, **11**, 297–324.
Schelten, J. and Hossfeld R. (1971). *J. Appl. Crystallogr.*, **4**, 210–223.
Schelten, J., Shlecht, P., Schmatz, W., and Mayer, A. (1972). *J. Biol. Chem.*, **247**, 5436–5441.
Schmatz, W. (1975). *Riv. Nuovo Cimento*, **65**, 398–422.
Schmidt, P. W. (1965). *Acta Crystallogr.*, **19**, 938–942.
Schmidt, P. W. (1967). *J. Math. Phys.*, **8**, 475–477.
Schmidt, P. W. and Fedorov, B. A. (1978). *J. Appl. Crystallogr.*, **11**, 411–416.
Schwann, D. and Schmatz, W. (1978). *Acta Metall.*, **26**, 1571–1578.
Schwartz, S., Cain, J. E., Dratz, E. A., and Blasie, J. K. (1975). *Biophys. J.*, **15**, 1201–1233.
Seeger, A. and Kröner, E. (1959). *Z. Naturforsch.*, **14A**, 74–80.
Serdyuk, I. N. (1974). *Doklady AN SSSR*, **217**, 232–234 (in Russian).
Serdyuk, I. N. and Fedorov, B. A. (1973). *J. Polym. Sci., Polym. Lett. Ed.*, **11**, 645–647.
Serdyuk, I. N. and Grenader, A. K. (1974). *Makromol. Chem.*, **175**, 1881–1892.
Serdyuk, I. N. and Grenader, A. K. (1975). *FEBS Lett.*, **59**, 133–136.
Serdyuk, I. N., Zaccai, G., and Spirin, A. S. (1978). *FEBS Lett.*, **94**, 349–352.
Serdyuk, I. N., Agalarov, S. Ch, Sedelnikova, S. E., Spirin, A. S., and May, R. P. (1983), *J. Mol. Biol.*, **169**, 409–425.
Serwer, P. (1976). *J. Mol. Biol.*, **107**, 271–291.
Shaffer, L. B. and Beeman, W. W. (1970). *J. Appl. Crystallogr.*, **3**, 379–384.

Shannon, C. E. and Weaver, W. (1949). *The Mathematical Theory of Communication*, University of Illinois Press, Urbana.
Sharp, P. and Bloomfield, V. A. (1968). *Biopolymers*, **6**, 1201–1211.
Shull, C. G. and Roess, L. C. (1947). *J. Appl. Phys.*, **18**, 295–307; 308–313.
Sjöberg, B. (1974). *J. Appl. Crystallogr.*, **7**, 192–199.
Sneddon, I. (1951). *Fourier Transforms*, McGraw-Hill, New York.
Soler, J. (1975). *Kristallografia*, **29**, 1175–1177. [Engl. transl. (1975). *Sov. Phys. Crystallogr.*, **20**, 713–714.]
Soler, J. (1976). Ph.D thesis, Institute of Macromolecular Chemistry, Prague.
Sosfenov, N. I. and Feigin, L. A. (1970). *Appar. Metody Rentgenovskogo Anal.*, **8**, 15–22 (in Russian).
Sosfenov, N. I., Feigin, L. A., Bondarenko, K. P., and Mirensky, A. V. (1969). *Appar. Metody Rentgenovskogo Anal.*, **5**, 53–72 (in Russian).
Spirin, A. S., Serdiuk, I. N., Shpungin, I. L., and Vasiliev, V. D. (1979). *Proc. Natl. Acad. Sci. USA*, **76**, 4867–4871.
Stabinger, H. and Kratky, O. (1978). *Makromol. Chem.*, **179**, 1655–1659.
Steeb, S. and Lamparter, P. (1984). *J. Non-Cryst. Solids*, **61–62**, 237–248.
Stribeck, N. and Ruland, W. (1978). *J. Appl. Crystallogr.*, **11**, 536–539.
Strobl, G. R. (1970). *Acta Crystallogr., Sect A*, **26**, 367–375.
Stuhrmann, H. B. (1970a). *Acta Crystallogr., Sect. A*, **26**, 297–306.
Stuhrmann, H. B. (1970b). *Z. Phys. Chem. (Frankfurt am Main)*, **72**, 177–184.
Stuhrmann, H. B. (1970c). *Z. Phys. Chem. (Frankfurt am Main)*, **72**, 185–198.
Stuhrmann, H. B. (1981a). *Kristallografia*, **26**, 956–964. [Engl. transl. (1981). *Sov. Phys. Crystallogr.*, **26**, 544–548.]
Stuhrmann, H. B. (1981b). *Q. Rev. Biophys.*, **14**, 433–460.
Stuhrmann, H. B., ed. (1982). *Uses of Synchrotron Radiation in Biology*, Academic Press, London.
Stuhrmann, H. B. (1985). *Adv. Polym. Sci.*, **67**, 123–163.
Stuhrmann, H. B. and Fuess, H. (1976). *Acta Crystallogr., Sect. A*, **32**, 67–74.
Stuhrmann, H. B. and Gabriel, A. (1983). *J. Appl. Crystallogr.*, **16**, 563–571.
Stuhrmann, H. B. and Kirste, R. G. (1965). *Z. Phys. Chem. (Frankfurt am Main)*, **46**, 247–250.
Stuhrmann, H. B. and Kirste, R. G. (1967). *Z. Phys. Chem. (Frankfurt am Main)*, **56**, 333–337.
Stuhrmann, H. B. and Miller, A. (1978). *J. Appl. Crystallogr.*, **11**, 325–345.
Stuhrmann, H. B. and Notbohm, H. (1981). *Proc. Natl. Acad. Sci. USA*, **78**, 6216–6220.
Stuhrmann, H. B., Haas, J., Ibel, K., Dewoif, B., Koch, M. H. J., Parfait, R., and Crichton, R. R. (1976a). *Proc. Natl. Acad. Sci. USA*, **73**, 2379–2383.
Stuhrmann, H. B., Haas, J., Ibel., K., Koch, M. H. J., and Crichton, R. R. (1976b). *J. Mol. Biol.*, **100**, 399–413.
Stuhrmann, H. B., Koch, M. H. J., Parfait, J., Hass, J., Ibel, K., and Crichton, R. R. (1975a). *Proc. Natl. Acad. Sci. USA*, **72**, 2316–2320.
Stuhrmann, H. B., Koch, M. H. J., Parfait, J., Haas, S., Ibel, K., and Crichton, R. R. (1978). *J. Mol. Biol.*, **119**, 203–212.
Stuhrmann, H. B., Tardieu, A., Mateu, L., Sardet, C., Luzzati, V., Aggrebeck, L., and Scanu, A. M. (1975b). *Proc. Natl. Acad. Sci. USA*, **72**, 2270–2273.
Svergun, D. I. and Semenyuk, A. V. (1985). *Doklady AN SSSR*, **284**, 621–626 (in Russian).
Svergun, D. I. and Semenyuk, A. V. (1986). *Studia Biophys.*, **112**, 255–262.
Svergun, D. I. and Semenyuk, A. V. (1987). *Doklady AN SSSR* (in press).
Svergun, D. I., Feigin, L. A., and Schedrin, B. M. (1981). *Doklady AN SSSR*, **261**, 878–882 (in Russian).
Svergun, D. I., Feigin, L. A., and Schedrin, B. M. (1982). *Acta Crystallogr., Sect. A*, **38**, 827–835.

Svergun, D. I., Feigin, L. A., and Schedrin, B. M. (1983). *Kristallografia*, **28**, 252–259 (in Russian). [Engl. transl. (1984). *Sov. Pys. Crystallogr.*, **28**, 147–150.]
Svergun, D. I., Feigin, L. A., and Schedrin, B. M. (1984). *Acta Crystallogr., Sect A*, **40**, 137–142.
Svergun, D. I., Kayushina, R. L., Izotova, T. D., and Khurgin, Yu. I. (1985). *Doklady AN SSSR*, **282**, 103–108 (in Russian).
Tardieu, A., Mateu, L., Sardet, C., Weiss, B., Luzzati, V., Aggerbeck, L., and Scanu, A. M. (1976). *J. Mol. Biol.*, **101**, 129–153.
Taupin, D. and Luzzati, V. (1982). *J. Appl. Crystallogr.*, **15**, 289–300.
Taylor, T. R. and Schmidt, P. W. (1967). *Acta Phys. Austriaca*, **25**, 293–296.
Templeton, D. H., Templeton, L. K., Phillips, J. C., and Hodgson, K. O. (1980). *Acta Crystallogr., Sect. A*, **36**, 436–442.
Tikhonov, A. N. and Arsenin, V. Ya. (1977) *Solution of Ill-Posed Problems*, Wiley, New York.
Timothy, J. G. and Madden, R. P. (1983). In: *Handbook on Synchrotron Radiation*, Vol. 1A (Koch, E.-E., ed.), pp. 315–366, North-Holland, Amsterdam.
Tsvankin, D. Ya. (1964). *Vysokomol. Soed.*, **6**, 2078–2082, 2083–2130 (in Russian).
Vachette, P. (1985). Private communication.
Vagov, V. A., Kunchenko, A. B., Ostanevich, Yu. M., and Salamatin, I. M. (1983). Preprint of Joint Institute of Nuclear Research No. 14–83–898, JINR, Dubna (in Russian).
Vainshtein, B. K. (1966). *Diffraction of X-Rays by Chain Molecules*, Elsevier, Amsterdam.
Vainshtein, B. K. (1981). *Modern Crystallography*, Vol. 1, Springer-Verlag, Berlin.
Vainshtein, B. K. and Kayushina, R. L. (1966). *Kristallografia*, **11**, 526–535. [Engl. transl. (1966). *Sov. Phys. Crystallogr.*, **10**, 698–706.]
Vainshtein, B. K., Sosfenov, N. I., and Feigin, L. A. (1970). *Doklady AN SSSR*, **190**, 574–577 (in Russian).
Vainshtein, B. K., Feigin, L. A., Lvov, Yu. M., Gvozdev, R. I., Marakushev, S. A., and Likhtenshtein, G. I. (1970). *FEBS Lett.*, **116**, 107–110.
Vasilevskaya, T. N., Golubkov, V. V., and Porai-Koshits, E. A. (1980). *Fiz. Khim. Stekla*, **6**, 51–58 (in Russian).
Vasilevskaya, T. N., Golubkov, V. V., and Porai-Koshits, E. A. (1981). *Fiz. Khim. Stekla*, **7**, 31–34 (in Russian).
Vasiliev, V. D., Selivanova, O. M., and Kotelyansky, V. E. (1978). *FEBS Lett.*, **95**, 273–276.
Vazina, A. A., Gerasimov, V. S., Zeleznaya, L. A., Matyushin, A. M., Sorokin, B. A., Skrebnitskaya, L. N., Shelestov, V. N., Frank, G. M., Avakyan, Sh. M., and Alikhanyan, A. I. (1975). *Biofizika*, **20**, 801–806 (in Russian).
Vilkov, L. V., Mastrukov, V. S., and Sadova, N. I. (1978). *The Geometrical Structure of Free Molecules*, Khimia, Leningrad (in Russian).
Vintaikin, E. Z., Barkalaya, A. A., Belyatskaya, I. S., and Sahno, V. M. (1977). *Fiz. Met. Metalloved.*, **43**, 734–742 (in Russian).
Vintaikin, E. Z., Dmitriev, V. B., and Udovenko, V. A. (1978). *Fiz. Met. Metalloved.*, **46**, 790–795 (in Russian).
Volkenshtein, M. V. (1963). *Configurational Statistics of Polymer Chains*, Interscience, New York.
Vonk, C. G. (1971). *J. Appl. Crystallogr.*, **4**, 340–342.
Vonk, C. G. (1976). *J. Appl. Crystallogr.*, **9**, 433–440.
Vonk, C. G. (1978). *J. Appl. Crystallogr.*, **11**, 540–546.
Vonk, C. G. (1982). In: *Small-Angle X-Ray Scattering* (Glatter, O. and Kratky, O., eds.), pp. 433–466, Academic Press, London.
Vonk, C. G. and Kortleve, G. (1967). *Kolloid-Z. Z. Polym.*, **220**, 19–24.
Walter, G., Gerber, Th., and Kranold, R. (1983). *Stud. Biophys.*, **97**, 129–134.

Walter, G., Kranold, R., Göeke, V., and Müller, J. J. (1977). *Kristallografia*, **22**, 951–961. [Engl. transl. (1977). *Sov. Phys. Crystallogr.*, **22**, 543–549.]
Walter, G., Kranold, R., Müller, J. J., and Damaschun, G. (1975). *Crystallographic Computing Techniques* (Ahmed, F. R., ed.), pp. 383–388, Munksgaard, Copenhagen.
Walther, A. (1962). *Opt. Acta*, **10**, 41–49.
Waser, J. and Schomaker, V. (1953). *Rev. Mod. Phys.*, **25**, 671–690.
Wendorff, J. H. and Fisher, E. W. (1973). *Koloid-Z. Z. Polym.*, **251**, 876–883.
White, S. W., Hulmes, D. J. S., Miller, A., and Timmins, P. (1977). *Nature (London)*, **266**, 421–425.
Wiegand, W. and Ruland, W. (1979). *Prog. Colloid Polym. Sci.*, **66**, 355–366.
Wilhelm, P., Pilz, I., Palm, W., and Bauer, K. (1978). *Eur. J. Biochem.*, **84**, 457–463.
Wilson, A. J. C. (1949). *X-Ray Optics*, Wiley, New York.
Worthington, C. R. (1973). *Exp. Eye Res.*, **17**, 487–501.
Worthington, C. R. (1981). *J. Appl. Crystallogr.*, **14**, 383–386.
Wu, H. and Schmidt, P. W. (1971). *J. Appl. Crystallogr.*, **4**, 224–231.
Yoda, O. (1984). *J. Appl. Crystallogr.*, **17**, 337–343.
Yoon, D. Y. (1978). *J. Appl. Crystallogr.*, **11**, 531–535.
Yoon, D. Y. and Flory, P. J. (1977). *Polymer*, **18**, 509–513.
Zaiss, W., Steeb, S., and Bauer, G. S. (1976). *Phys. Chem. Liq.*, **6**, 21–41.
Zernike, F. and Prins, J. A. (1927). *Z. Phys.*, **41**, 184–194.
Zimm, B. N. and Stockmayer, W. H. (1949). *J. Chem. Phys.*, **17**, 1301–1314.
Zipper, P. (1969). *Acta Phys. Austriaca*, **30**, 143–151.
Zipper, P. (1972). *Acta Phys. Austriaca*, **36**, 27–38.

Index

Absolute measurements, 73–76
Absorption
  edge, 16, 143
  of neutrons, 23
  of X rays, 22
Amorphous solids, 240

Babinet principle, 52
Bacteriophages, 81, 100, 112, 177
Basic functions, 115
Bessel functions, 28, 157
  separation, 167–176
Box-function refinement, 153–155
Breit–Wigner formula, 18

Calibrated sample, 75
Characteristic function, 41, 116
Chord distribution, 44
Clusters, 221, 243
Collimation corrections, 295–303
  systems, 250, 257–262, 269
    Bonse–Hart, 261
    Kratky, 260
    point, 257, 266, 270
    slit, 258–261
Combined use of radiations, 139
Concentration effects, 69
Contrast, 35
  variation, 115–131
    complex, 141
Convolution, 8
  square root, 152
Correlation function, 39

Debye formula, 39, 77, 95
Defects, 220–224
Delta function, 8
Desmearing, 281
Detectors
  neutron, 269
  position-sensitive, 256, 270
  X-ray, 256
Deuteration, 118, 125, 135
Direct methods, 147, 150, 164
Dislocations, 221
Distance distribution function, 40, 135

Errors
  propagation, 67, 287
  statistical, 280
  systematic, 276

Fluctuations
  of concentration, 243
  statistical, 50, 240
Focusing, 264
Form factor, 13, 34
Fourier transform, 5–10
Frequency filtering, 291

Glasses, 240–243
Guinier approximation, 69

Hankel transformation, 157–158
H–D exchange, 120–122, 130
Heavy-atom labels, 131, 202
Homogeneous approximation, 76–81

Incorrectly stated problems, 236, 309
Indirect methods, 64, 310–320
  Fourier transformation, 311
Information content, 63–68
Interference effects, 38, 69
Invariants, 59–63
Isomorphous replacement, 131–133
Isotopic replacement, 118–122
Iteration methods, 309

Laplace transformation, 300
Large periods, 212, 216
Largest dimension, 62, 82
Lipoproteins, 122–125
Lorentz factor, 216

Matrix inversion technique, 301
Mellin transformation, 304
Membrane, 214
Miscibility gap, 229
Modeling, 94, 98–104
  cube method, 97
  spheres method, 95
    in real space, 98
  subparticle method, 95
Molecular mass, 73
  distribution, 195
Monochromatization
  neutron, 269
  X rays, 263–265
Multipole theory, 156–164
Muscle, 212

Ornstein–Zernike formula, 244

Particle
  axially symmetric, 174
  isometric, 100, 164
  lamellar, 49, 63, 149
  rodlike, 47, 63, 149
  spherically symmetric, 47, 112, 148
  uniform, 40–44, 165–167
  volume, 46, 79
Patterson function, 9
Phase separation, 225–229, 241
Polychromaticity, 279
  corrections, 303
Polymers
  amorphous, 201–203, 210
  chain, Gaussian, 189
    persistent, 191
    perturbed, 194
  crystalline, 203–209
Polynomial smoothing, 284
Porod invariant, 46, 79, 111
  law, 45, 77, 307
Proteins, 99, 126, 132, 141, 143, 180

Radius of gyration, 60, 68–73, 117
Regularization, 304, 317
Resolution
  instrumental, 252
  space, 27, 33
Ribosome, 103, 128, 134–136

Sampling theorem, 65, 314–317
Scattering
  additive components, 276
  amplitude, 4, 10–14, 148
    atomic, 20
  anomalous, 16, 140–143
  cross section, 6, 15, 19
  density, 4, 120
    excess, 35, 109
  from shape, 115
  incoherent, 19, 75
  intensity, 6
  large-angle, 112–115
  length, 4
  light, 139, 217
  magnetic, 21, 228
  simple bodies, 10, 90–94
Self-convolution, 9, 40
Shape determination, 165
  of reflections, 213
Sign problem, 150–155, 216
Size distribution, 34, 195, 231–239
Solvent influence, 108, 112
Sources
  neutron, 268
  X-ray, 255, 262
Spherical harmonics, 156–164
Spinoidal decomposition, 225, 241
Splines, 288–291, 310–313

Surface
  of particle, 43, 46, 61, 79, 85
  of phase separation, 53
Synchrotron radiation, 262
Systems
  anisotropic, 210–217
  binary, 52, 229
  fibrillar, 211
  lamellar, 203, 214
  monodisperse, 34
  nonparticulate, 50–55, 240
  polydisperse, 34, 195, 230

Temperature factor, 308
Termination effects, 305–309
Theta solvent, 194
Time-resolved measurements, 130, 212
Titchmarsh transform, 233
Triangulation methods, 132–136

Virial coefficients, 194, 198
Voids, 221

Zernike–Prince formula, 32
Zimm plot, 198